ACOUSTIC SIGNAL PROCESSING FOR TELECOMMUNICATION

THE KLUWER INTERNATIONAL SERIES IN ENGINEERING AND COMPUTER SCIENCE

ACOUSTIC SIGNAL PROCESSING FOR TELECOMMUNICATION

Edited by
STEVEN L. GAY
Bell Laboratories, Lucent Technologies

JACOB BENESTY
Bell Laboratories, Lucent Technologies

Kluwer Academic Publishers
Boston/Dordrecht/London

Distributors for North, Central and South America:
Kluwer Academic Publishers
101 Philip Drive
Assinippi Park
Norwell, Massachusetts 02061 USA
Telephone (781) 871-6600
Fax (781) 871-6528
E-Mail <kluwer@wkap.com>

Distributors for all other countries:
Kluwer Academic Publishers Group
Distribution Centre
Post Office Box 322
3300 AH Dordrecht, THE NETHERLANDS
Telephone 31 78 6392 392
Fax 31 78 6546 474
E-Mail services@wkap.nl>

Electronic Services <http://www.wkap.nl>

Library of Congress Cataloging-in-Publication

Acoustic signal processing for telecommunication / edited by Steven L. Gay, Jacob Benesty.
p. cm. -- (Kluwer international series in engineering and computer science ; SECS 551)
Includes bibliographical references and index.
ISBN 0-7923-7814-8
1. Signal processing--Digital techniques. 2. Algorithms. 3. Adaptive signal processing.
4. Noise control. I. Gay, Steven L. II. Benesty, Jacob. III. Series.

TK5102.9. A27 2000
621.382'2--dc21

00-022051

Printed on acid-free paper.

Printed in the United States of America

Contents

List of Figures

List of Tables

Preface

The overriding goal of acoustic signal processing for telecommunication systems is to promote the feeling of "telepresence" among users. That is, make users feel they are in the actual physical presence of each other even though they may be separated into many groups over large distances. Unfortunately, there are many obstacles which prevent system designers from easily attaining this goal. These include the user's acoustic environments, the physical and architectural aspects of modern telecommunication systems, and even the human auditory perceptual system itself.

Telepresence implies the use of hands-free communication which give rise to problems that are almost nonexistent when handsets are used. These difficulties have motivated a considerable body of research in signal processing algorithms. Technologies such as noise reduction and dereverberation algorithms using one or more microphones (Parts III and IV), camera tracking (Chapter 11), echo control algorithms (Parts I and II), virtual sound (Part V), and blind source separation (Part VI) have arisen to stabilize audio connections, eliminate echo, and improve audio transmission and rendering.

Researchers are now endeavoring to enhance the telepresence experience by using multi-channel audio streams between locations to increase spatial realism, signal separation, and talker localization and identification, by taking advantage of our binaural hearing system. While stereo and surround-sound are common examples of one-way free space multi-channel audio, realizing these technologies in the full duplex telecommunications realm has raised a set of new fundamental problems that have only recently been addressed in a satisfactory manner. Furthermore, multi-channel duplex communications enabled by multi-channel echo cancellation and control algorithms will allow participants of point-to-point and even multi-point teleconferences to instinctively know who is talking and from where, simply by using the normal auditory cues that have evolved in humans over millennia.

Acoustic signal processing also plays an important role in enhancing the visual aspect of multi-media telecommunication. Algorithms which localize and identify the nature of sound sources allow cameras to be steered automatically to the active participants of a teleconference, allowing participants to concentrate on the issues at hand rather than cumbersome camera manipulation.

Our strategy for selecting the chapters for this book has been to present digital signal processing techniques for telecommunications acoustics that are both cutting edge and practical. Each chapter presents material that has not appeared in book form before and yet is easily realizable in today's technology. To this end, those chapters that do not explicitly discuss implementation are followed by those that discuss implementation aspects on the same subject. The end result is a book that, we hope, is interesting to both researchers and developers.

STEVEN L. GAY

JACOB BENESTY

Contributing Authors

Jacob Benesty
Bell Laboratories, Lucent Technologies

Michael S. Brandstein
Division of Engineering and Applied Sciences, Harvard University

Benoit Champagne
Department of Electrical and Computer Engineering, McGill University

Jiashu Chen
Lucent Technologies

Eric J. Diethorn
Microelectronics and Communications Technologies, Lucent Technologies

Gary W. Elko
Bell Laboratories, Lucent Technologies

Peter Eneroth
Department of Applied Electronics, Lund University

Steven L. Gay
Bell Laboratories, Lucent Technologies

Mohamed Ghanassi
EXFO Fiber Optic Test Equipment

Scott M. Griebel
Division of Engineering and Applied Sciences, Harvard University

Tomas Gänsler
Bell Laboratories, Lucent Technologies

Yiteng (Arden) Huang
Georgia Institute of Technology

Dennis R. Morgan
Bell Laboratories, Lucent Technologies

Bernhard H. Nitsch
Fachgebiet Theorie der Signale, Darmstadt University of Technology

Darren B. Ward
University College, The University of New South Wales

Chapter 1

AN INTRODUCTION TO ACOUSTIC ECHO AND NOISE CONTROL

Steven L. Gay
Bell Laboratories, Lucent Technologies
slg@bell-labs.com

Jacob Benesty
Bell Laboratories, Lucent Technologies
jbenesty@bell-labs.com

Abstract This chapter introduces the acoustic echo and noise problems that occur in hands-free communications. These problems (and possible solutions) are a good background of acoustic signal processing for telecommunication in general. Moreover, the network echo canceler which was invented at Bell Labs in the 1960's is thoroughly explained here since this solution fully inspired researchers in the derivation of the most widely used solution to the acoustic echo problem: the acoustic echo canceler.

Keywords: Acoustic Echo Cancellation, Suppression, Noise Reduction, Spectral Subtraction, Adaptive Algorithms, LMS, NLMS, RLS

1. HUMAN PERCEPTION OF ECHOES

Conversations rarely occur in anechoic environments. The acoustics of rooms almost always consists of reflections from walls, floors, and ceilings. Sound reflections with short round trip delays, on the order of a few milliseconds, are perceived as spectral coloration or reverberation and are generally preferred over anechoic environments [1]. When a sound reflection's round-trip delay exceeds a few tens of milliseconds and is unattenuated or only slightly attenuated, it is perceived as a distinct echo, and when the round-trip delay approaches a quarter of a second most people find it difficult to carry on a normal conversation.

Table 1.1 Subjective reaction to echo delay.

Round-Trip Delay (ms)	Mean Required Loss (dB)
0	1.4
20	11.1
40	17.7
60	22.7
80	27.2
100	30.9

In the early days of satellite telecommunication design there was concern that the long propagation delay between ground stations and geosynchronous satellites would make the technology unusable. Even if perfect echo control were possible, there was the question of whether the 480 ms round-trip delay alone would inhibit normal conversation patterns.

This and other echo associated issues of terrestrial telecommunication origins motivated investigations on the effects of echo round-trip delay and attenuation on conversations [2, 3, 4, 5].

Table 1.1 shows the attenuation required to prevent echoes from becoming objectionable by the average person. The mean required loss is shown for various amounts of round-trip delay. The table shows that as the round-trip delay increases, so does the mean required loss.

The results of another study showing the same effect are shown in Table 1.2. Here, two measures, the decrease of mean opinion score (MOS) and the percent of subjects experiencing difficulty communicating are shown as a function of round-trip delay when the return loss is fixed at 15 dB.

MOS is a subjective evaluation system with a five step scale; "unsatisfactory," "poor," "fair," "good," and "excellent" corresponding to numerical values of 1 through 5 respectively. Subjects judge the quality of speech using this scale and the results are averaged yielding the *mean-opinion score*.

The effect of the increasing round-trip delay is to lower the mean opinion score and increase the difficulty of the subjects in carrying on a conversation.

In order to determine the efficacy of perfect echo control, several of these studies [6, 4, 3] examined the effect of round-trip delay in the absence of echo on subjects conducting conversations. More specifically, researchers were interested in the effect on the protocol of one speaker interrupting another. With no delay, interruptions occur in a conversation at the rate of about two per minute. If the round-trip delay is increased to 600 ms the rate doubles to about 4 per minute. However, when compared to the no-delay/no-echo condition, round-

Table 1.2 Subjective effect of 15 dB echo return loss.

Round-Trip Delay (ms)	Decrease in MOS	Percent Difficulty
0	0	0
300	1.3	30
600	2.0	60
1200	2.0	60

trip delays of up to 600 ms resulted in only insignificant decreases in MOS and increases in percent communication difficulty.

The overall conclusion is that as round-trip delay increases, increasing amounts of attenuation of the reflected signal is required for speaker comfort and that if echoes can be eliminated entirely, quite long round-trip delays can be tolerated.

2. THE NETWORK ECHO PROBLEM

Most of the acoustic control techniques currently employed were first developed to address the network echo problem.

This section discusses the network, or *electrical* echo problem. Figure 1.1 shows a simplified long distance connection. Talker A is connected to talker B via two 2-wire circuits (also known as local loops) and a 4-wire long distance connection. The 2-wire connections which connect the talkers to their respective local central offices carry both the receive and transmit signals on a single pair of wires. This results in considerable savings in wire and other facilities over a 4-wire arrangement. The circuits in the 4-wire connection insert delay that is unavoidable. Part of it is caused by the transport media: The finite speed of electrical waves through copper wire, or light through fiber, or radio waves through space. The other part is caused by switching and/or multiplexing equipment.

The 2-wire circuit is connected to the 4-wire long distance circuit by means of a device called a hybrid. It is a passive bridge circuit which passes the output of the 4-wire circuit into the 2-wire circuit and the output of the 2-wire circuit to input of the 4-wire circuit without attenuation, yet provides attenuation between the output and input of the 4-wire circuit.

A given 4-wire circuit terminated by a hybrid can be switched to any number of 2-wire circuits. As a result, a compromise balance circuit is used. The result is that some of the 4-wire out signal is coupled back to the 4-wire input, resulting in echo.

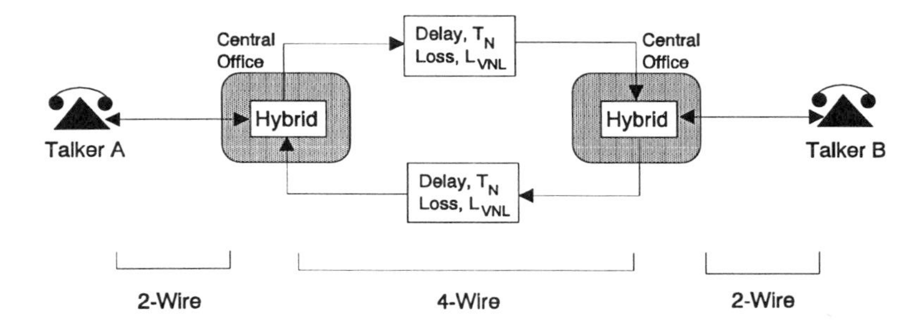

Figure 1.1 A simplified long distance connection.

In the 1920's Clark and Mathes [7] introduced the echo *suppressor*. Nominally, this is a voice activated switch at the 4-wire termination at each end of the long distance connection.

The echo suppressor operates two switches, the receive switch and the transmit switch. The receive switch has the ability to switch in 6 to 12 dB of loss in the path of the far-end speech while the transmit switch may break the path of the near-end speech (and echo) in the 4-wire out direction to the far-end user. When the far-end user is talking and the near-end speaker is silent, the receive switch is in the no-loss position and the transmit switch is in the open position. When the far-end signal goes silent, the transmit switch transitions to the closed position following a hold period of about 25 ms. This allows the response of the far-end speech to die out before switching, preventing echo from being heard by the far-end user. When the far-end speaker is interrupted by the near-end talker, the transmit switch transitions back to the closed position while the receive switch transitions to the attenuation position. In order to bridge brief silent periods in the near-end speech, an interruption hangover period of about 200 ms must expire after cessation of near-end speech before the transmit and receive switches transition back to the far-end talker only state.

Echo suppressors that do not perform optimally, can cause the following degradations as observed by the far-end user:

1. Echo caused by slow initial far-end speech detection or false near-end speech detection.

2. Clipping of the near-end speech due to slow initial interruption switching.

3. Chopping of the near-end speech due to early release of the interruption state.

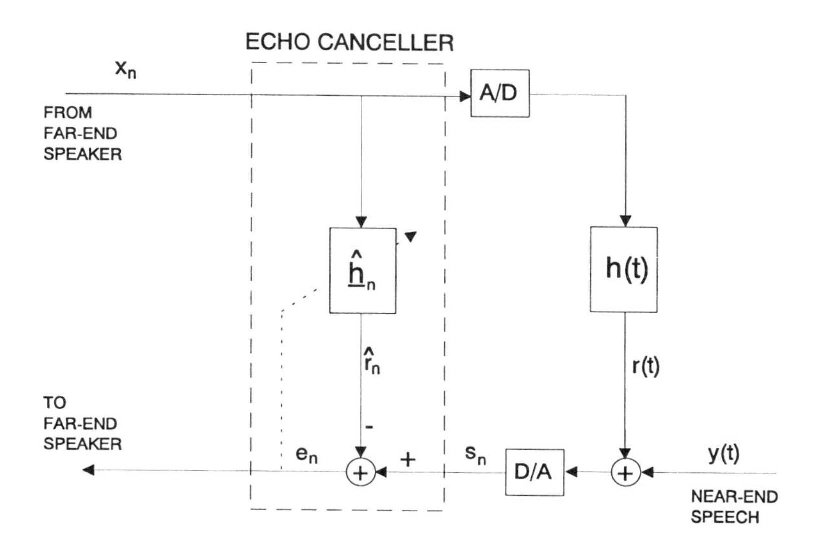

Figure 1.2 A simplified network echo canceler.

The most serious of these problems is the clipping and chopping of speech in the long round-trip delay type connections such as satellite links.

In the 1960's a new echo control device, the echo canceler [8, 9, 10] was introduced. It addresses many of the problems associated with the echo suppressor. A digital network echo canceler is shown in Fig. 1.2. The basic idea is to build a model of the impulse response of the echo path and then excite this model with the same signal that is sent into the hybrid and local loop, that is, the far-end speaker's speech.

The echo path impulse response is represented by $h(t)$ and its model in the canceler is represented by the vector $\hat{\mathbf{h}}_n$. The signal x_n is the sampled far-end signal and $y(t)$ is the near-end signal. The response of the model $\hat{r}_n$, is subtracted from the combination of the echo and the speech of the near-end speaker s_n leaving only the sampled speech of the near-end speaker y_n to be sent to the far-end user. The problem, of course, is in building (and maintaining) the model and, to some extent, obtaining the response of the model to the excitation signal.

Echo cancelers as in Fig. 1.2 are predominantly used to terminate long distance 4-wire circuits. As with the echo suppressor and hybrid, they are commonly switched to different circuits on a per call basis, each circuit having a different impulse response. Therefore, the echo path model $\hat{\mathbf{h}}_n$ must have the ability to learn and adapt to the new echo path impulse response at the beginning of each call. To accomplish this, the echo canceler uses an adaptive filter to construct the echo impulse response model. The adaptive filter is usually based on the normalized least mean square (NLMS) finite impulse response (FIR) adaptive filter. This adaptive filter attempts to build the echo impulse response model by adjusting its filter coefficients (or taps) in such a way as to drive e_n to zero. This is fine if s_n consists only of the echo of the far-end speech. In that case, the correlation of x_n and s_n contains valuable information about the echo impulse response. If, on the other hand, s_n also contains significant amounts of near-end signal y_n, then the echo impulse response information is corrupted by any extraneous correlation between x_n and y_n. For this reason, practical echo cancelers need to inhibit adaptation of the filter taps when significant near-end signal is present. This is accomplished by a function known as a near-end speech detector (see Chapter 5). Only the adaptation of the filter taps is inhibited during near-end speech periods, not the generation of the echo replica $\hat{r}_n$. This means that even during double talk periods echo will be canceled.

The impulse response of the echo path as seen from the 4-wire termination is typically characterized by a pure delay followed by the impulse response of the hybrid and two-wire circuits whose length is about 8 ms. Adaptive filters of 32 ms in length are usually sufficient to cope with most circuits of this nature.

3. THE ACOUSTIC ECHO PROBLEM

The previous section dealt with echoes in the telecommunications network which mainly arise from impedance mismatches at hybrids. The second major source of echoes in telecommunications today is due to acoustic coupling between the microphones and loudspeakers as in speakerphones.

Loudspeaker telephony has become increasingly popular in the past several years. Its two major advantages are hands-free operation and convenient conferencing. Hands-free telephony is critical for safety in applications such as mobile telephone and is a convenience in the white collar office environment for efficient multitasking that includes workers communicating with colleagues.

In the mobile telephone, loudspeaker telephony allows the mobile subscriber to drive with both hands while conversing over the cellular phone. In many countries hands free operation of mobile telephones may soon be mandated by law.

The inherent conferencing feature provides business people the ability to hold meetings in two or more remote locations; saving both time and travel expenses. The modern conference call typically includes at least one location where several people are gathered in a room using a speakerphone so that all present may converse with those at remote locations.

Video teleconferencing is an extension to the audio-only conference call. The audio problems are similar, although video conference rooms are often carefully engineered acoustically to minimize the possibilities of singing and the effects of echoes. Without the application of acoustic echo cancellation, the location and amplification levels of microphones and speakers must be given careful attention. An additional requirement to the video conference room application is that often the speech must be delayed to synchronize with the video signal due to the long delay in the video coder/decoder.

The ultimate goal of acoustic echo control is to provide true full-duplex hands-free communication so that normal speech patterns, including polite interjections, can be made without disruption of speech. The apparent solution is echo cancellation. Unfortunately, the acoustic environment is more hostile than the network environment to echo cancelers. The problems associated with the acoustic environment are as follows:

1. The echo path is extremely long, on the order of 125 ms.

2. The echo path may rapidly change at any time during the connection.

3. The background noise in the near-end signal is stronger.

The Acoustic Echo Path:

The excessive length of the acoustic echo path in time, especially when compared to the network echo path, is mainly due to the slow speed of sound through air. Multiple reflections off walls, chairs, desks, and other objects in the room also serve to lengthen the response. The energy of a reverberating signal in a room is a function of the room's size and the materials inside it, different materials having different reflection coefficients. The average sound intensity due to room reverberation decays exponentially for the typical room. The reverberation time, T_{60}, is the time it takes for the reverberation level to drop by 60 dB. For a typical office, the reverberation time is about 200 to 300 ms. Therefore, to reduce the acoustic echo of the typical office by 30 dB, a 100 to 150 ms length echo canceler is required. At an 8 KHz sample rate, this implies an adaptive filter on the order of 1000 taps is needed.

To complicate matters more, the impulse response of the room is not static over time. It varies with ambient temperature, pressure, and humidity. In addition, movement of objects, such as human bodies, doors, and the location of microphones and speakers can all dramatically and rapidly modify the acoustic impulse response.

The Near-End Noise:

The magnitude of sound pressure level decreases proportionally as the distance from the source increases. A talker 18 to 21 inches from a speakerphone microphone requires 25 dB of gain to bring it to the level comparable to the signal level of a talker speaking directly into the microphone of a handset. In

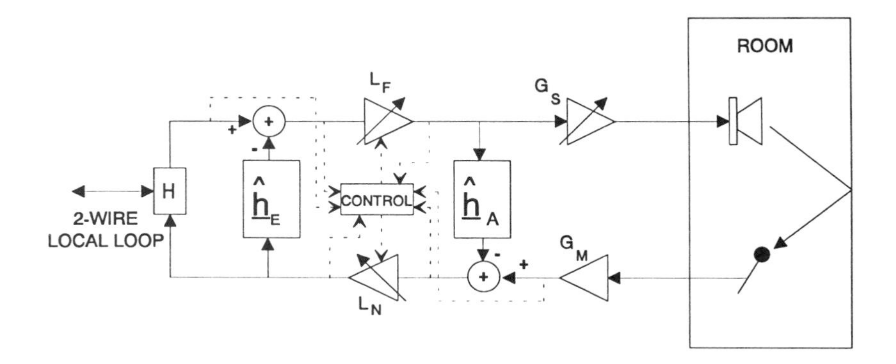

Figure 1.3 Speakerphone with suppression and echo cancellation.

addition to increasing the level of the signal coming directly from the near-end speaker, the speakerphone's microphone amplifier increases the level of the reverberation and the background noise of the room as well.

The speech level at the microphone from a speaker 18 inches away is about 70 dB SPL [11], while noise level in a moderately noisy environment is less than 60 dB SPL. These levels guarantee intelligibility of the speech received at the microphone [12], however, as the near-end speaker moves farther away from the microphone, intelligibility will decrease.

Figure 1.3 shows a simplified block diagram of a 2-wire speakerphone with echo suppression and both acoustic and electrical echo cancelers. The electrical echo cancelers enhance the loss across the hybrid, providing singing margin in closed loop within the speakerphone. Once the electrical echo cancelers have converged they will more or less stay converged since the electrical echo path does not change significantly during the call.

It is desirable to avoid echo suppression and, as much as possible, rely on echo cancellation for singing and echo control. However, some echo suppression must be used in practical speakerphones to provide a minimal amount of singing margin at the beginning of a call, before the speech of the near and far-end talkers have had a chance to train the echo cancelers. In the speakerphone-to-speakerphone connection, suppression also prevents station-to-station singing during periods when the acoustic echo paths of both sides have changed and the acoustic echo cancelers have not yet had a chance to reconverge.

4. ADAPTIVE FILTERS FOR ECHO CANCELLATION

This section briefly reviews methods of adaptive filtering as applicable to the problem of echo cancellation.

As the name implies, adaptive filters are filters with coefficients which are modified periodically in order to attempt to meet some performance criterion. That performance criterion is usually in the form of some error or cost function minimization. Most fundamental is the maximum a posteriori (MAP) performance measure, which attempts to find the most likely set of inputs (echo path impulse response) given the observed signals (excitation signal and echo plus noise). Unfortunately MAP requires a priori knowledge of the probabilities of the inputs which are not available. A related method is the maximum likelihood (ML) criterion which selects the set of inputs that maximize the probability of obtaining the observed outputs given the trial set of inputs. ML is derived by expanding MAP using Baye's rule and simply ignoring the a priori input probabilities. Hence, ML's one advantage over MAP – it does not require foreknowledge of the input probabilities. ML and MAP are equivalent when the input probability density function is uniform. When the excitation signal and noise corrupting the echo are Gaussian (a typical simplifying assumption) and uncorrelated, it can be shown that minimizing the magnitude square of the error vector between the predicted echo (from the estimated coefficients) and the observed echo, the method of least squares (LS) is equivalent to optimizing the ML criterion. Finally, if the first and second order statistics of the excitation and echo are known, then minimizing the *expected* error yields yet another estimate of the system impulse response known as the *Wiener* solution. If the involved signals are ergodic then the LS solution is asymptotically convergent to the Wiener solution as the order of the error vector grows.

In the following, we discuss the least mean square (LMS), normalized least mean square (NLMS), and recursive least squares (RLS) algorithms.

4.1 THE LMS AND NLMS ALGORITHMS

The least mean square (LMS) algorithm [13] is the most commonly used adaptive filtering algorithm in the world today. It is used in system identification problems (including, of course, echo cancellation), channel equalization problems, and speech coding. Though its speed of convergence is rather slow, it remains popular because of its robust performance, simplicity, and low cost.

Let us make the following definitions:

1. $\mathbf{x}_n = [x_n, x_{n-1}, \ldots, x_{n-L+1}]^t$ is the *excitation* or *tap-delay line* vector.
2. s_n is the *desired* or the *system output* signal. It is the echo plus any additional signal added in the echo path such as *near-end* speech or background noise.
3. $\mathbf{h}_n = [h_{0,n}, h_{1,n}, \ldots, h_{L-1,n}]^t$ where $h_{i,n}$ is the ith adaptive *tap weight* or *coefficient* at time n .
4. e_n is the *a priori* error (or simply, the "error") or the *residual echo.*

The error signal at time n is

$$e_n = s_n - \mathbf{x}_n^t \mathbf{h}_{n-1}. \tag{1.1}$$

The performance index for LMS is the expectation of the squared error,

$$J(\mathbf{h}) = E\{e_n^2\} = E\{(s_n - \mathbf{x}_n^t \mathbf{h})^2\}. \tag{1.2}$$

Setting the derivative of this with respect to the coefficients $\mathbf{h}$ to zero yields the equations:

$$\mathbf{R}_{xx}\mathbf{h} = \mathbf{r}_{sx}, \tag{1.3}$$

where

$$\mathbf{R}_{xx} = E\{\mathbf{x}_n \mathbf{x}_n^t\} \tag{1.4}$$

and

$$\mathbf{r}_{sx} = E\{s_n \mathbf{x}_n\}. \tag{1.5}$$

Assuming that the inverse of $\mathbf{R}_{xx}$ exists, the optimal Wiener solution is:

$$\mathbf{h}_{\text{opt}} = \mathbf{R}_{xx}^{-1} \mathbf{r}_{sx}. \tag{1.6}$$

Now, given an L-length coefficient vector at sample period $n-1$, $\mathbf{h}_{n-1}$ we want to update this vector with another L-length vector, $\tilde{\mathbf{h}}_n$,

$$\mathbf{h}_n = \mathbf{h}_{n-1} + \tilde{\mathbf{h}}_n, \tag{1.7}$$

where $\tilde{\mathbf{h}}_n$ is chosen so as to decrease the value of $J(\mathbf{h})$. That is, we want that

$$J(\mathbf{h}_n) = J(\mathbf{h}_{n-1} + \tilde{\mathbf{h}}_n) \leq J(\mathbf{h}_{n-1}). \tag{1.8}$$

If the update vector $\tilde{\mathbf{h}}_n$ is chosen small enough, then using a Taylor series expansion we have:

$$J(\mathbf{h}_{n-1} + \tilde{\mathbf{h}}_n) \approx J(\mathbf{h}_{n-1}) + \tilde{\mathbf{h}}_n^t \left. \frac{\partial J(\mathbf{h})}{\partial \mathbf{h}} \right|_{\mathbf{h}=\mathbf{h}_{n-1}}. \tag{1.9}$$

Using this in the inequality of (1.8) (and keeping in mind that $J(\mathbf{h}_n)$ is never negative) yields the desired inequalities

$$0 \leq J(\mathbf{h}_{n-1}) + \tilde{\mathbf{h}}_n^t \left. \frac{\partial J(\mathbf{h})}{\partial \mathbf{h}} \right|_{\mathbf{h}=\mathbf{h}_{n-1}} \leq J(\mathbf{h}_{n-1}) \tag{1.10}$$

which are satisfied if $\tilde{\mathbf{h}}_n$ is chosen to be proportional to the negative gradient of $J(\mathbf{h})$ evaluated at $\mathbf{h}_{n-1}$,

$$\tilde{\mathbf{h}}_n = -\mu \left. \frac{\partial J(\mathbf{h})}{\partial \mathbf{h}} \right|_{\mathbf{h}=\mathbf{h}_{n-1}}, \tag{1.11}$$

where the constant of proportionality, μ is called the *stepsize*. Then, from (1.2) and (1.11) we have

$$\begin{aligned} \tilde{\mathbf{h}}_n &= 2\mu E\{\mathbf{x}_n e_n\} & (1.12) \\ &= 2\mu(\mathbf{r}_{sx} - \mathbf{R}_{xx}\mathbf{h}_{n-1}), & (1.13) \end{aligned}$$

where we have used (1.1) in the last step. In general, however, a priori knowledge of $\mathbf{R}_{xx}$ and $\mathbf{r}_{sx}$ is not available, so, for LMS, the *stochastic approximation* is made, that is, the expectation operation in (1.12) is simply ignored. Using this in (1.7) yields the LMS coefficient update

$$\mathbf{h}_n = \mathbf{h}_{n-1} + 2\mu \mathbf{x}_n e_n. \tag{1.14}$$

The average coefficient vector convergence is easily derived [14] by first restoring the expectations. (The derivation of the coefficient vector correlation matrix convergence [15] is quite a bit more complicated and is not essential for this brief introduction.) So, using (1.13) in (1.7), solving the resulting difference equation and applying (1.6), yields,

$$\mathbf{h}_n = \mathbf{h}_{\text{opt}} + [\mathbf{I} - 2\mu\mathbf{R}_{xx}]^n[\mathbf{h}_0 - \mathbf{h}_{\text{opt}}]. \tag{1.15}$$

Let $\lambda_{xx,i}$ represent the eigenvalues of $\mathbf{R}_{xx}$. Then from (1.15) one can see that $\mathbf{h}_n - \mathbf{h}_{\text{opt}}$ tends to zero as do the eigen-modes of (1.15), that is, as,

$$[1 - 2\mu\lambda_{xx,i}]^n \longrightarrow 0. \tag{1.16}$$

So, for $\mathbf{h}_n$ to converge to $\mathbf{h}_{\text{opt}}$, we must have, for $i = 1, \ldots, L$,

$$0 < \mu < 1/\lambda_{xx,i}. \tag{1.17}$$

Since $\mathbf{R}_{xx}$ is positive definite, all its eigenvalues are positive. So, the maximum eigenvalue of $\mathbf{R}_{xx}$ is less than the sum of its eigenvalues. The sum of the eigenvalues is equal to the trace which in this case is just L times the variance of x_n, σ_x^2. Therefore, if we choose μ such that

$$0 < \mu < \frac{1}{L\sigma_x^2} \tag{1.18}$$

then the convergence of the average of $\mathbf{h}_n$ is guaranteed.

From (1.16) we note that once the value of μ is determined, modes with smaller eigenvalues will converge slower than those with large ones. This is why LMS converges slowly when excited by highly colored signals like speech.

Further analysis indicates that when there is very little noise, LMS converges fastest when μ is in the middle of the range indicated by (1.18). To effectively use this fact, the step size must be a function of σ_x^2. The result is called the *normalized* LMS coefficient update:

$$\mathbf{h}_n = \mathbf{h}_{n-1} + \frac{\alpha}{L\sigma_x^2}\mathbf{x}_n e_n. \tag{1.19}$$

Comparing (1.14) to (1.19) we have changed the constant terms in the coefficient update from 2μ to $\frac{\alpha}{L\sigma_x^2}$. The 2 in the LMS update has been subsumed into the constant α and the factors $L\sigma_x^2$ appear in the denominator. If the above stability analysis is carried out for (1.19) rather than for (1.14) we find that (1.15) becomes

$$\mathbf{h}_n = \mathbf{h}_{\text{opt}} + [\mathbf{I} - \frac{\alpha}{L\sigma_x^2}\mathbf{R}_{xx}]^n[\mathbf{h}_0 - \mathbf{h}_{\text{opt}}] \tag{1.20}$$

and (1.18) becomes

$$0 < \alpha < 2. \tag{1.21}$$

From (1.20) and (1.21) it is apparent that both the speed of convergence and the range of stability for α are independent of the excitation signal's power for NLMS. This, of course, does not fix the slow convergence of those modes with small eigenvalues.

For non-stationary signals such as speech, the estimate of σ_x^2 must be made time varying. A good way of doing this is to replace $L\sigma_x^2$ with $\mathbf{x}_n^t\mathbf{x}_n$. This is easily updated each sample period using the recursion

$$\mathbf{x}_n^t\mathbf{x}_n = \mathbf{x}_{n-1}^t\mathbf{x}_{n-1} + x_n^2 - x_{n-L}^2. \tag{1.22}$$

So far, we have developed NLMS as a modification of the LMS algorithm made to accommodate non-stationary signals. While this is a perfectly valid interpretation, another one that lends itself to generalization (and also improved performance) is NLMS as an affine projection. This is discussed in Chapter 2.

4.2 LEAST SQUARES AND RECURSIVE LEAST SQUARES ALGORITHMS

In the method of least squares (LS) we start with the *a posteriori* error vector,

$$\mathbf{e}_{1,n} = \mathbf{s}_n - \mathbf{X}_n^t\mathbf{h}, \tag{1.23}$$

where for the moment we leave undefined the exact bounds on the samples comprising the matrices and vectors except to say that, at least after the first N sample periods, N, the length of the error vector is greater than, L, the length of the filter. Note that the 1 in the subscript of $\mathbf{e}_{1,n}$ denotes the a posteriori error as opposed to the a priori error.

The LS performance index is

$$J_{\text{LS}} = \mathbf{e}_{1,n}^t \mathbf{W} \mathbf{e}_{1,n}, \tag{1.24}$$

where $\mathbf{W}$ serves to weight the a posteriori errors and is diagonal with elements $w_n, \ldots, w_{n-N+1}$. Using (1.23) in (1.24) and setting the derivative of J_{LS} to zero we get the *deterministic* normal equations,

$$\mathbf{X}_n \mathbf{W} \mathbf{X}_n^t \mathbf{h} = \mathbf{X}_n \mathbf{W} \mathbf{s}_n, \tag{1.25}$$

whose solution is:

$$\mathbf{h}_{\text{LS}} = [\mathbf{X}_n \mathbf{W} \mathbf{X}_n^t]^{-1} \mathbf{X}_n \mathbf{W} \mathbf{s}_n. \tag{1.26}$$

The L by L matrix $\mathbf{X}_n \mathbf{W} \mathbf{X}_n^t$ is a scaled time averaged version of the covariance matrix $\mathbf{R}_{xx}$. Similarly, the L length vector $\mathbf{X}_n \mathbf{W} \mathbf{s}_n$ is a scaled time average version of the cross covariance vector $\mathbf{r}_{sx}$. As mentioned earlier, if the excitation signal x_n and the noise in the echo path y_n are gaussian and uncorrelated, then as N grows, the solution $\mathbf{h}_{\text{LS}}$ approaches the Wiener solution, $\mathbf{h}_{\text{opt}}$, asymptotically.

We now address more specifically the bounds on the vectors and matrices of (1.23). The definitions of the principal vectors and matrices are:

$$\mathbf{x}_0 = [x_0, x_{-1}, \ldots, x_{-L+1}]^t, \tag{1.27}$$

$$\ldots$$

$$\mathbf{x}_n = [x_n, x_{n-1}, \ldots, x_{n-L+1}]^t, \tag{1.28}$$

$$\mathbf{X}_n = [\mathbf{x}_n, \mathbf{x}_{n-1}, \ldots, \mathbf{x}_{n-N+1}], \tag{1.29}$$

$$\mathbf{s}_n = [s_n, s_{n-1}, \ldots, s_{n-N+1}]^t, \tag{1.30}$$

$$\mathbf{e}_{1,n} = [e_{1,n}, e_{1,n-1}, \ldots, e_{1,n-N+1}]^t. \tag{1.31}$$

Different assumptions about data windows and initializations of the LS algorithm lead to different *recursive* LS methods which we will shortly address. A partial list of these methods is as follows:

1. The *prewindowing* method assumes that $x_n = 0$ for $n < 0$, $N = n + 1$, and $w_{n-i} = 1$ for $0 \leq i \leq n$.

2. The *covariance* method allows x_n to take on non-zero values for $n < 0$, sets $N = n + 1$, and $w_{n-i} = 1$ for $0 \leq i \leq n$.

3. The *exponentially windowed* method assumes that $x_n = 0$ for $n < 0$, sets $N = n + 1$, and $w_{n-i} = \lambda^{n-i}$ for $0 \leq i \leq n$.

4. The *sliding windowed* method assumes that $x_n = 0$ for $n < 0$, fixes N to some value greater than L, and $w_{n-i} = 1$ for $0 \leq i \leq N - 1$.

When $n < L - 1$ the LS solution as defined in (1.26) is undefined since the rank of $\mathbf{X}_n\mathbf{X}_n^t$ is at most $n + 1$. There are a couple of ways around this problem. One is to set $L = n + 1$ while in this initial transient phase and only fix L when it reaches the desired value. This is called *exact* initialization.

Another approach is to use the method of *soft constraints*. Here, the performance index of (1.24) is replaced with

$$J_{\mathrm{LS}} = \delta[\mathbf{h}_n - \mathbf{h}_0]^t\hat{\mathbf{W}}[\mathbf{h}_n - \mathbf{h}_0] + \mathbf{e}_n^t\mathbf{W}\mathbf{e}_n, \tag{1.32}$$

where $\hat{\mathbf{W}} = \mathrm{diag}\{1, \lambda, ..., \lambda^{L-1}\}$. The minimization of (1.32) leads to the solution

$$\mathbf{h}_{\mathrm{opt}} = [\mathbf{X}_n\mathbf{W}\mathbf{X}_n^t + \delta\hat{\mathbf{W}}]^{-1}[\mathbf{X}_n\mathbf{W}\mathbf{s}_n + \delta\hat{\mathbf{W}}\mathbf{h}_0]. \tag{1.33}$$

The parameter δ is a soft constraint on the performance index which gives more (when δ is *large*) or less (when δ is *small*) weight to the initial values of the coefficients $\mathbf{h}_0$. It also serves to *regularize* the inverse in (1.33) by making the matrix $\mathbf{X}_n\mathbf{W}\mathbf{X}_n^t + \delta\hat{\mathbf{W}}$ non-singular.

The prewindowed method is preferred in cases where the excitation signal is stationary and the unknown system is fixed. In cases where neither of these assumptions is valid, e.g. voice excited acoustic echo cancellation, the exponential or sliding windowed methods allow the filter to forget or eliminate consideration of errors that occurred too far in the past. The price for this forgetfulness is a decrease in fidelity in the estimation of $\mathbf{h}_{\mathrm{LS}}$.

The direct solution of (1.33) has $O(L^3)$ complexity, making it an impractical method for echo path lengths of any reasonable size. This is particularly true for adaptive filters where the coefficients are updated each sample period. In recursive least squares, RLS, the complexity is reduced to $O(L^2)$ operations per sample period. RLS exploits the following observations:

1. The matrix

$$\mathbf{R}_n = \mathbf{X}_n\mathbf{W}\mathbf{X}_n^t \tag{1.34}$$

can be updated from sample to sample period with two rank one matrix additions (or equivalently one rank-two addition),

$$\mathbf{R}_n = \lambda \mathbf{R}_{n-1} + \mathbf{x}_n \mathbf{x}_n^t - \lambda^N \mathbf{x}_{n-N} \mathbf{x}_{n-N}^t. \tag{1.35}$$

2. The vector

$$\mathbf{r}_n = \mathbf{X}_n \mathbf{W} \mathbf{s}_n \tag{1.36}$$

can be similarly updated,

$$\mathbf{r}_n = \lambda \mathbf{r}_{n-1} + \mathbf{x}_n \mathbf{s}_n - \lambda^N \mathbf{x}_{n-N} \mathbf{s}_{n-N}. \tag{1.37}$$

Depending on the choice of N and λ the updates of R_n and r_n can satisfy those required for any of the above methods.

1. For the prewindowed and covariance case, $\lambda = 1$ and $N = n + 1$, and since $x_n = 0$ for $n < 0$, updates (1.35) and (1.37) become rank one updates.
2. For the exponential windowed case, $0 < \lambda < 1$ and $N = n + 1$, once again resulting in rank-one updates.
3. For the sliding windowed case, $\lambda = 1$ and N is fixed, resulting in a rank-two update.

For now, we concentrate on the exponentially windowed case. Then, (1.35) and (1.37) become

$$\mathbf{R}_n = \lambda \mathbf{R}_{n-1} + \mathbf{x}_n \mathbf{x}_n^t \tag{1.38}$$

and

$$\mathbf{r}_n = \lambda \mathbf{r}_{n-1} + \mathbf{x}_n \mathbf{s}_n \tag{1.39}$$

respectively.

The matrix inversion (Woodbury's) lemma states

$$\mathbf{A}^{-1} = \mathbf{B}^{-1} - \mathbf{B}^{-1}\mathbf{C}\left[\mathbf{I} + \mathbf{D}\mathbf{B}^{-1}\mathbf{C}\right]^{-1}\mathbf{D}\mathbf{B}^{-1}, \tag{1.40}$$

where $\mathbf{A} = \mathbf{B} + \mathbf{CD}$ and $\mathbf{B}$ are assumed to be non-singular. Applying this to (1.37) yields

$$\mathbf{R}_n^{-1} = \lambda^{-1}\mathbf{R}_{n-1}^{-1} - \lambda^{-2}\mathbf{R}_{n-1}^{-1}\mathbf{x}_n \left[1 + \lambda^{-1}\mathbf{x}_n^t \mathbf{R}_{n-1}^{-1}\mathbf{x}_n\right]^{-1} \mathbf{x}_n^t \mathbf{R}_{n-1}^{-1}. \tag{1.41}$$

We now define the *a priori Kalman gain vector*,

$$\mathbf{k}_{0,n} = \mathbf{R}_{n-1}^{-1}\mathbf{x}_n, \tag{1.42}$$

the *a posteriori Kalman gain vector*,

$$\mathbf{k}_{1,n} = \mathbf{R}_n^{-1}\mathbf{x}_n, \tag{1.43}$$

and the *likelihood variable*,

$$\gamma_n = \left[1 + \frac{1}{\lambda}\mathbf{x}_n^t\mathbf{R}_{n-1}^{-1}\mathbf{x}_n\right]^{-1} = \left[1 + \frac{1}{\lambda}\mathbf{k}_{0,n}^t\mathbf{x}_n\right]^{-1}. \tag{1.44}$$

Using these definitions we can rewrite (1.41) as

$$\mathbf{R}_n^{-1} = \frac{1}{\lambda}\mathbf{R}_{n-1}^{-1} - \frac{\gamma_n}{\lambda^2}\mathbf{k}_{0,n}\mathbf{k}_{0,n}^t. \tag{1.45}$$

Using (1.44) and (1.45) in (1.43) we can express the a posteriori Kalman gain vector in terms of the a priori Kalman gain vector and the likelihood variable as

$$\begin{aligned}\mathbf{k}_{1,n} &= \frac{\gamma_n}{\lambda}\mathbf{k}_{0,n} \\ &= \frac{\lambda^{-1}\mathbf{R}_{n-1}^{-1}\mathbf{x}_n}{1 + \lambda^{-1}\mathbf{x}_n^t\mathbf{R}_{n-1}^{-1}\mathbf{x}_n}.\end{aligned} \tag{1.46}$$

Incidentally, applying (1.46) to (1.42) allows us to express $\mathbf{R}_n^{-1}$ in yet another manner,

$$\mathbf{R}_n^{-1} = \frac{1}{\lambda}\mathbf{R}_{n-1}^{-1} - \frac{1}{\lambda}\mathbf{k}_{1,n}\mathbf{x}_n^t\mathbf{R}_{n-1}^{-1}. \tag{1.47}$$

The best least squares filter coefficients, $\mathbf{h}_{\mathrm{LS,n}}$ can be found each sample period, assuming $\delta = 0$, by using (1.33) and definitions (1.34) and (1.36) expanded using (1.45) and (1.42), respectively,

$$\mathbf{h}_{\mathrm{LS,n}} = \mathbf{h}_{\mathrm{LS,n-1}} + \mathbf{k}_{1,n}e_{0,n}, \tag{1.48}$$

where $e_{0,n}$ is defined as the *a priori error*,

$$e_{0,n} = s_n - \mathbf{x}_n^t\mathbf{h}_{\mathrm{LS,n-1}}. \tag{1.49}$$

Bringing equations (1.46), (1.49), (1.48), and (1.47) together we have the RLS algorithm.

Finally, the RLS coefficient update can be expressed as

$$\mathbf{h}_{\mathrm{LS,n}} = \mathbf{h}_{\mathrm{LS,n-1}} + \mathbf{R}_n^{-1}\mathbf{x}_n e_{0,n}. \tag{1.50}$$

This is similar to the coefficient update of LMS as expressed in (1.14) except that the constant 2μ has been generalized to the matrix $\mathbf{R}_n^{-1}$. Assuming stationarity, from (1.38) the steady state, we can write:

$$\mathbf{R}_n = \frac{1}{1-\lambda}\mathbf{R}_{xx}. \tag{1.51}$$

So, substituting $\left[\frac{1}{1-\lambda}\mathbf{R}_{xx}\right]^{-1}$ for 2μ we have, as an expression of the first order convergence of the RLS coefficients,

$$\begin{aligned} \mathbf{h}_{\mathrm{LS,n}} &= \mathbf{h}_{\mathrm{opt}} + \left[\mathbf{I} - (1-\lambda)\mathbf{R}_{xx}^{-1}\mathbf{R}_{xx}\right]\left[\mathbf{h}_0 - \mathbf{h}_{\mathrm{opt}}\right] \quad (1.52) \\ &= \mathbf{h}_{\mathrm{opt}} + \lambda^n \left[\mathbf{h}_0 - \mathbf{h}_{\mathrm{opt}}\right]. \quad (1.53) \end{aligned}$$

So, the first order convergence of $\mathbf{h}_{\mathrm{LS,n}}$ in completely independent of the eigenstructure of $\mathbf{R}_{xx}$ and all modes of $\mathbf{h}_{\mathrm{LS,n}}$ converge at the rate of λ^n, a considerable improvement over LMS.

The complexity of RLS is in $O(L^2)$. Fast RLS (FRLS) algorithms yield the same convergence with only $O(L)$ complexity. These algorithms will be covered in Chapter 6.

5. NOISE REDUCTION

The enhancement of speech signals disturbed by additive noise is particularly important in hands-free communication (the higher the distance between the microphone and the talker, the worse the signal-to-noise ratio – SNR), especially in cars where the level of noise is very high. The most common methods for noise reduction suppose that the acoustic noise and speech are picked up by one microphone. They are mostly based on spectral magnitude subtraction where the short-time spectral amplitude of noise is estimated during speech pauses and subtracted from the noisy microphone signal [16]. This implies the use of a voice activity detector (VAD) to determine at every frame whether there is speech present in that frame. It is clear that the performance of the methods depend a great deal on the efficacy of the VAD. There are, however, new methods based on minimum statistics where a VAD is not required anymore [17].

Consider the disturbed signal:

$$\mathbf{x}_n = \mathbf{s}_n + \mathbf{v}_n, \quad (1.54)$$

where $\mathbf{s}_n$ and $\mathbf{v}_n$ denote respectively, the speech signal and the noise. Using the discrete Fourier transform, an estimation of the noise-free speech spectrum at the frame number m is:

$$\hat{S}_m(\omega) = [|X_m(\omega)| - |\hat{V}_m(\omega)|]e^{j\varphi_m(\omega)}, \quad (1.55)$$

where $|X_m(\omega)|$ and $|\hat{V}_m(\omega)|$ are respectively the spectral magnitude of the noisy speech and a spectral magnitude estimate of the noise, ω is the frequency index, and $\varphi_m(\omega)$ is the phase of the disturbed speech signal. Note that the phase of the signal is not processed in general and the original noisy signal phase is used to reconstruct the time domain enhanced speech signal. Mathematically, the principles of spectral subtraction based methods are rather simple. However, they introduce a distortion known as *musical tones*. (See Chapter 9 for more details.)

6. CONCLUSIONS

As we have seen and as will be shown in the next chapters, the estimation of the impulse response between a source (a loudspeaker or a human talker) and a sensor (a microphone or a human ear) plays a key role in problems and solutions of most hands-free applications. For example, in the acoustic echo problem where a reference signal is accessible, a *good* estimation of the impulse response between the loudspeaker and the microphone is *relatively* easy; as a result, a satisfactory solution to this problem can be derived. But for noise suppression and dereverberation problems, no reference signal is available in general and the estimation of the impulse response of interest is much more challenging. As a result, an adequate solution is hard to obtain.

This chapter has provided a brief introduction to two very important problems that occur in hands-free communications and some of the background information of acoustic signal processing for telecommunication. The subsequent chapters will deal with these issues in greater detail and discuss many more problems and new innovative solutions.

References

[1] M. M. Sondhi and D. A. Berkley, "Silencing echoes on the telephone network," *Proc. of the IEEE*, vol. 68, pp. 948-963, Aug. 1980.

[2] H. R. Huntly, "Transmission design of intertoll telephone trunks," *Bell Syst. Tech. J.*, vol. 32, pp. 1019-1036, Sep. 1953.

[3] G. Williams and L. S. Moye, "Subjective evaluation of unsuppressed echo in simulated long-delay telephone communications," *Proc Inst. Elec. Eng.*, vol. 118, pp. 401-408, 1971.

[4] R. R. Riesz and E. T. Klemmer, "Subjective evaluation of delay and echo suppressors in telephone communications," *Bell Syst. Tech. J.*, vol. 42, pp. 2919-2943, 1963.

[5] H. Kuttruff, *Room Acoustics*. Applied Science Publishers Ltd., London, UK, 2nd edition, 1973.

[6] R. G. Gould and G. K. Helder, "Transmission delay and echo suppression," *IEEE Spectrum*, pp. 47-54, Apr. 1970.

[7] A. B. Clark and R. C. Mathes, "Echo suppressors for Long telephone circuits," *Proc. AIEE*, vol. 44, pp. 481-490, Apr. 1925.

[8] M. M. Sondhi and A. J. Presti, "A self-adaptive echo canceler," *Bell Syst. Tech. J.*, vol. 45, pp. 1851-1854, 1966.

[9] M. M. Sondhi, "An adaptive echo canceler," *Bell Syst. Tech. J.*, vol. 46, pp. 497-511, Mar. 1967.

[10] F. K. Becker and H. R. Rudin, "Application of automatic transversal filters to the problem of echo suppression," *Bell Syst. Tech. J.*, vol. 45, pp. 1847-1850, 1966.

[11] W. F. Clemency and W. D. Goodale, Jr., "Functional design of a voice-switched speakerphone," *Bell Syst. Tech. J.*, vol. XL, pp. 649-668, May 1961.

[12] J. P. A. Lochner and J. F. Burger, "The intelligibility of speech under reverberant conditions," *Acustica*, vol. 11, pp. 195-200, 1961.

[13] B. Widrow and S. D. Stearns, *Adaptive Signal Processing*. Prentice-Hall Inc., Englewood Cliffs, N.J., 1985.

[14] S. J. Orfanidis, *Optimum Signal Processing: An Introduction*. Second Edition, Macmillan Publishing Co., New York, N.Y., 1988.

[15] S. Haykin, *Adaptive Filter Theory*. Third Edition, Prentice Hall, Englewood Cliffs, N.J., 1996.

[16] J. S. Lim and A. V. Oppenheim, "Enhancement and bandwidth compression of noisy speech," *Proc. of the IEEE*, vol. 67, pp. 1586-1604, Dec. 1979.

[17] R. Martin, "Spectral subtraction based on minimum statistics," in *Proc. EUSIPCO*, 1994, pp. 1182-1185.

I

MONO-CHANNEL ACOUSTIC ECHO CANCELLATION

Chapter 2

THE FAST AFFINE PROJECTION ALGORITHM

Steven L. Gay
Bell Labs, Lucent Technologies
slg@bell-labs.com

Abstract This chapter discusses an adaptive filtering algorithm called fast affine projections (FAP). FAP's key features include LMS like complexity and memory requirements (low), and RLS like convergence (fast) for the important case where the excitation signal is speech. Another of FAP's important features is that it causes no delay in the input or output signals. In addition, the algorithm is easily regularized resulting in robust performance even for highly colored excitation signals. The combination of these features make FAP an excellent candidate for the adaptive filter in the acoustic echo cancellation problem. A simple, low complexity numerical stabilization method for the algorithm is also introduced.

Keywords: LMS, NLMS, RLS, FRLS, FAP, Affine Projection, Acoustic Echo, Cancellation

1. INTRODUCTION

The affine projection algorithm (APA) [1] is a generalization of the well known normalized least mean square (NLMS) adaptive filtering algorithm [2]. Under this interpretation, each tap weight vector update of NLMS is viewed as a one dimensional affine projection. In APA the projections are made in multiple dimensions. As the projection dimension increases, so does the convergence speed of the tap weight vector, and unfortunately, the algorithm's computational complexity. Using techniques similar to those which led to fast (i.e., computationally efficient) recursive least squares (FRLS) [3] from recursive least squares (RLS) [4] a fast version of APA, fast affine projections (FAP) may be derived [5, 6, 7].

As with RLS and FRLS, FAP requires the solution to a system of equations involving the implicit inverse of the excitation signal's covariance matrix. Although, with FAP the dimension of the covariance matrix is the dimension of the projection, N, not the length of the joint process estimation, L. This is advantageous because usually N is much smaller than L. FAP uses a sliding windowed FRLS [8] to assist in a recursive calculation of the solution. Since sliding windowed FRLS algorithms easily incorporate regularization of the covariance matrix inverse, FAP is regularized as well, making it robust to measurement noise. The complexity of FAP is roughly $2L + 20N$ multiplications per sample period. For applications like acoustic echo cancellation, FAP's complexity is comparable to NLMS's ($2L$ multiplications per sample period). Moreover, FAP does not require significantly greater memory than NLMS.

2. THE AFFINE PROJECTION ALGORITHM

The affine projection algorithm, in a general form, is defined by the following two equations:

$$\mathbf{e}_n = \mathbf{s}_n - \mathbf{X}_n^t \mathbf{h}_{n-1}, \tag{2.1}$$

$$\mathbf{h}_n = \mathbf{h}_{n-1} + \mu \mathbf{X}_n \left[\mathbf{X}_n^t \mathbf{X}_n\right]^{-1} \mathbf{e}_n, \tag{2.2}$$

where the superscript t denotes transpose and the following definitions are made:

1. x_n is the *excitation* signal and n is the time index.

2. $$\mathbf{x}_n = [x_n, x_{n-1}, \dots, x_{n-L+1}]^t \tag{2.3}$$

   is the L length excitation or *tap-delay line* vector.

3. $$\alpha_n = [x_n, x_{n-1}, \dots, x_{n-N+1}]^t \tag{2.4}$$

   is the N length excitation vector.

4. $$\mathbf{X}_n = \left[\mathbf{x}_n, \mathbf{x}_{n-1}, \dots, \mathbf{x}_{n-N+1}\right] = \begin{bmatrix} \alpha_n^t \\ \alpha_{n-1}^t \\ \vdots \\ \alpha_{n-L+1}^t \end{bmatrix} \tag{2.5}$$

   is the L by N excitation matrix.

5. $$\mathbf{h}_n = [h_{0,n}, h_{1,n}, \dots, h_{L-1,n}]^t \tag{2.6}$$

is the L length adaptive coefficient vector where $h_{i,n}$ is the ith adaptive *tap weight* or *coefficient* at time n.

6.

$$\mathbf{h}_{ep} = [h_{0,ep}, h_{1,ep}, \ldots, h_{L-1,ep}]^t \tag{2.7}$$

is the L length echo path impulse response vector where $h_{i,ep}$ is the ith tap weight or coefficient.

7. y_n is the measurement noise signal. In the language of echo cancellation it is the *near-end signal* which consists of the near-end talker's voice and/or back ground noise.

8.

$$\mathbf{y}_n = [y_n, y_{n-1}, \ldots, y_{n-N+1}]^t \tag{2.8}$$

is the N length near-end signal vector.

9.

$$\mathbf{s}_n = [s_n, s_{n-1}, \ldots, s_{n-N+1}]^t = \mathbf{X}_n^t \mathbf{h}_{ep} + \mathbf{y}_n \tag{2.9}$$

is the N length *desired* or the *system output* vector. Its elements consist of the echo plus any additional signal added in the echo path.

10.

$$e_n = s_n - \mathbf{x}_n^t \mathbf{h}_n \tag{2.10}$$

is the *a priori* error signal or the *residual echo.*

11.

$$\mathbf{e}_n = [e_n, e_{n-1}, \ldots, e_{n-N+1}]^t \tag{2.11}$$

is the N length *a priori* error vector.

12. μ is the *relaxation* or *step-size* parameter.

If we set N to one in (2.1) and (2.2) we get

$$e_n = s_n - \mathbf{x}_n^t \mathbf{h}_{n-1}, \tag{2.12}$$

$$\mathbf{h}_n = \mathbf{h}_{n-1} + \mu \mathbf{x}_n \left[\mathbf{x}_n^t \mathbf{x}_n\right]^{-1} e_n, \tag{2.13}$$

which is the familiar NLMS algorithm. Thus, we see that APA is a generalization of NLMS.

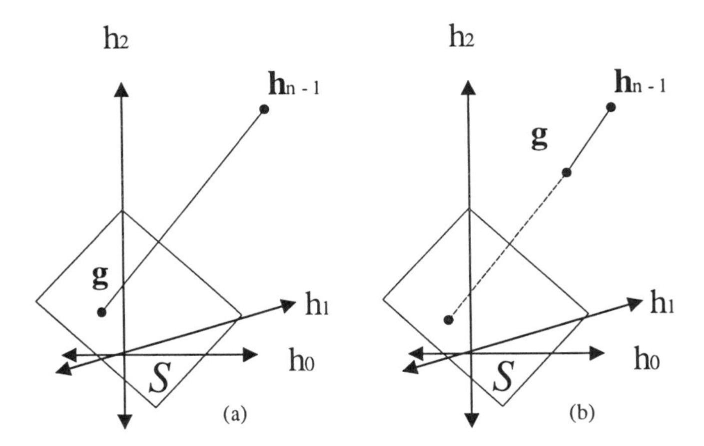

Figure 2.1 (a) Projection onto a linear subspace. (b) Relaxed projection onto a linear subspace.

2.1 PROJECTIONS ONTO AN AFFINE SUBSPACE

We now show that the APA as expressed in (2.1) and (2.2) indeed represents a projection onto an affine subspace. In Fig. 2.1(a) we show the projection of a vector, $\mathbf{h}_{n-1}$ onto a linear subspace, S, where we have a space dimension of $L = 3$ and a subspace dimension of $L - N = 2$. Note that an $L - N$ dimensional linear subspace is a subspace spanned by any linear combination of $L - N$ vectors. One of those combinations is where all of the coefficients are zero; so a linear subspace always includes the origin. Algebraically, we represent the projection as,

$$\mathbf{g} = \mathbf{Q}\mathbf{h}_{n-1}, \tag{2.14}$$

where $\mathbf{Q}$ is a projection matrix of the form

$$\mathbf{Q} = \mathbf{U} \begin{bmatrix} 0 & 0 & 0 \\ 0 & 1 & 0 \\ 0 & 0 & 1 \end{bmatrix} \mathbf{U}^t \tag{2.15}$$

and $\mathbf{U}$ is a unitary matrix (i.e. a rotation matrix). In general, the diagonal matrix in (2.15) has N zeros and $L - N$ ones along its diagonal.

Figure 2.1(b) shows a relaxed projection. Here, $\mathbf{g}$ ends up only part way between $\mathbf{h}_{n-1}$ and S. The relaxed projection is still represented by (2.14), but

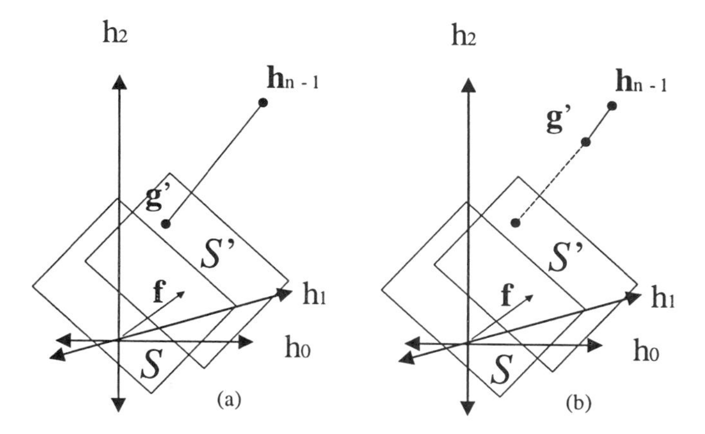

Figure 2.2 (a) Projection onto an affine subspace. (b) Relaxed projection onto an affine subspace.

with

$$\mathbf{Q} = \mathbf{U}\begin{bmatrix} 1-\mu & 0 & 0 \\ 0 & 1 & 0 \\ 0 & 0 & 1 \end{bmatrix}\mathbf{U}^t. \tag{2.16}$$

An affine subspace, S', as shown in Fig. 2.2(a), is defined as a subspace parallel to a linear subspace, offset by a perpendicular vector, $\mathbf{f}$. Note that the affine subspace does not include the origin. Algebraically, the projection onto the affine subspace is represented as

$$\mathbf{g}' = \mathbf{Q}\mathbf{h}_{n-1} + \mathbf{f}, \tag{2.17}$$

where $\mathbf{f}$ is in the null space of $\mathbf{Q}$; that is, $\mathbf{Qf}$ equal an all zero vector. Figure 2.2(b) shows a relaxed projection onto the affine subspace. As before, $\mu = 1/3$.

Manipulating equations (2.1), (2.2), and (2.9) and assuming $\mathbf{y}_n = \mathbf{0}$, we can express the APA tap update as,

$$\mathbf{h}_n = \left[\mathbf{I} - \mu\mathbf{X}_n\left[\mathbf{X}_n^t\mathbf{X}_n\right]^{-1}\mathbf{X}_n^t\right]\mathbf{h}_{n-1} + \mu\mathbf{X}_n\left[\mathbf{X}_n^t\mathbf{X}_n\right]^{-1}\mathbf{X}_n^t\mathbf{h}_{ep}. \tag{2.18}$$

Define,

$$\mathbf{Q}_n = \mathbf{I} - \mu \mathbf{X}_n \left[\mathbf{X}_n^t \mathbf{X}_n\right]^{-1} \mathbf{X}_n^t = \mathbf{U}_n \begin{bmatrix} 1-\mu & & & & & \\ & \ddots & & & & \\ & & 1-\mu & & & \\ & & & 1 & & \\ & & & & \ddots & \\ & & & & & 1 \end{bmatrix} \mathbf{U}_n^t, \tag{2.19}$$

and if $\mu = 1$,

$$\mathbf{Q}_n = \mathbf{U}_n \begin{bmatrix} 0 & & & & & \\ & \ddots & & & & \\ & & 0 & & & \\ & & & 1 & & \\ & & & & \ddots & \\ & & & & & 1 \end{bmatrix} \mathbf{U}_n^t, \tag{2.20}$$

where there are N zeros and $L - N$ ones in the diagonal matrix. Similarly, define,

$$\mathbf{P}_n = \mu \mathbf{X}_n \left[\mathbf{X}_n^t \mathbf{X}_n\right]^{-1} \mathbf{X}_n^t = \mathbf{U}_n \begin{bmatrix} \mu & & & & & \\ & \ddots & & & & \\ & & \mu & & & \\ & & & 0 & & \\ & & & & \ddots & \\ & & & & & 0 \end{bmatrix} \mathbf{U}_n^t, \tag{2.21}$$

and if $\mu = 1$,

$$\mathbf{P}_n = \mathbf{U}_n \begin{bmatrix} 1 & & & & & \\ & \ddots & & & & \\ & & 1 & & & \\ & & & 0 & & \\ & & & & \ddots & \\ & & & & & 0 \end{bmatrix} \mathbf{U}_n^t, \tag{2.22}$$

where there are N ones and $L - N$ zeros in the diagonal matrix. That is, $\mathbf{Q}_n$ and $\mathbf{P}_n$ represent projection matrices onto orthogonal subspaces when $\mu = 1$ and relaxed projection matrices when $0 < \mu < 1$. The matrix $\mathbf{Q}_n$ has the same form as in equation (2.15). Using (2.19) and (2.21) in (2.18), the APA coefficient vector update becomes

$$\mathbf{h}_n = \mathbf{Q}_n \mathbf{h}_{n-1} + \mathbf{P}_n \mathbf{h}_{ep}, \tag{2.23}$$

which is the same form as the affine projection defined in equation (2.17), where now $\mathbf{Q} = \mathbf{Q}_n$ and $\mathbf{f} = \mathbf{P}_n \mathbf{h}_{ep}$. Thus, equations (2.1) and (2.2) represent the relaxed projection of the system impulse response estimate onto an affine subspace which is determined by: 1) the excitation matrix $\mathbf{X}_n$ [according to (2.19) and (2.21)] and 2) the true system impulse response, $\mathbf{h}_{ep}$ [according to (2.23)].

2.2 CONVERGENCE AND REGULARIZATION

Equation (2.23) gives us an intuitive feel for the convergence of $\mathbf{h}_n$ to $\mathbf{h}_{ep}$. Let us assume that $\mu = 1$. We see that as N increases from 1 toward L, the contribution to $\mathbf{h}_n$ from $\mathbf{h}_{n-1}$ decreases because the nullity of $\mathbf{Q}_n$ is increasing, while the contribution from $\mathbf{h}_{ep}$ increases because the rank of $\mathbf{P}_n$ is increasing. In principle, when $N = L$, $\mathbf{h}_n$ should converge to $\mathbf{h}_{ep}$ in one step, since $\mathbf{Q}_n$ has a rank of zero and $\mathbf{P}_n$ a rank of L. In practice however, we find that as N approaches L the condition number of the matrix, $\mathbf{X}_n^t \mathbf{X}_n$ begins to grow. As a result, the inverse of $\mathbf{X}_n^t \mathbf{X}_n$ becomes more and more dubious and must be replaced with either a regularized or pseudo-inverse. Either way, $\mathbf{P}_n$ ends up with a rank somewhat less than L. Still, for moderate values of N, even when the inverse of $\mathbf{X}_n^t \mathbf{X}_n$ is regularized, the convergence of $\mathbf{h}_n$ is quite impressive, as we shall demonstrate.

The inverse of $\mathbf{X}_n^t \mathbf{X}_n$ can be regularized by adding the matrix, $\delta \mathbf{I}$ prior to taking the inverse. The matrix, $\mathbf{I}$ is the $N \times N$ identity matrix and δ is a small positive scalar. Where $\mathbf{X}_n^t \mathbf{X}_n$ may have eigenvalues close to zero, creating problems for the inverse, $\mathbf{X}_n^t \mathbf{X}_n + \delta \mathbf{I}$ has δ as its smallest possible eigenvalue which if large enough, yields a well behaved inverse. The regularized APA tap update is then

$$\mathbf{h}_n = \mathbf{h}_{n-1} + \mu \mathbf{X}_n \left[\mathbf{X}_n^t \mathbf{X}_n + \delta \mathbf{I} \right]^{-1} \mathbf{e}_n. \tag{2.24}$$

2.3 THE CONNECTION BETWEEN APA AND RECURSIVE LEAST SQUARES

Using the matrix inversion lemma we can show the connection between APA and RLS. The matrix inversion lemma states that if the non-singular matrix $\mathbf{A}$ can be written as,

$$\mathbf{A} = \mathbf{B} + \mathbf{C}\mathbf{D}, \tag{2.25}$$

where $\mathbf{B}$ is also non-singular. Then its inverse can be written as

$$\mathbf{A}^{-1} = \mathbf{B}^{-1} - \mathbf{B}^{-1}\mathbf{C}\left[\mathbf{I} + \mathbf{D}\mathbf{B}^{-1}\mathbf{C}\right]^{-1}\mathbf{D}\mathbf{B}^{-1}. \tag{2.26}$$

From this we see that,

$$\left[\mathbf{X}_n\mathbf{X}_n^t + \delta\mathbf{I}\right]^{-1}\mathbf{X}_n = \mathbf{X}_n\left[\mathbf{X}_n^t\mathbf{X}_n + \delta\mathbf{I}\right]^{-1}. \tag{2.27}$$

Using (2.27) in (2.24) we see that regularized APA can be written as

$$\mathbf{h}_n = \mathbf{h}_{n-1} + \mu\left[\mathbf{X}_n\mathbf{X}_n^t + \delta\mathbf{I}\right]^{-1}\mathbf{X}_n\mathbf{e}_n. \tag{2.28}$$

Consider the vector $\mathbf{e}_n$. By definition,

$$\mathbf{e}_n = \mathbf{s}_n - \mathbf{X}_n^t\mathbf{h}_{n-1} = \left[\begin{array}{c} s_n - \mathbf{x}_n^t\mathbf{h}_{n-1} \\ \bar{\mathbf{s}}_{n-1} - \bar{\mathbf{X}}_{n-1}^t\mathbf{h}_{n-1}\end{array}\right], \tag{2.29}$$

where the matrix $\bar{\mathbf{X}}_{n-1}$ has dimension $L \times (N-1)$ and consists of the $N-1$ left-most (newest) columns of $\mathbf{X}_{n-1}$ and the $N-1$ length vector $\bar{\mathbf{s}}_{n-1}$ consists of the $N-1$ upper (newest) elements of the vector $\mathbf{s}_{n-1}$.

We first address the lower $N-1$ elements of (2.29). Define the *a posteriori* residual echo vector for sample period $n-1$, $\mathbf{e}_{1,n-1}$ as

$$\begin{aligned} \mathbf{e}_{1,n-1} &= \mathbf{s}_{n-1} - \mathbf{X}_{n-1}^t\mathbf{h}_{n-1} & (2.30)\\ &= \mathbf{s}_{n-1} - \mathbf{X}_{n-1}^t\left[\mathbf{h}_{n-2} + \mu\mathbf{X}_{n-1}\left[\mathbf{X}_{n-1}^t\mathbf{X}_{n-1} + \delta\mathbf{I}\right]^{-1}\mathbf{e}_{n-1}\right] & \\ &= \mathbf{e}_{n-1} - \mu\mathbf{X}_{n-1}^t\mathbf{X}_{n-1}\left[\mathbf{X}_{n-1}^t\mathbf{X}_{n-1} + \delta\mathbf{I}\right]^{-1}\mathbf{e}_{n-1}. & (2.31)\end{aligned}$$

We now make the approximation, $\mathbf{X}_{n-1}^t\mathbf{X}_{n-1} + \delta\mathbf{I} \approx \mathbf{X}_{n-1}^t\mathbf{X}_{n-1}$. Which is valid as long as δ is significantly smaller than the eigenvalues of $\mathbf{X}_{n-1}^t\mathbf{X}_{n-1}$. Using this approximation we have,

$$\mathbf{e}_{1,n-1} \approx (1-\mu)\mathbf{e}_{n-1}. \tag{2.32}$$

Recognizing that the lower $N-1$ elements of (2.29) are the same as the upper $N-1$ elements of (2.30), we see that we can use (2.32) to express $\mathbf{e}_n$ as

$$\mathbf{e}_n = \mathbf{s}_n - \mathbf{X}_n^t\mathbf{h}_{n-1} = \left[\begin{array}{c} s_n - \mathbf{x}_n^t\mathbf{h}_{n-1} \\ (1-\mu)\bar{\mathbf{e}}_{n-1}\end{array}\right] = \left[\begin{array}{c} e_n \\ (1-\mu)\bar{\mathbf{e}}_{n-1}\end{array}\right]. \tag{2.33}$$

Then, for $\mu = 1$,

$$\mathbf{e}_n = \mathbf{s}_n - \mathbf{X}_n^t\mathbf{h}_{n-1} = \left[\begin{array}{c} e_n \\ \mathbf{0}\end{array}\right]. \tag{2.34}$$

Using (2.34) in (2.28) we see that,

$$\mathbf{h}_n = \mathbf{h}_{n-1} + \left[\mathbf{X}_n\mathbf{X}_n^t + \delta\mathbf{I}\right]^{-1} \mathbf{x}_n e_n. \tag{2.35}$$

Equation (2.35) is very similar to RLS. The difference is that the matrix which is inverted is a regularized, rank deficient form of the usual estimated autocorrelation matrix. If w let $\delta = 0$ and $N = n$, (2.35) becomes the growing windowed RLS.

3. FAST AFFINE PROJECTIONS

First, we write the relaxed and regularized affine projection algorithm in a slightly different form,

$$\mathbf{e}_n = \mathbf{s}_n - \mathbf{X}_n^t\mathbf{h}_{n-1}, \tag{2.36}$$

$$\epsilon_n = \left[\mathbf{X}_n^t\mathbf{X}_n + \delta\mathbf{I}\right]^{-1} \mathbf{e}_n, \tag{2.37}$$

$$\mathbf{h}_n = \mathbf{h}_{n-1} + \mu\mathbf{X}_n\, \epsilon_n, \tag{2.38}$$

where we have defined the N length *normalized residual echo vector,* $\epsilon_n = [\epsilon_{0,n}, \cdots, \epsilon_{N-1,n}]^t$.

The complexity of APA is $2LN + K_{inv}N^2$ multiplies per sample period, where K_{inv} is a constant associated with the complexity of the inverse required in (2.37). If a generalized Levinson algorithm is used to solve the systems of equations in (2.37), K_{inv} is about 7. One way to mitigate this computational complexity is to only update the coefficients once every N sample periods [9] reducing the average complexity (over N sample periods) to $2L + K_{inv}N$ multiplies per sample period. This is known as PRA, the partial rank algorithm. Simulations indicate that when very highly colored excitation signals are used, the convergence of PRA is somewhat inferior to APA. For speech excitation, however, we have found that PRA achieves close to the same convergence as APA. The main disadvantage of PRA is that its computational complexity is bursty. So, depending on the speed of the implementing technology, there is often a delay in the generation of the error vector, $\mathbf{e}_n$. As will be shown below, FAP performs a complete N dimensional APA update each sample period with $2L + O(N)$ multiplies per sample without delay.

3.1 FAST RESIDUAL ECHO VECTOR CALCULATION

Earlier, we justified the approximation in relation (2.32) on the assumption that the regularization factor δ would be much smaller than the smallest eigenvalue in $\mathbf{X}_n^t\mathbf{X}_n$. In this section we examine the situation where that assumption does not hold, yet we would like to use relation (2.32) anyway. This case arises for instance when N is selected to be in the neighborhood of 50, speech is the excitation signal, and the near-end background noise signal energy is larger than the smaller eigenvalues of $\mathbf{X}_n^t\mathbf{X}_n$.

We begin by rewriting (2.31) slightly,

$$\mathbf{e}_{1,n-1} = \left[\mathbf{I} - \mu\mathbf{X}_{n-1}^t\mathbf{X}_{n-1}\left[\mathbf{X}_{n-1}^t\mathbf{X}_{n-1} + \delta\mathbf{I}\right]^{-1}\right]\mathbf{e}_{n-1}. \tag{2.39}$$

The matrix $\mathbf{X}_{n-1}^t\mathbf{X}_{n-1}$ has the similarity decomposition,

$$\mathbf{X}_{n-1}^t\mathbf{X}_{n-1} = \mathbf{V}_{n-1}\Lambda_{n-1}\mathbf{V}_{n-1}^t, \tag{2.40}$$

where $\mathbf{V}_{n-1}$ is an N by N unitary matrix and Λ_{n-1} is a N by N diagonal matrix with its ith diagonal element being the ith eigenvalue of $\mathbf{X}_{n-1}^t\mathbf{X}_{n-1}$, $\lambda_{i,n-1}$.

Defining the a priori and a posteriori modal error vectors,

$$\mathbf{e}'_{n-1} = \mathbf{V}_{n-1}^t\mathbf{e}_{n-1} \tag{2.41}$$

and

$$\mathbf{e}'_{1,n-1} = \mathbf{V}_{n-1}^t\mathbf{e}_{1,n-1} \tag{2.42}$$

respectively; we can multiply (2.39) from the left by $\mathbf{V}_{n-1}^t$ and show that the ith a posteriori modal error vector element, $e'_{1,i,n-1}$, can be found from the ith a priori modal error vector element, $e'_{i,n-1}$, by,

$$e'_{1,i,n-1} = \left[1 - \frac{\mu\lambda_{i,n-1}}{\delta + \lambda_{i,n-1}}\right]e'_{i,n-1}. \tag{2.43}$$

From (2.43) it can be shown that

$$e'_{1,i,n-1} \approx \begin{cases} (1-\mu)e'_{i,n-1} & \lambda_{i,n-1} \gg \delta \\ e'_{i,n-1} & \lambda_{i,n-1} \ll \delta. \end{cases} \tag{2.44}$$

Assume that δ is chosen to be approximately equal to the power of y_n. Then, for those modes where, $\lambda_{i,n-1} \ll \delta$, $e'_{i,n-1}$ is mainly dominated by the background noise and little can be learned from it about $\mathbf{h}_{ep}$. So, suppressing these modes by multiplying them by $1-\mu$ will attenuate somewhat the background noise's effect on the overall echo path estimate. Applying this to (2.44) and multiplying from the left by $\mathbf{V}_{n-1}$ we have

$$\mathbf{e}_{1,n-1} \approx (1-\mu)\mathbf{e}_{n-1} \tag{2.45}$$

and from this (2.33). From (2.39) we see that approximation (2.33) becomes an equality when $\delta = 0$, but then, the inverse in (2.39) is not regularized. Simulations show that by making adjustments in δ the convergence performance of APA with and without the approximation (2.39) can be equated.

The complexity of (2.39) is L operations to calculate e_n and $N-1$ operations to update $(1-\mu)\bar{\mathbf{e}}_{n-1}$. For the case where $\mu = 1$ the $N-1$ operations are obviously unnecessary.

3.2 FAST ADAPTIVE COEFFICIENT VECTOR CALCULATION

In the problem of echo cancellation, as in other problems, the overall system output that is observed by the user is the residual echo. Therefore, it is permissible to maintain any form of $\mathbf{h}_n$ that is convenient in the algorithm, as long as the first sample of $\mathbf{e}_n$ is not modified in any way. This is the basis of FAP. The fidelity of $\mathbf{e}_n$ is maintained at each sample period, but $\mathbf{h}_n$ is not. Another vector, $\hat{\mathbf{h}}_n$ is maintained, where only the last column of $\mathbf{X}_n$ is weighted and accumulated into $\hat{\mathbf{h}}_n$ each sample period [10]. Thus, the computational complexity of the tap weight-update process is no more complex than NLMS, L multiplications.

One can express the current echo path estimate, $\mathbf{h}_n$, in terms of the original echo path estimate, $\mathbf{h}_0$, and the subsequent $\mathbf{X}_i$'s and ϵ_i's,

$$\mathbf{h}_n = \mathbf{h}_0 + \mu \sum_{i=0}^{n-1} \mathbf{X}_{n-i}\, \epsilon_{n-i}. \tag{2.46}$$

Expanding the vector/matrix multiplication, we have,

$$\mathbf{h}_n = \mathbf{h}_0 + \mu \sum_{i=0}^{n-1} \sum_{j=0}^{N-1} \mathbf{x}_{n-j-i} \epsilon_{j,n-i}, \tag{2.47}$$

where $\mathbf{x}_{n-j-i}$ is the jth column of matrix $\mathbf{X}_{n-i}$ and $\epsilon_{j,n-i}$ is the jth element of vector ϵ_{n-i} . If we assume that the excitation vectors, $\mathbf{x}_k$'s are only non-zero for $1 \leq k \leq n$, then we can apply a window to (2.47) without changing the result,

$$\mathbf{h}_n = \mathbf{h}_0 + \mu \sum_{i=0}^{n-1} \sum_{j=0}^{N-1} \mathbf{x}_{n-j-i} w_{1,j+i} \epsilon_{j,n-i}, \tag{2.48}$$

where

$$w_{1,j+i} = \begin{cases} 1 & 0 \leq j+i \leq n-1 \\ 0 & \text{elsewhere.} \end{cases} \tag{2.49}$$

Changing the order of summations and applying the change of variables, $i = k - j$ to (2.48) yields,

$$\mathbf{h}_n = \mathbf{h}_0 + \mu \sum_{j=0}^{N-1} \sum_{k=j}^{n-1+j} \mathbf{x}_{n-k} w_{1,k} \epsilon_{j,n-k+j}. \tag{2.50}$$

Then, applying the definition of $w_{1,k}$ in (2.50) we can modify the second summation to

$$\mathbf{h}_n = \mathbf{h}_0 + \mu \sum_{j=0}^{N-1} \sum_{k=j}^{n-1} \mathbf{x}_{n-k} \epsilon_{j,n-k+j}. \tag{2.51}$$

Now, we break the second summation into two parts, one from $k = j$ to $k = N - 1$ and one from $k = N$ to $k = n - 1$ with the result,

$$\mathbf{h}_n = \mathbf{h}_0 + \mu \sum_{j=0}^{N-1} \sum_{k=j}^{N-1} \mathbf{x}_{n-k} \epsilon_{j,n-k+j} + \mu \sum_{k=N}^{n-1} \sum_{j=0}^{N-1} \mathbf{x}_{n-k} \epsilon_{j,n-k+j}, \tag{2.52}$$

where we have also changed the order of summations in the second double sum.

Directing our attention to the first double sum, let us define a second window as

$$w_{2,k-j} = \begin{cases} 1 & 0 \leq k - j \\ 0 & \text{elsewhere.} \end{cases} \tag{2.53}$$

Without altering the result we can apply this window to the first double sum by beginning the second summation in it at $k = 0$ rather than $k = j$,

$$\sum_{j=0}^{N-1} \sum_{k=0}^{N-1} \mathbf{x}_{n-k} w_{2,k-j} \epsilon_{j,n-k+j} = \sum_{j=0}^{N-1} \sum_{k=j}^{N-1} \mathbf{x}_{n-k} \epsilon_{j,n-k+j}. \tag{2.54}$$

Now we can exchange the order of summations and re-apply the window, $w_{2,k-j}$ to change the end of the second summation at $j = k$ rather than $j = N - 1$,

$$\sum_{k=0}^{N-1} \sum_{j=0}^{k} \mathbf{x}_{n-k} \epsilon_{j,n-k+j} = \sum_{k=0}^{N-1} \sum_{j=0}^{N-1} \mathbf{x}_{n-k} w_{2,k-j} \epsilon_{j,n-k+j}. \tag{2.55}$$

Applying (2.54) and (2.55) to (2.52) we finally arrive at,

$$\mathbf{h}_n = \mathbf{h}_0 + \mu \sum_{k=0}^{N-1} \mathbf{x}_{n-k} \sum_{j=0}^{k} \epsilon_{j,n-k+j} + \mu \sum_{k=N}^{n-1} \mathbf{x}_{n-k} \sum_{j=0}^{N-1} \epsilon_{j,n-k+j}. \tag{2.56}$$

If we define the first term and the second pair of summations on the right side of (2.56) as

$$\hat{\mathbf{h}}_n = \mathbf{h}_0 + \mu \sum_{k=N}^{n-1} \mathbf{x}_{n-k} \sum_{j=0}^{N-1} \epsilon_{j,n-k+j} \tag{2.57}$$

and recognize the first pair of summations in (2.56) as a vector-matrix multiplication,

$$\mathbf{X}_n \mathbf{E}_n = \sum_{k=0}^{N-1} \mathbf{x}_{n-k} \sum_{j=0}^{k} \epsilon_{j,n-k+j}, \tag{2.58}$$

where

$$\mathbf{E}_n = \begin{bmatrix} \epsilon_{0,n} \\ \epsilon_{1,n} + \epsilon_{0,n-1} \\ \vdots \\ \epsilon_{N-1,n} + \epsilon_{N-2,n-1} + \epsilon_{0,n-N+1} \end{bmatrix}. \tag{2.59}$$

If we define $\bar{\mathbf{E}}_n$ as an $N-1$ length vector consisting of the upper most $N-1$ elements of $\mathbf{E}_n$, can write the recursion,

$$\mathbf{E}_n = \epsilon_n + \begin{bmatrix} 0 \\ \bar{\mathbf{E}}_{n-1} \end{bmatrix}. \tag{2.60}$$

Then we can express (2.56) as

$$\mathbf{h}_n = \hat{\mathbf{h}}_{n-1} + \mu \mathbf{X}_n \mathbf{E}_n. \tag{2.61}$$

It is easily seen from (2.57) that

$$\hat{\mathbf{h}}_n = \hat{\mathbf{h}}_{n-1} + \mu \mathbf{x}_{n-N+1} \sum_{k=0}^{N-1} \epsilon_{j,n-1+j} \tag{2.62}$$

$$= \hat{\mathbf{h}}_{n-1} + \mu \mathbf{x}_{n-N+1} E_{N-1,n}. \tag{2.63}$$

Using (2.63) in (2.61) we see that we can alternately express the current echo path estimate as

$$\mathbf{h}_n = \hat{\mathbf{h}}_n + \mu \bar{\mathbf{X}}_n \bar{\mathbf{E}}_n, \tag{2.64}$$

where $\bar{\mathbf{X}}_n$ is an L by $N-1$ matrix consisting of the $N-1$ left most columns in $\mathbf{X}_n$.

We now address the relationship between $\mathbf{e}_n$ and $\mathbf{e}_{n-1}$. From (2.33) we have,

$$\mathbf{e}_n = \mathbf{s}_n - \mathbf{X}_n^t \mathbf{h}_{n-1} \approx \begin{bmatrix} s_n - \mathbf{x}_n^t \mathbf{h}_{n-1} \\ (1-\mu)\bar{\mathbf{e}}_{n-1} \end{bmatrix} = \begin{bmatrix} e_n \\ (1-\mu)\bar{\mathbf{e}}_{n-1} \end{bmatrix}. \tag{2.65}$$

Unfortunately, $\mathbf{h}_{n-1}$ is not readily available to us. But, we can use equation (2.64) in the first element of (2.65) to get

$$e_n = s_n - \mathbf{x}_n^t \hat{\mathbf{h}}_{n-1} + \mu \mathbf{x}_n^t \bar{\mathbf{X}}_{n-1} \bar{\mathbf{E}}_{n-1} \tag{2.66}$$

$$= \hat{e}_n - \mu \tilde{\mathbf{r}}_n^t \bar{\mathbf{E}}_{n-1}, \tag{2.67}$$

where

$$\hat{e}_n = s_n - \mathbf{x}_n^t \hat{\mathbf{h}}_{n-1} \tag{2.68}$$

and

$$\tilde{\mathbf{r}}_n = \mathbf{x}_n^t \bar{\mathbf{X}}_{n-1} = \tilde{\mathbf{r}}_{n-1} + x_n \, \tilde{\alpha}_n - x_{n-L} \, \tilde{\alpha}_{n-L}, \tag{2.69}$$

where $\tilde{\alpha}_n$ is an $N-1$ length vector consisting of the last $N-1$ elements of α_n.

3.3 FAST NORMALIZED RESIDUAL ECHO VECTOR CALCULATION

To efficiently compute (2.37) we need to find a recursion for the vector

$$\epsilon_n = \left[\mathbf{X}_n^t \mathbf{X}_n + \delta \mathbf{I}\right]^{-1} \mathbf{e}_n. \tag{2.70}$$

Define,

$$\mathbf{R}_n = \mathbf{X}_n^t \mathbf{X}_n + \delta \mathbf{I}, \tag{2.71}$$

and let $\mathbf{a}_n$ and $\mathbf{b}_n$ denote the optimum forward and backward linear predictors for $\mathbf{R}_n$ and let $E_{a,n}$ and $E_{b,n}$ denote their respective prediction error energies. Also define $\bar{\mathbf{R}}_n$ and $\tilde{\mathbf{R}}_n$ as the upper left and lower right $N-1$ by $N-1$ matrices within $\mathbf{R}_n$, respectively. Then, given the identities:

$$\mathbf{R}_n^{-1} = \begin{bmatrix} 0 & \mathbf{0}^t \\ \mathbf{0} & \tilde{\mathbf{R}}_n^{-1} \end{bmatrix} + \frac{1}{E_{a,n}} \mathbf{a}_n \mathbf{a}_n^t \tag{2.72}$$

$$= \begin{bmatrix} \bar{\mathbf{R}}_n^{-1} & \mathbf{0} \\ \mathbf{0}^t & 0 \end{bmatrix} + \frac{1}{E_{b,n}} \mathbf{b}_n \mathbf{b}_n^t \tag{2.73}$$

and the definitions,

$$\tilde{\epsilon}_n = \tilde{\mathbf{R}}_n^{-1} \tilde{\mathbf{e}}_n, \tag{2.74}$$

$$\bar{\epsilon}_n = \bar{\mathbf{R}}_n^{-1} \bar{\mathbf{e}}_n, \tag{2.75}$$

(where $\bar{\mathbf{e}}_n$ and $\tilde{\mathbf{e}}_n$ consist of the upper and lower $N-1$ elements of $\mathbf{e}_n$, respectively) we can multiply (2.72) from the right by $\mathbf{e}_n$ and using (2.70) and (2.74)

$$\epsilon_n = \begin{bmatrix} 0 \\ \tilde{\epsilon}_n \end{bmatrix} + \frac{1}{E_{a,n}} \mathbf{a}_n \mathbf{a}_n^t \mathbf{e}_n. \tag{2.76}$$

Similarly, multiplying (2.73) from the right by $\mathbf{e}_n$ and using (2.70) and (2.75)

$$\epsilon_n = \begin{bmatrix} \bar{\epsilon}_n \\ 0 \end{bmatrix} + \frac{1}{E_{b,n}} \mathbf{b}_n \mathbf{b}_n^t \mathbf{e}_n. \tag{2.77}$$

Solving for $\begin{bmatrix} \bar{\epsilon}_n \\ 0 \end{bmatrix}$ we have

$$\begin{bmatrix} \bar{\epsilon}_n \\ 0 \end{bmatrix} = \epsilon_n - \frac{1}{E_{b,n}} \mathbf{b}_n \mathbf{b}_n^t \mathbf{e}_n. \tag{2.78}$$

The quantities, $E_{a,n}$, $E_{b,n}$, $\mathbf{a}_n$, and $\mathbf{b}_n$ can be calculated efficiently (complexity $10N$) using a sliding windowed FRLS algorithm (see the Appendix).

The relationship between $\tilde{\epsilon}_n$ and $\tilde{\epsilon}_{n-1}$ is now investigated. It can easily be shown that

$$\tilde{\mathbf{R}}_n = \bar{\mathbf{R}}_{n-1}. \tag{2.79}$$

Using (2.79), the definition of $\tilde{\mathbf{e}}_n$, $\bar{\mathbf{e}}_n$, (2.79), and (2.33) we have,

$$\tilde{\epsilon}_n = \tilde{\mathbf{R}}_n^{-1}\tilde{\mathbf{e}}_n = \bar{\mathbf{R}}_{n-1}^{-1}(1-\mu)\bar{\mathbf{e}}_n = (1-\mu)\,\bar{\epsilon}_{n-1}. \tag{2.80}$$

3.4 THE FAP ALGORITHM

The FAP algorithm with regularization and relaxation is as follows:

1. Initialization: $E_{a,0} = E_{b,0} = \delta$ and $\mathbf{a}_0 = [1, \mathbf{0}]^t$, $\mathbf{b}_0 = [\mathbf{0}, 1]^t$,
2. Use sliding windowed FRLS to update $E_{a,n}$, $E_{b,n}$, $\mathbf{a}_n$, and $\mathbf{b}_n$,
3. $\tilde{\mathbf{r}}_n = \tilde{\mathbf{r}}_{n-1} + x_n\,\tilde{\alpha}_n - x_{n-L}\,\tilde{\alpha}_{n-L}$,
4. $\hat{e}_n = s_n - \mathbf{x}_n^t\hat{\mathbf{h}}_{n-1}$,
5. $e_n = \hat{e}_n - \mu\tilde{\mathbf{r}}_n^t\bar{\mathbf{E}}_{n-1}$,
6. $\mathbf{e}_n = \begin{bmatrix} e_n \\ (1-\mu)\bar{\mathbf{e}}_{n-1} \end{bmatrix}$,
7. $\epsilon_n = \begin{bmatrix} 0 \\ \tilde{\epsilon}_n \end{bmatrix} + \frac{1}{E_{a,n}}\mathbf{a}_n\mathbf{a}_n^t\mathbf{e}_n$,
8. $\begin{bmatrix} \bar{\epsilon}_n \\ 0 \end{bmatrix} = \epsilon_n - \frac{1}{E_{b,n}}\mathbf{b}_n\mathbf{b}_n^t\mathbf{e}_n$,
9. $\mathbf{E}_n = \epsilon_n + \begin{bmatrix} 0 \\ \bar{\mathbf{E}}_{n-1} \end{bmatrix}$,
10. $\hat{\mathbf{h}}_n = \hat{\mathbf{h}}_{n-1} + \mu\mathbf{x}_{n-N+1}E_{N-1,n}$,
11. $\tilde{\epsilon}_{n+1} = (1-\mu)\,\bar{\epsilon}_n$.

Step 2 is of complexity $10N$ when FTF (fast transversal filter) is used. Steps 4 and 10 are both of complexity L, steps 3, 7, and 8 are each of complexity $2N$, and steps 5, 6, 9, and 11 are of complexity N. This gives us an overall complexity of $2L + 20N$.

If we eliminate relaxation, that is, set μ to one, we can realize considerable savings in complexity. For example, in step 11 we can see that $\tilde{\epsilon}_n$ will always be zero. Therefore $\bar{\epsilon}_n$ need not be calculated. Thus steps 8 and 11 may be eliminated. Furthermore, step 6 is no longer needed since only the first element in $\mathbf{e}_n$ is non-zero. Steps 7 and 9 may also be combined into a single complexity

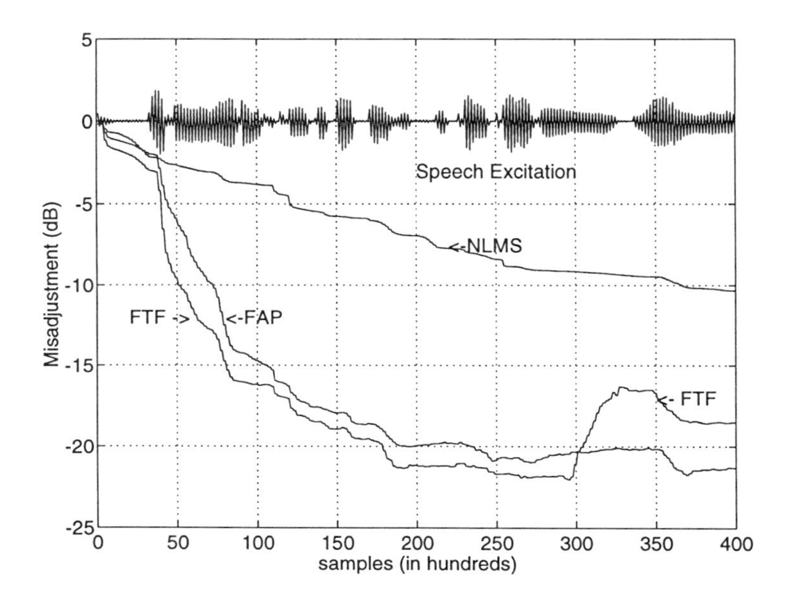

Figure 2.3 Comparison of coefficient error for FAP, FTF, and NLMS with speech as excitation.

N calculation, $\mathbf{E}_n = \begin{bmatrix} 0 \\ \bar{\mathbf{E}}_{n-1} \end{bmatrix} + \frac{e_n}{E_{a,n}} \mathbf{a}_n$. Thus, FAP without relaxation may be written as follows:

1. Initialization: $E_{a,0} = E_{b,0} = \delta$ and $\mathbf{a}_0 = [1, \mathbf{0}]^t$, $\mathbf{b}_0 = [\mathbf{0}, 1]^t$,

2. Use sliding windowed FRLS to update $E_{a,n}$, $E_{b,n}$, $\mathbf{a}_n$, and $\mathbf{b}_n$,

3. $\tilde{\mathbf{r}}_n = \mathbf{x}_n^t \bar{\mathbf{X}}_{n-1} = \tilde{\mathbf{r}}_{n-1} + x_n \, \tilde{\alpha}_n - x_{n-L} \, \tilde{\alpha}_{n-L}$,

4. $\hat{e}_n = s_n - \mathbf{x}_n^t \hat{\mathbf{h}}_{n-1}$,

5. $e_n = \hat{e}_n - \mu \tilde{\mathbf{r}}_n^t \bar{\mathbf{E}}_{n-1}$,

6. $\mathbf{E}_n = \begin{bmatrix} 0 \\ \bar{\mathbf{E}}_{n-1} \end{bmatrix} + \frac{e_n}{E_{a,n}} \mathbf{a}_n$,

7. $\hat{\mathbf{h}}_n = \hat{\mathbf{h}}_{n-1} + \mu \mathbf{x}_{n-N+1} E_{N-1,n}$.

Here, steps 4 and 7 are still of complexity L, step 3 is of complexity $2N$, and steps 5 and 6 are of complexity N. Taking into account the sliding windowed FTF, we now have a total complexity of $2L + 14N$.

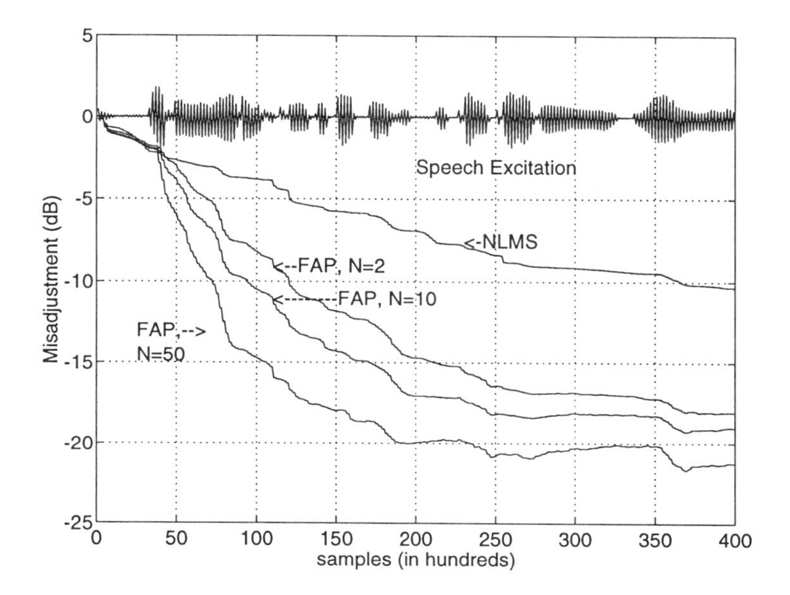

Figure 2.4 Comparison of FAP for different orders of projection, N, with speech as excitation.

4. SIMULATIONS

Figure 2.3 shows a comparison of the convergence of NLMS, FTF (Fast Transversal Filter, an FRLS technique), and FAP coefficient error magnitudes. The excitation signal was speech sampled at 8 kHz, the echo path of length, $L = 1000$, was fixed and the white gaussian additive noise, y_n, was 30 dB down from the echo. Soft initialization was used for both algorithms. For FTF, $E_{a,0}$ and $E_{b,0}$ were both set to $2\sigma_x^2$ (where σ_x^2 is the average power of x_n) and λ, the forgetting factor was set to $\frac{3L-1}{3L}$. For FAP, $E_{a,0}$ and $E_{b,0}$ were set to $\delta = 20\sigma_x^2$ and N was 50. FAP converges at roughly the same rate as FTF with about $2L$ complexity versus $7L$ complexity, respectively. Both FAP and FTF converge faster than NLMS.

Another important implementation consideration is memory. FAP requires about the same amount of memory as NLMS, about $2L$ locations where FTF requires about $5L$ locations.

In Fig. 2.4, we show the convergence of NLMS and FAP with various orders of projections. Once again, speech was the excitation, the length of the filter was 1000 samples, and the signal to noise ratio was 30 dB. We see that quite a bit of improvement is gained with just $N = 2$ and that increasing N to 10 does not improve the speed of convergence significantly. However, if N is further increased to 50, there is again a significant gain in the speed of convergence. Note that for FAP, the increase from $N = 2$ to $N = 50$ does not significantly increase the computational complexity. Thus, very perceptible increases in convergence are realized with only moderate increases in computational complexity.

5. NUMERICAL CONSIDERATIONS

FAP uses the sliding window technique to update and down-date data in its implicit regularized sample correlation matrix and cross correlation vector. Errors introduced by finite arithmetic in practical implementations of the algorithm therefore cause the correlation matrix and cross correlation vector to take random walks with respect to their infinite precision counterparts. A stabilized sliding windowed FRLS algorithm [11] has been introduced, with complexity $14N$ multiplications per sample period (rather than $10N$ for non-stabilized versions). However, even this algorithm is stable only for stationary signals, a class of signals which certainly does not include speech. Another approach, which is very straightforward and rather elegant for FAP, is to periodically start a new sliding window in parallel with the old sliding window, and when the data is the same in both processes, replace the old sliding window based parameters with the new ones. Although this increases the sliding window based parameter calculations by about 50 percent on average (assuming the restarting is done every $L + N$ sample periods), the overall cost is small since only those parameters with computational complexity proportional to N are affected. The overall complexity is only $2L + 21N$ for FAP without relaxation and $2L + 30N$ for FAP with relaxation. Since this approach is basically a periodic restart, it is numerically stable for all signals.

6. CONCLUSIONS

This chapter has discussed a fast version of the affine projection algorithm, FAP. When the length of the adaptive filter is L and the dimension of the affine projection (performed each sample period) is N, FAP's complexity is either $2L + 14N$ or $2L + 20N$ depending on whether the relaxation parameter is one or smaller, respectively. Usually $N \ll L$. Simulations demonstrate that FAP converges as fast as the more complex and memory intensive FRLS methods when the excitation signal is speech. The implicit correlation matrix inverse of FAP is regularized, so the algorithm is easily stabilized for even highly colored excitation. Finally, a simple, low complexity numerical stabilization method for the algorithm was also introduced.

References

[1] K. Ozeki and T. Umeda, "An adaptive filtering algorithm using an orthogonal projection to an affine subspace and its properties," *Elec. Comm. Japan*, vol. J67-A, pp. 126-132, Feb. 1984.

[2] B. Widrow and S. D. Stearns, *Adaptive Signal Processing*. Prentice-Hall, Inc. Englewood Cliffs, N.J., 1985.

[3] J. M. Cioffi and T. Kailath, "Fast, recursive-least-squares transversal filters for adaptive filtering," *IEEE Trans. Acoust., Speech, Signal Processing*, vol. ASSP-32, Apr. 1984.

[4] S. J. Orfanidis, *Optimum Signal Processing, An Introduction*. MacMillan, New York, 1985.

[5] S. L. Gay, "A fast converging, low complexity adaptive filtering algorithm," Third Intl. Workshop on Acoustic Echo Control, Sept. 1993, France.

[6] S. L. Gay, *Fast Projection Algorithms with Application to Voice Excited Echo Cancellers*. Ph.D. Dissertation, Rutgers University, Piscataway, New Jersey, 1994.

[7] M. Tanaka, Y. Kaneda, and S. Makino, "Reduction of computation for high-order projection algorithm," 1993 Electronics Information Communication Society Autumn Seminar, Tokyo, Japan (in Japanese).

[8] J. M. Cioffi and T. Kailath, "Windowed fast transversal filters adaptive algorithms with normalization," *IEEE Trans. Acoust., Speech, Signal Processing*, vol. ASSP-33, June 1985.

[9] S. G. Kratzer and D. R. Morgan, "The partial-rank algorithm for adaptive beamforming," in *Proc. SPIE Int. Soc. Optic. Eng.*, 1985, vol. 564, pp. 9-14.

[10] Y. Maruyama "A fast method of projection algorithm," in *Proc. IEICE Spring Conf.*, 1990, B-744.

[11] D. T. M. Slock and T. Kailath, "Numerically stable transversal filters for recursive least squares adaptive filtering," *IEEE Trans. Signal Processing*, vol. 39, Jan. 1991.

Appendix: Sliding Windowed Fast Recursive Least Squares

This section describes, without derivation, the sliding windowed versions of the fast transversal filter and fast Kalman algorithms.

We begin by making the following definitions:

1. $e_{0,n,a}$ is the a priori forward prediction error from the N length data vector α_n,

$$e_{0,n,a} = \alpha_n^t \mathbf{a}_{n-1}. \tag{2.A.1}$$

   The first subscript, 0 indicates that it is an *a priori* variable, i.e. it is calculated from a forward prediction vector derived in the previous sample period. A "1" would indicate an *a posteriori* variable, i.e. the error calculated from sample period n's forward prediction coefficient vector,

$$e_{1,n,a} = \alpha_n^t \mathbf{a}_n. \tag{2.A.2}$$

   The second subscript denotes the data related time index (for instance, $e_{0,n-L,a}$ is calculated using the data vector, α_{n-L}),

$$e_{0,n-L,a} = \alpha_{n-L}^t \mathbf{a}_{n-1}. \tag{2.A.3}$$

   The third subscript, a, indicates that it is a *forward prediction* error.

2. Similarly, $e_{0,n,b}$ is the a priori backward prediction error from the N length data vector α_n.

3. $\mathbf{a}_n$ is the N length forward prediction vector of time n. Its first element is always 1.

4. $E_{n,a}$ is the expected forward error energy of time n.

5. $\mathbf{b}_n$ is the N length backward prediction vector of time n. Its last element is always 1.

6. $E_{n,b}$ is the expected backward error energy of time n.

7. $\mathbf{k}_{0,n}$ is the N length a priori kalman gain update vector at time n. It is defined as

$$\mathbf{k}_{0,n} = \mathbf{R}_n^{-1}\, \alpha_{n-1}, \tag{2.A.4}$$

   where $\mathbf{R}_n$ is defined in equation (2.71).

8. $\mathbf{k}_{0,n-L}$ is the N length delayed a priori kalman gain down-date vector at time n. It is defined as

$$\mathbf{k}_{0,n-L} = \mathbf{R}_n^{-1}\, \alpha_{n-L-1}. \tag{2.A.5}$$

9. Similarly, $\mathbf{k}_{1,n}$ is the N length a posteriori kalman gain update vector defined as

$$\mathbf{k}_{1,n} = \mathbf{R}_n^{-1}\, \alpha_n. \tag{2.A.6}$$

10. $\mathbf{k}_{1,n-L}$ is the N length delayed a priori kalman gain down-date vector at time n. It is defined as

$$\mathbf{k}_{1,n-L} = \mathbf{R}_n^{-1}\, \alpha_{n-L}. \tag{2.A.7}$$

11. Both $\mathbf{k}_{0,n}$ and $\mathbf{k}_{1,n}$ and their corresponding down-dates have corresponding *tilde* and *bar* versions of length $N-1$ defined as:

$$\tilde{\mathbf{k}}_{0,n} = \tilde{\mathbf{R}}_n\, \tilde{\alpha}_{n-1}, \tag{2.A.8}$$

$$\bar{\mathbf{k}}_{0,n} = \bar{\mathbf{R}}_n\, \bar{\alpha}_{n-1}, \tag{2.A.9}$$

$$\tilde{\mathbf{k}}_{1,n} = \tilde{\mathbf{R}}_n\, \tilde{\alpha}_n, \tag{2.A.10}$$

$$\bar{\mathbf{k}}_{1,n} = \bar{\mathbf{R}}_n\, \bar{\alpha}_n, \tag{2.A.11}$$

where $\tilde{\mathbf{R}}_n$ and $\bar{\mathbf{R}}_n$ are defined in (2.72) and (2.73), respectively. The down-date vectors are defined analogously.

Using the above definitions, the sliding windowed Fast Kalman algorithm is:

1. $\begin{bmatrix} e_{0,n,a} \\ -e_{0,n-L,a} \end{bmatrix} = \begin{bmatrix} \alpha_n^t \\ -\alpha_{n-L}^t \end{bmatrix} \mathbf{a}_{n-1},$

2. $\mathbf{a}_n = \mathbf{a}_{n-1} - \begin{bmatrix} 0 & 0 \\ \tilde{\mathbf{k}}_{1,n} & \tilde{\mathbf{k}}_{1,n-L} \end{bmatrix} \begin{bmatrix} e_{1,n,a} \\ -e_{1,n-L,a} \end{bmatrix},$

3. $\begin{bmatrix} e_{1,n,a} \\ -e_{1,n-L,a} \end{bmatrix} = \begin{bmatrix} \alpha_n^t \\ -\alpha_{n-L}^t \end{bmatrix} \mathbf{a}_n,$

4. $E_{n,a} = E_{n-1,a} + e_{0,n,a}e_{1,n,a} - e_{0,n-L,a}e_{1,n-L,a},$

5. $k_{1,n,1} = \frac{e_{1,n,a}}{E_{n,a}},\ k_{1,n-L,1} = \frac{e_{1,n-L,a}}{E_{n,a}},$

6. $\left[\mathbf{k}_{1,n}, \mathbf{k}_{1,n-L}\right] = \begin{bmatrix} 0 & 0 \\ \tilde{\mathbf{k}}_{1,n} & \tilde{\mathbf{k}}_{1,n-L} \end{bmatrix} \mathbf{a}_n \left[k_{1,n,1}, k_{1,n-L,1}\right]$ extract last coefficients, $k_{1,n,N}$ and $k_{1,n-L,N}$,

7. $\begin{bmatrix} e_{0,n,b} \\ -e_{0,n-L,b} \end{bmatrix} = \begin{bmatrix} \alpha_n^t \\ -\alpha_{n-L}^t \end{bmatrix} \mathbf{b}_{n-1},$

8. $\mathbf{b}_n = \frac{\mathbf{b}_{n-1} - \mathbf{k}_{1,n} e_{0,n,b} + \mathbf{k}_{1,n-L} e_{0,n-L,b}}{1 - k_{1,n,N} e_{0,n,b} + k_{1,n-L,N} e_{0,n-L,b}},$

9. $\begin{bmatrix} \bar{\mathbf{k}}_{1,n} & \bar{\mathbf{k}}_{1,n-L} \\ 0 & 0 \end{bmatrix} = \left[\mathbf{k}_{1,n}, \mathbf{k}_{1,n-L}\right] - \mathbf{b}_n \left[k_{1,n,N}, k_{1,n-L,N}\right],$

10. $\begin{bmatrix} e_{0,n} \\ -e_{0,n-L} \end{bmatrix} = \begin{bmatrix} s_n \\ -s_{n-L} \end{bmatrix} - \begin{bmatrix} \alpha_n^t \\ -\alpha_{n-L}^t \end{bmatrix} \mathbf{h}_{n-1},$

11. $\mathbf{h}_n = \mathbf{h}_{n-1} + \left[\mathbf{k}_{1,n}, \mathbf{k}_{1,n-L}\right] \begin{bmatrix} e_{0,n} \\ -e_{0,n-L} \end{bmatrix}.$

The sliding windowed Fast Transversal filter is:

1. $\begin{bmatrix} e_{0,n,a} \\ -e_{0,n-L,a} \end{bmatrix} = \begin{bmatrix} \alpha_n^t \\ -\alpha_{n-L}^t \end{bmatrix} \mathbf{a}_{n-1},$

2. $\begin{bmatrix} e_{1,n,a} \\ -e_{1,n-L,a} \end{bmatrix} = \tilde{\mathbf{U}}_n \begin{bmatrix} e_{0,n,a} \\ -e_{0,n-L,a} \end{bmatrix},$

3. $\left[k_{1,n,1}, k_{1,n-L,1}\right] = \left[\frac{e_{0,n,a}}{E_{n-1,a}}, \frac{e_{0,n-L,a}}{E_{n-1,a}}\right],$

4. $E_{n,a} = E_{n-1,a} + e_{0,n,a} e_{1,n,a} - e_{0,n-L,a} e_{1,n-L,a},$

5. $\left[\mathbf{k}_{0,n}, \mathbf{k}_{0,n-L}\right] = \begin{bmatrix} 0 & 0 \\ \tilde{\mathbf{k}}_{0,n} & \tilde{\mathbf{k}}_{0,n-L} \end{bmatrix} \mathbf{a}_{n-1} \left[k_{0,n,1}, k_{0,n-L,1}\right]$ extract last coefficients, $k_{0,n,N}$ and $k_{0,n-L,N}$,

6. $\begin{bmatrix} e_{0,n,b} \\ -e_{0,n-L,b} \end{bmatrix} = E_{n-1,b} \begin{bmatrix} \mathbf{k}_{0,n,N} \\ -\mathbf{k}_{0,n-L,N} \end{bmatrix},$

7. $\begin{bmatrix} \bar{\mathbf{k}}_{0,n} & \bar{\mathbf{k}}_{0,n-L} \\ 0 & 0 \end{bmatrix} = \left[\mathbf{k}_{0,n}, \mathbf{k}_{0,n-L}\right] - \mathbf{b}_{n-1} \left[k_{0,n,N}, k_{0,n-L,N}\right],$

8. $\mathbf{U}_n = \tilde{\mathbf{U}}_n - \frac{1}{E_{n,a}} \begin{bmatrix} e_{1,n,a} \\ e_{1,n-L,a} \end{bmatrix} \left[e_{1,n,a}, e_{1,n-L,a}\right],$

9. $\bar{\mathbf{U}}_n = \mathbf{U}_n \left[\mathbf{I} - \frac{1}{E_{n-1,b}} \begin{bmatrix} e_{0,n,b} \\ -e_{0,n-L,b} \end{bmatrix} \left[e_{0,n,b}, e_{0,n-L,b}\right] \mathbf{U}_n\right]^{-1},$

10. $\begin{bmatrix} e_{1,n,b} \\ -e_{1,n-L,b} \end{bmatrix} = \bar{\mathbf{U}}_n \begin{bmatrix} e_{0,n,b} \\ -e_{0,n-L,b} \end{bmatrix},$

11. $E_{n,b} = E_{n-1,b} + e_{0,n,b} e_{1,n,b} - e_{0,n-L,b} e_{1,n-L,b},$

12. $\mathbf{a}_n = \mathbf{a}_{n-1} - \begin{bmatrix} 0 & 0 \\ \tilde{\mathbf{k}}_{0,n} & \tilde{\mathbf{k}}_{0,n-L} \end{bmatrix} \begin{bmatrix} e_{1,n,a} \\ -e_{1,n-L,a} \end{bmatrix}$,

13. $\mathbf{b}_n = \mathbf{b}_{n-1} - \begin{bmatrix} \bar{\mathbf{k}}_{0,n} & \bar{\mathbf{k}}_{0,n-L} \\ 0 & 0 \end{bmatrix} \begin{bmatrix} e_{1,n,b} \\ -e_{1,n-L,b} \end{bmatrix}$,

14. $\begin{bmatrix} e_{0,n} \\ e_{0,n-L} \end{bmatrix} = \begin{bmatrix} s_n \\ s_{n-L} \end{bmatrix} - \begin{bmatrix} \alpha_n^t \\ \alpha_{n-L}^t \end{bmatrix} \mathbf{h}_{n-1}$,

15. $\begin{bmatrix} e_{1,n} \\ e_{1,n-L} \end{bmatrix} = \tilde{\mathbf{U}}_n^t \begin{bmatrix} e_{0,n} \\ e_{0,n-L} \end{bmatrix}$,

16. $\mathbf{h}_n = \mathbf{h}_{n-1} + \left[\mathbf{k}_{0,n}, \mathbf{k}_{0,n-L}\right] \begin{bmatrix} e_{1,n} \\ -e_{1,n-L} \end{bmatrix}$.

Chapter 3

SUBBAND ACOUSTIC ECHO CANCELLATION USING THE FAP-RLS ALGORITHM: FIXED-POINT IMPLEMENTATION ISSUES

Mohamed Ghanassi
EXFO Fiber Optic Test Equipment
mghanassi@exfo.com

Benoit Champagne
Department of Electrical and Computer Engineering, McGill University
champagne@ece.mcgill.ca

Abstract In this chapter, we investigate quantization effects in the fixed-point implementation of a subband AEC system based on a modified form of the FAP algorithm, referred to as FAP-RLS. The latter is similar in concept to the standard FAP, except for two modifications: use of sliding window RLS-type approach (instead of FRLS) to compute the normalized residual echo vector, for improved robustness in implementation; and simplified update of the residual echo vector. Subband decomposition is performed with modified uniform DFT filter banks, realized efficiently via the weighted overlap-add (WOA) technique for flexibility in oversampling. We characterize the main sources of errors in both the FAP-RLS and the DFT filter banks, and propose simple and effective solutions for stable operation of the subband AEC system. Our findings are supported experimentally.

Keywords: Acoustic Echo Cancellation, Subband Adaptive Filtering, Fast Affine Projection Algorithm, Recursive Least-squares, Fixed-point Implementation, Uniform DFT Filter Banks, WOA Method

1. INTRODUCTION

Acoustic echo cancellation (AEC), combined with proper voice activity detection algorithms, is recognized as one of the best solutions to the control of acoustic echoes generated by hands-free audio terminals in offices and conference rooms [8]. In this approach, the acoustic echo path between the terminal's loudspeaker and microphone (also called room impulse response) is identified by an adaptive filter; the filter output, which provides an electronic replica of the acoustic echo, is subtracted from the microphone signal to cancel the echo.

Fundamental difficulties facing the design of an adaptive filter for AEC include the long duration and time-varying nature of the acoustic echo path as well as the highly non-stationarity character of the excitation signal (i.e. speech). As as result, instantaneous stochastic gradient algorithms such as the normalized least-mean-square (NLMS), often favored for their low computational complexity and robustness, exhibit slow convergence and tracking behaviors and cannot usually achieve a level of performance which is adequate for practical AEC applications, as set forth for example by the International Telecommunication Union (ITU) [9]. Furthermore, and despite the significant and rapid progress in digital signal prossessor (DSP) technology, even the NLMS algorithm is considered too costly by equipment manufacturers for use in commercial products, In the last ten years or so, these cost/performance considerations have driven the search of improved adaptive filtering structures and algorithms for AEC applications.

Among the various structures that have been investigated, subband adaptive filtering still remains one of the most attractive and effective. In subband AEC, echo estimation and cancellation is realized via a set of parallel, independent adaptive filters operating on subband versions of the loudspeaker and microphone signals. Significant computational savings result from operating the subband adaptive filters at a reduced sampling rate, which depends on the number of subbands and the properties of the subband analysis/synthesis filters [11, 7]. Some recent works addressing the problem of filter bank design for AEC applications include [4, 13]. In addition to reduced complexity, improved convergence performance of the adaptive filtering algorithm, as compared to a fullband scheme, may usually be achieved with a subband structure by individually optimizing the algorithm parameters in each subband.

In parallel to these developments, several new adaptive filtering algorithms have been proposed for AEC applications, with the aim of achieving faster and signal-independent convergence while preserving the low complexity of NLMS. Among these, the fast affine projection (FAP) [5, 15] has been receiving considerable attention recently. FAP is a (computationally) fast version of the affine projection algorithm (APA) originally proposed in [14]. It uses efficient schemes for updating the residual error vector and filter weight vector, and fast recursive least-squares (FRLS) predictors for the implicit computation of the inverse data covariance matrix required in APA. Because of well-known difficulties associated to the implementation of FRLS algorithms in practice, i.e. numerical instabilities, memory requirements and code overhead, alternate forms of the standard FAP that make use of a sliding-window RLS-type approach, instead of FRLS, have been proposed [10, 12]. These versions remain attractive for the small values of projection order typically needed in AEC applications.

In this chapter, we investigate quantization effects in the fixed-point implementation of a subband AEC system based on one such alternate FAP algorithm, simply referred to here as the FAP-RLS. In addition to using a sliding-window RLS-type approach as described above, the FAP-RLS under study makes an additional simplification in the computation of the normalized residual echo vector [12]. The FAP-RLS is used in a subband filtering structure based on modified uniform DFT filter banks, realized efficiently via the weighted overlap-add (WOA) technique for flexibility in oversampling [13]. We characterize the main sources of errors in both the FAP-RLS and the DFT filter banks, and propose simple and effective solutions for stable operation of the FAP-based subband AEC system. Our findings are supported experimentally.

This chapter is organized as follows: Section 2 provides an overview of the subband AEC system under consideration, including the FAP-RLS algorithm and the uniform DFT filter banks. Section 3 describes the scope of the study, specifying the objectives and methodology. The quantization effects of fixed-point arithmetic on the FAP-RLS algorithm and the uniform DFT filter banks are studied individually in Section 4 and 5, respectively. The behavior of the complete FAP-based subband AEC system is addressed in Section 6. Section 7 contains some concluding remarks.

2. OVERVIEW OF FAP-BASED SUBBAND AEC SYSTEM

In subband AEC (see Fig. 3.1), the loudspeaker signal, $x(k)$, and the microphone signal, $y(k)$, where k denotes the time index at the original sampling rate F_s, are fed to analysis banks where they are split into K adjacent frequency subbands and downsampled by an integer factor $D \leq K$. The resulting subband signals are denoted as $x_i(m)$ and $y_i(m)$, respectively, where $i = 0, \ldots, K-1$ is the subband index and m is the time index at the reduced rate F_s/D. In each subband, an adaptive filter (AF) with input $x_i(m)$ is used to cancel the echo component in $y_i(m)$, yielding a subband residual echo $e_i(m)$. Finally, the subband residuals are upsampled by D and combined into a single fullband residual $e(k)$ by a synthesis filter bank, for transmission over a communication network. Note that when the near-end user is active, filter adaptation must be disabled to avoid undesirable distortion to his/her signal; this requires the use of a voice activity detector (not discussed here).

As a result of subband downsampling within this approach, important computational savings of the order of D^2/K may be achieved. Another benefit is the improved performance of the adaptive filtering scheme that results from op-

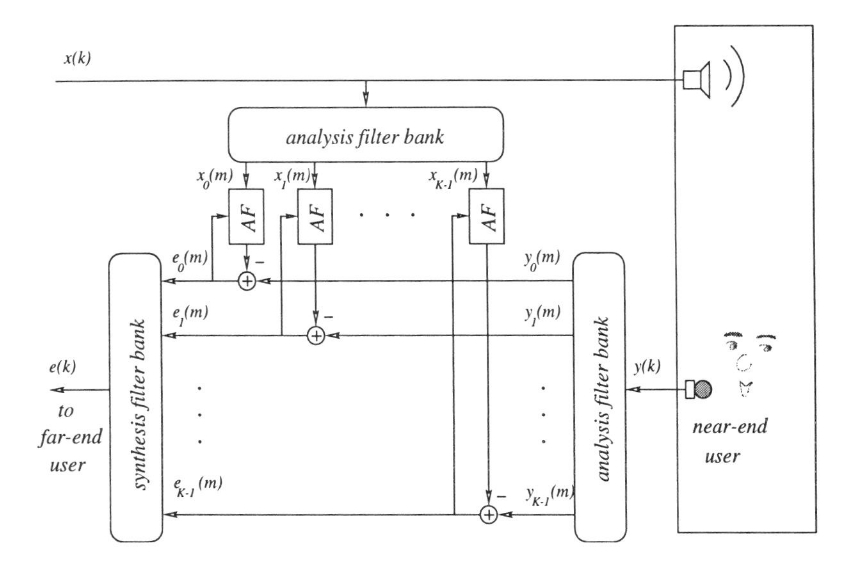

Figure 3.1 Block diagram of generic subband AEC system.

timizing its parameters individually within each subband. The design of filter banks for subband AEC is subject to conflicting requirements: low processing delay, near-perfect reconstruction (NPR) and low complexity (see e.g. [4]). The delay constraint limits the length, and thus the selective properties of the analysis/synthesis filters. To avoid the associated decimation aliasing in the subbands, oversampling (i.e. $D < K$) is now commonly used in practical systems.

The subband AEC system under consideration in this study may be characterized as follows: echo cancellation in the subband is realized with adaptive transversal filters updated separately with so-called FAP-RLS algorithms [12]; subband analysis/synthesis is achieved via oversampled, modified uniform DFT filter banks designed and realized as in [13] to meet the above requirements. Additional details follow.

2.1 FAP-RLS ALGORITHM

In [14], the affine projection algorithm (APA) is proposed as a generalization of NLMS. APA uses an improved direction of update for its transversal filter weight vector, based on a concept of projections on affine subspaces. APA has complexity $2NM$, where N is the length of the weight vector an M is the projection order. For $M = 1$, APA reduces to the NLMS; as M increases, the convergence performance of APA improves as the result of the decorrelation effect of the projections.

FAP is a kind of (computationally) fast version of APA (see Chapter 2). Three crucial steps are invoked in its derivation from APA, namely [5, 15]: (1) fast update of the residual echo vector via a kind of push-down approach; (2) use, and fast adaptation, of an alternate filter weight vector; and (3) use of the sliding-window FRLS algorithm [1] for the implicit computation of the inverse data covariance matrix of size $M \times M$ needed in APA. The complexity of FAP is roughly $2N + 20M$, compared to $O(MN)$ for APA. When using FAP in the AEC context, where large values of N in the thousands are typical, significant improvements in convergence speed over NLMS are observed with only small values of M, that is $M \ll N$.

The main disadvantage of FAP is its use of FRLS in step (3). Despite the low complexity of $10M$ for this step, as compared to $O(M^2)$ for a standard RLS, FRLS is quite sensitive to errors introduced by finite arithmetic, which rapidly lead to its instability. Even the so-called stabilized versions of FRLS suffer from instability when the data covariance matrix is poorly conditioned [6]. This is a serious problem in a fixed-point implementation, and particularly in subband AEC where multiple adaptive filters are run in parallel and the risk of instability is thus higher. In addition, the code overhead and memory requirements of FRLS algorithms are relatively large, so that they do not necessarily represent the most economical solution.

Based on such considerations, modified or alternate versions of FAP have been proposed for AEC applications to further reduce its complexity and/or improve its numerical behavior [10, 3, 12]. In the present study, we focus our attention on a modified FAP algorithm, conveniently called FAP-RLS [12]. There are two main differences between the standard FAP and the FAP-RLS algorithms. Firstly, as in [10], FAP-RLS uses a standard sliding-window RLS-type of approach based on the matrix inversion lemma to update the inverse data covariance matrix (see step (3) above). The main advantage of this modification is a reduced sensitivity to, and easier control of, quantization effects in finite arithmetic. Secondly, a simplification is made in the update of the residual echo vector, as explained below.

To simplify the presentation, fullband AEC notation will be used. In the APA context, the residual echo or error vector is defined as

$$\mathbf{e}(k) = \mathbf{y}(k) - \mathbf{X}^t(k)\mathbf{w}(k) \tag{3.1}$$

where $\mathbf{y}(k) = [y(k), y(k-1), ..., y(k-M+1)]^t$, $\mathbf{X}(k) = [\mathbf{x}(k), \mathbf{x}(k-1), ..., \mathbf{x}(k-M+1)]$, $\mathbf{x}(k) = [x(k), x(k-1), ..., x(k-N+1)]^t$, $\mathbf{w}(k)$ is the APA weight vector of size $N \times 1$ and the superscript ^{t} denotes transposition.

Table 3.1 FAP-RLS algorithm (complex version).

Equation
$\mathbf{a}(k) = \mathbf{R}^{-1}(k-1)\mathbf{s}^*(k)$
$\alpha(k) = [1 + \mathbf{s}^t(k)\mathbf{a}(k)]^{-1}$
$\mathbf{Q}^{-1}(k) = \mathbf{R}^{-1}(k-1) - \alpha(k)\mathbf{a}(k)\mathbf{a}^h(k)$
$\mathbf{b}(k) = \mathbf{Q}^{-1}(k)\mathbf{s}^*(k-N)$
$\beta(k) = [1 - \mathbf{s}^t(k-N)\mathbf{b}(k)]^{-1}$
$\mathbf{R}^{-1}(k) = \mathbf{Q}^{-1}(k) + \beta(k)\mathbf{b}(k)\mathbf{b}^h(k)$
$\mathbf{r}(k) = \mathbf{r}(k-1) + x(k)\overline{\mathbf{s}}^*(k-1) - x(k-N)\overline{\mathbf{s}}^*(k-N-1)$
$e(k) = y(k) - \hat{\mathbf{w}}^h(k-1)\mathbf{x}(k) - \mu\overline{\eta}^h(k-1)\mathbf{r}(k)$
$\epsilon(k) = e^*(k)\mathbf{p}(k)$, where $\mathbf{p}(k) =$ 1st column of $\mathbf{R}^{-1}(k)$
$\eta(k) = \epsilon(k) + [0, \overline{\eta}(k-1)^t]^t$
$\hat{\mathbf{w}}(k) = \hat{\mathbf{w}}(k-1) + \mu\mathbf{x}(N-(M-1))\eta_{M-1}(k)$

In [5, 15], vector $\mathbf{e}(k)$ is updated via

$$\mathbf{e}(k) = \begin{bmatrix} e(k) \\ (1-\mu)\overline{\mathbf{e}}(k-1) \end{bmatrix} \tag{3.2}$$

where $\overline{\mathbf{e}}(k)$ is made up of the first (i.e. top) $M-1$ entries of $\mathbf{e}(k)$ and μ is the relaxation factor (i.e. step size) in APA. We note that in practical applications of FAP, μ is relatively close to one, say $.5 < \mu \leq 1$, so that the last $M-1$ entries in $\mathbf{e}(k)$ (3.2) are scaled down relative to the first entry. It has been observed that for such values of μ, setting the last $M-1$ entries of $\mathbf{e}(k)$ to zero, i.e. forcing

$$\mathbf{e}(k) = \begin{bmatrix} e(k) \\ \mathbf{0} \end{bmatrix} \tag{3.3}$$

has little effects on the FAP's behavior [12]. The FAP-RLS algorithm under consideration here makes use of this simplification, equivalent to setting $\mu = 1$ in (3.2), without changing μ in the remaining FAP equations.

The FAP-RLS algorithm is summarized in its complex version in Table 3.1. For the real version, simply drop the complex conjugate symbol * and replace hermitian transpose ^{h} with plain transpose ^{t}. Note that both the real and complex versions are needed in our particular subband AEC structure (see below). Referring to Table 3.1: $\mathbf{s}(k) = [x(k), ..., x(k-M+1)]^t$, $\mathbf{R}(k)$ is a $M \times M$

sample data covariance matrix, $\mathbf{r}(k)$ is a correlation vector of size $M - 1$, $e(k)$ is the residual echo, $\epsilon(k)$ is the normalized residual echo vector and $\hat{\mathbf{w}}(k)$ is the alternate weight vector. Vector $\epsilon(k)$ is given an interpretation as a decorrelation filter in [15]. Under the approximation (3.3), $\epsilon(k)$ in FAP-RLS is in effect a least-squares forward error prediction filter.

For initialization of the FAP-RLS, we may use $\mathbf{P}(0) = \mathbf{I}/\delta$ where δ is a regularization parameter, $\mathbf{r}(0) = \mathbf{0}$, $\eta(0) = \mathbf{0}$, and $\hat{\mathbf{w}}(0) = \mathbf{0}$. The computational complexity of FAP-RLS in its complex version is about $2N + 3M^2 + 6.5M$ complex multiplies plus 2 divisions per iteration. The complexity of the real version is $2N + 3M^2 + 8M$ real multiplies plus 2 divisions per iteration. For the small values of M used in our application, typically $M \leq 5$, the computational complexity of FAP-RLS is comparable to (or less than) the original FAP.

2.2 UNIFORM DFT FILTER BANKS

We now briefly review the approach of [13] for the design and realization of the oversampled uniform DFT filter banks.

In the analysis bank, the input signal $x(k)$ is passed through a set of K parallel branches in which it is complex modulated by W_K^{-ik}, where $W_K = \exp(j2\pi/K)$ and $i \in \{0, ..., K-1\}$ is the subband index, convolved with a common low-pass prototype filter $h(n)$ (with cut-off at π/K), and downsampled by integer factor D. The digital spectrum $[0, 2\pi]$ is thus divided into K uniform subbands of width $\Delta\omega = 2\pi/K$ and center frequency $i\Delta\omega$. In the synthesis bank, the input subband signals are upsampled by D, low-pass filtered by $g(k)$, modulated by $W_K^{i(k+1)}$ and summed. Aliasing distortion is made acceptably small by using oversampling, i.e. $D < K$. Because of the special choice of modulating functions in the synthesis bank, which differ from those in conventional uniform DFT filter banks, phase distortion may be eliminated by setting $g(k) = h(L - 1 - k)$. Finally, $h(k)$ is obtained as an FIR filter of length $L = n_b K$, where n_b is an integer, by interpolation of a 2-channel quadrature mirror filter (QMF) so that it approximately minimizes a composite measure of amplitude distortion and out-of-band energy.

An efficient weighted-overlap-add (WOA) approach is used for the implementation of the above uniform DFT filter banks [2]. This approach allows the use of arbitrary (integer) values of downsampling D for optimum performance of the subband AEC system, i.e. trade-off between complexity reduction and subband aliasing. The main steps in the WOA implementation of the analysis bank for an arbitrary input signal $x(k)$ are summarized below. At discrete-time m (reduced sampling rate):

1. Compute $y_m(r) = h(r)x(mD - r)$, $r = 0, 1, ..., L - 1$.

2. Partition $\{y_m(r)\}_{r=0}^{L-1}$ into $n_b = L/K$ blocks of K consecutive samples.

3. Compute $\xi_m(t), t = 0, 1, ..., K-1$, by adding the corresponding samples of each block.

4. Compute $x_m(t) = \xi_m((t - mD) \bmod K), t = 0, 1, ..., K - 1$.

5. Compute $x_i(m), i = 0, 1, ..., K - 1$, as the K-point DFT of $\{x_m(t)\}_{t=0}^{K-1}$.

Except for a minor modification, the corresponding steps in the WOA implementation of the synthesis bank are the duals of the above ones. The output of the synthesis bank is scaled by D to achieve unit gain of the overall system.

In the present AEC application, the input signals to the analysis filter banks, i.e. $x(k)$ and $y(k)$ in Fig. 3.1, are real. As a result, subband outputs $x_0(m)$ and $x_{K/2}(m)$ are real while for $i = 1, ..., K/2 - 1$, $x_i(m) = x^*_{K-i}(m)$, with similar properties for the set $\{y_i(m)\}_{i=0}^{K-1}$. Thus, in the subband AEC system, real versions of FAP-RLS are used in subbands 0 and $K/2$, complex versions are used in subbands 1 to $K/2 - 1$, and no processing is required in the remaining subbands, the corresponding error signals being obtained from symmetry. The total complexity of the subband AEC system realized in this way is about

$$\frac{2(K-1)}{D}(2N_s + 3M^2 + 6.5M + 2k_d) + \frac{3}{D}(L + K \log_2 K) \qquad (3.4)$$

real multiplications per iteration at the rate F_s, where N_s is the length of the subband adaptive filters and k_d is the complexity of a real division.

3. SCOPE OF FIXED-POINT STUDY

In the past, we have extensively studied the performance of the above subband FAP-RLS based AEC system in floating-point arithmetic, using off-line C-language implementations running on computer workstations and a real-time assembler implementation running on a TMS320C40 DSP from Texas Instruments (TI). Compared to a fullband NLMS-based AEC system, the subband FAP-RLS may achieve much faster convergence at a fraction of the cost.

Due to cost considerations, equipment manufacturers generally prefer the use of fixed-point DSPs over floating-point ones in their AEC products. Quantization effects in the former result in a deviation of the adaptive filter performance from that predicted or observed in infinite precision. This deviation, which becomes more apparent as the number of representation bits is reduced, may take the form of an increased residual error after convergence (i.e. loss of precision), or more dramatically, of an unbounded accumulation of quantization errors over time (i.e. numerical instability). In practice, the cost of an implementation is an increasing function of the number of bits and there is a motivation to minimize this number (subject to DSP availability) while ensuring correct operation of the algorithm.

The main objective of this study is to develop a better understanding of quantization effects on the behavior of the subband FAP-RLS when implemented in fixed-point. This is judged important for minimizing implementation costs and avoiding undesirable numerical behaviors of the algorithm in commercial applications. Both issues of numerical stability and precision are addressed, with the aim of identifying potential problems and providing practical solutions for good and reliable operation of the algorithm in fixed-point.

We assume a b-bit fixed-point fractional representation extending from -1 to $1-\Delta$, where $\Delta = 2^{-(b-1)}$. MSB denotes the most significant bit (left) and LSB denotes the least significant bit (right). Two's complement is used for representing negative numbers. The product of two numbers may be quantized by either truncating or rounding. Experimental results not reported here show a much faster error accumulation with truncation. Accordingly, rounding is assumed in our analysis and used in the DSP experiments. The quantization error in a number x is defined as $q_x = Q(x) - x$, where $Q(.)$ is the quantization operator. Quantization errors are modeled as independent random variables uniformly distributed within $\pm\Delta/2$; the corresponding variance is $\Sigma_q^2 = \Delta^2/12$.

Overflow may occur as the result of addition or division. In most commercial fixed-point DSPs, overflow is minimized by the use of accumulators with additional guard bits to the left of the MSB. Scaling (i.e. shift) of a quantity may still be necessary to avoid overflow during memory transfer. Due to lack of space, only very essential details are provided on scaling, although it remains an essential programming aspect for optimal use of fixed-point resources.

The fixed-point behavior of the subband FAP-RLS is investigated analytically and by computer simulations. To this latter end, we have developed several programs that enable us to control the number of bits used for the representation of numbers, as well as other related aspects. The 16-bit representation is particularly important for applications. In this particular case, we have developed a software application written in C that emulates the functionality of the TMS320C54 DSP from TI.

4. FIXED-POINT IMPLEMENTATION OF FAP-RLS

In this section, we study the quantization effects introduced by fixed-point arithmetic on the behavior of the FAP-RLS. To simplify the presentation, full-band notation is assumed. However, in all experiments, the algorithm parameters are set to values that are representative of a subband AEC application. In particular, the filter length is set to $N = 128$ and only small values of M are used. The length of the acoustic echo path is also set to 128 and Gaussian white noise is added to the echo when appropriate.

4.1 UPDATE OF INVERSE DATA COVARIANCE MATRIX

In FAP-RLS, the update of the inverse data covariance matrix is achieved via a double application of a matrix inversion formula for rank-one additive modification (so-called matrix inversion lemma). As is well known, this approach is also subject to error accumulation and thus plays a key role in the overall stability aspect of the FAP-RLS.

Proceeding as in [16], we first investigate the propagation of a single quantization error in the computation of $\mathbf{R}^{-1}(k)$ in Table 3.1. For simplicity, first consider the case $M = 1$, where $\mathbf{R}(k)$ and $\mathbf{Q}(k)$ are scalar quantities. Assume that at time k_o, a quantization error $q(k_o)$ is introduced in $\mathbf{R}^{-1}(k_o)$. Using the equations in Table 3.1, one may show that the resulting error propagated by the algorithm at time $k \geq k_o$ is $q(k) = G(k)q(k_o)$ where

$$G(k) = \prod_{i=k_o+1}^{k} \left[1 + \mathbf{R}^{-1}(i-1)(|x(i)|^2 - |x(i-N)|^2)\right]^{-2}. \tag{3.5}$$

The behavior of the gain $G(k)$ has been investigated numerically. In the case of a stationary Gaussian white noise process $x(k)$, the results show a stationary behavior of $G(k)$, i.e. no decay of the quantization error with time. In light of (3.5), this is understandable since one may show that $E\{[1 + (|x(i)|^2 - |x(i-N)|^2)/\sigma_x^2]^{-2}\} \simeq 1$. In practice, i.e. where quantization errors would be made at every iteration, this behavior favors the accumulation of errors. In the case of a non-stationary signal such as speech, $G(k)$ is observed to increase significantly at the beginning of an interval of small signal amplitude. This behaviors, which is also predictable from (3.5), suggests that in a practical setting, quantization errors would build up very rapidly at the beginning of such intervals. Similar conclusions apply to the case $M > 1$, although the generalization and interpretation of (3.5) is more elaborate.

The accumulation effect of quantization errors in $\mathbf{R}^{-1}(k)$ is next investigated via computer experiments on the TMS320C54 simulator. We begin by noting that to avoid overflow, it is necessary to scale down (i.e. shift towards the LSB) the quantities $\mathbf{a}(k)$, $1 + \mathbf{s}^t(k)\mathbf{a}(k)$, $\mathbf{b}(k)$ and $\beta(k)$ that occur in the update of $\mathbf{R}^{-1}(k)$. More importantly, when the regularization parameter δ is smaller than one, which is not uncommon, the entries of $\mathbf{R}^{-1}(k)$ and $\mathbf{Q}^{-1}(k)$ must also be scaled down to avoid overflow. While these shifts may be compensated for in the algorithm, the ensuing loss of precision can not. In this context, the error accumulation phenomenon in the fixed-point computation of $\mathbf{R}^{-1}(k)$ is strongly influenced by the value of δ.

To support this claim, Fig. 3.2 illustrates the build-up of quantization errors in a 16-bit implementation of the $\mathbf{R}^{-1}(k)$ update, as observed experimentally for $\delta = 10\sigma_x^2$ and $50\sigma_x^2$. In these experiments, the input signal $x(k)$ is a

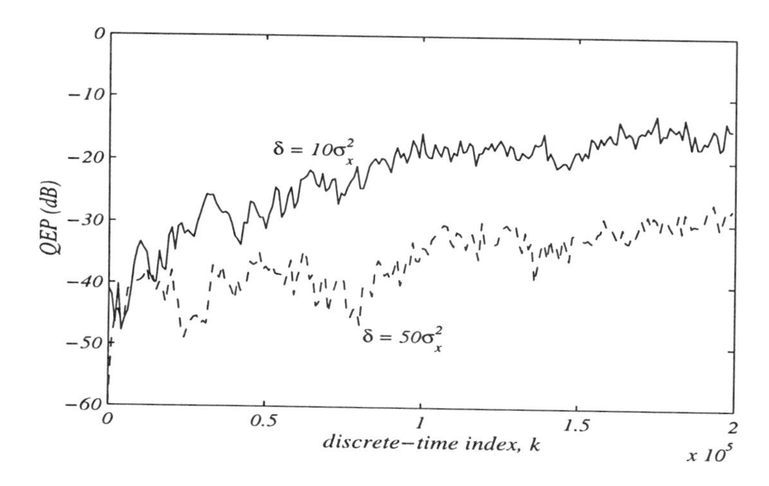

Figure 3.2 Quantization error power (QEP) in $[\mathbf{R}^{-1}(k)]_{11}$ versus time index k in 16-bit implementation of inverse data covariance matrix update for $\delta = 10\sigma_x^2$ and $50\sigma_x^2$.

stationary Gaussian white noise with variance $\sigma_x^2 = 0.025$; the value of the projection order is set to $M = 4$. For each value of δ, Fig. 3.2 shows the time evolution of the short-time normalized power of the quantization error, the latter being defined as $q(k) = [\mathbf{R}_q^{-1}(k) - \mathbf{R}^{-1}(k)]_{11}$, where $\mathbf{R}_q^{-1}(k)$ is computed in fixed-point with 16-bit precision and $\mathbf{R}^{-1}(k)$ is computed in floating-point with 64-bit precision (i.e. $\approx$ infinite precision). In both cases, the error increases with time; however, the rate of increase is faster for the smaller value of δ. In other experiments with non-stationary signals (e.g. speech), a net increase in the error power is observed at the beginning of an interval of small signal amplitude, in agreement with theoretical model (3.5).

In general, we observe a divergence of the FAP-RLS algorithm when the error level in Fig. 1 reaches about -10 dB. Based on such results, we conclude that 16-bit precision is not sufficient for the computation of $\mathbf{R}^{-1}(k)$. To extend the stable life of FAP-RLS, we increase the number of bits used in the computation of $\mathbf{R}^{-1}(k)$ only. Specifically, all the variables entering this computation are represented with 32 bits (i.e. two 16-bit words), except for $1 + \mathbf{s}^t(k)\mathbf{a}(k)$ and $1 - \mathbf{s}^t(k - N)\mathbf{b}(k)$ in Table 3.1, which are still represented with 16 bits to avoid costly 32-bit divisions. The evolution of the quantization error power for this 32/16-bit implementation is shown in Fig. 3.3 (curve labeled 32/16), along with the corresponding result for a 16-bit implementation (curve labeled 16-bit). In both cases, $\delta = 12\sigma_x^2$. The level of error in the 32/16-bit implementation is much lower than in the 16-bit one, by at least 40 dB. However, the computation of $\mathbf{R}^{-1}(k)$ with the 32/16-bit approach requires about 2.5 more cpu cycles. For small M, this is quite acceptable.

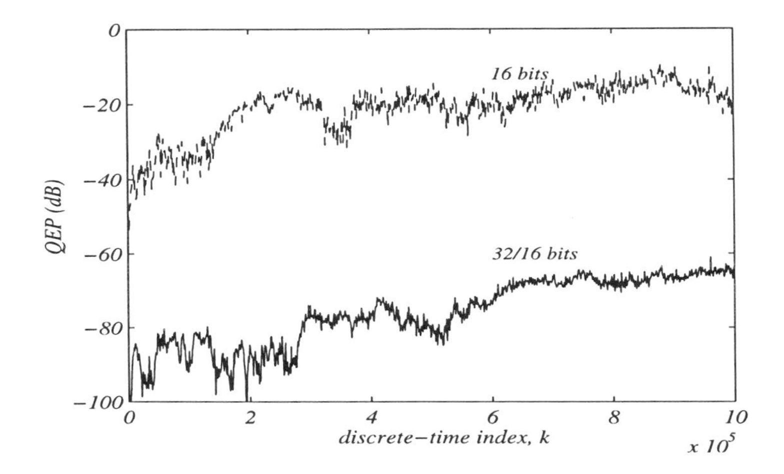

Figure 3.3 Quantization error power (QEP) in $[\mathbf{R}^{-1}(k)]_{11}$ versus time index k in 16-bit and 32/16-bit implementations of inverse data covariance matrix update ($\delta = 12\sigma_x^2$).

4.2 UPDATE OF CORRELATION VECTOR

Another potential source of error accumulation in FAP algorithms is in the computation of the correlation vector $\mathbf{r}(k)$ in Table 3.1 [5]. To minimize such accumulation in practice, one possibility is to represent the entries of $\mathbf{r}(k)$ in double precision, i.e. using $2b$ bits. However, we have found it simpler to use the following approach in updating $\mathbf{r}(k)$:

$$\mathbf{r}(k) = \mathbf{r}(k-1) + Q[x(k)\bar{\mathbf{s}}(k-1)] - Q[x(k-N)\bar{\mathbf{s}}(k-1-N)] \tag{3.6}$$

where $Q(.)$ denotes quantization by rounding. This approach introduces a small quantization error in the ith entry of $\mathbf{r}(k)$, equal to the sum of the N quantization errors of the products $x(t)x(t-1-i)$ $(t = k, ..., k-N+1)$, but completely eliminates the problem of error accumulation over time. This is exemplified in Fig. 3.4, which shows the evolution of the quantization error power in $[\mathbf{r}(k)]_1$ under the same experimental conditions as in Fig. 3.3.

4.3 FILTERING AND ADAPTATION

The computation of the residual error $e(k)$ and the filter weight vector $\hat{\mathbf{w}}(k)$ in FAP-RLS (see Table 3.1) is similar in nature to that in the NLMS algorithm, and is apparently not a source of instability. The computation of $\eta(k)$, for its part, does not permit error accumulation. However, these operations either explicitly or implicitly use $\mathbf{R}^{-1}(k)$, so that the complete FAP-RLS algorithm may exhibit unstable behavior.

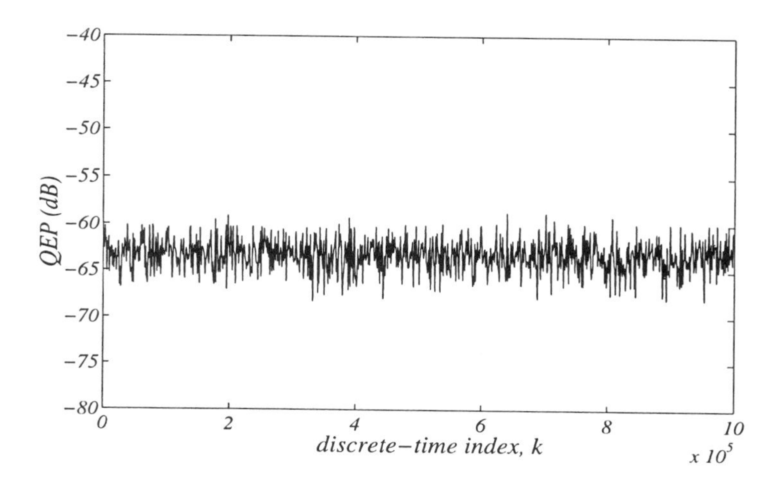

Figure 3.4 Quantization error power (QEP) in $[\mathbf{r}(k)]_1$ versus time index k in 16-bit implementation of (3.6).

As pointed out earlier, it has been observed experimentally that with stationary data, the accumulation of quantization errors in $\mathbf{R}^{-1}(k)$ eventually leads to instability of the FAP-RLS when the level of error reaches a certain threshold. This type of behavior was further investigated in the case of non-stationary signals using a composite source signal (CSS) as input $x(k)$. The CSS used here is made up of a succession of identical bursts, each burst being the concatenation of a voice sound, a pseudo-noise and a pause. The duration of a burst is about one third of a second. With 16-bit precision, we observe important instabilities of FAP-RLS occurring principally during silence periods, after a period of rapid error accumulation in $\mathbf{R}^{-1}(k)$. Reinitialization (see below) is not always effective to overcome this problem. However, when the FAP-RLS is implemented in 16-bit and the computation of $\mathbf{R}^{-1}(k)$ in 32/16-bit as described above, a robust performance is observed and the algorithm remains operational over extremely long periods of time.

To avoid potential problems related to the sudden and fast accumulation of quantization errors in $\mathbf{R}^{-1}(k)$ at the beginning of a silence period in practical situations involving speech signals, we have implemented and tested a reinitialization mechanism in which $\mathbf{R}^{-1}(k)$ is reset to $\mathbf{I}/\delta$ after detection of a silence period (i.e. signal power below predefined threshold) of duration N samples. In this case, previous signal samples $x(k-i)$, $i = 0, ..., N-1$, and vectors $\eta(k)$ and $\mathbf{r}(k)$ are also reset to zero. With this mechanism, instability were never observed in our experiments with the 32/16-bit implementation. This approach is simple, does not require significant extra computations, and has no effects on the observed convergence behavior of the FAP-RLS. Combined with the 32/16-bit implementation technique, it practically ensures the stability of the FAP-RLS algorithm.

4.4 ALGORITHM PRECISION

We finally investigate the effects of the number of representation bits, b, on the transient and steady-state behaviors of the algorithm.

To this end, we use a set of C-language programs allowing the emulation of a fixed-point DSP with a variable number of bits. As is typical for this type of experiments, the input signal $x(k)$ is a stationary Gaussian white noise with variance σ_x^2. To allow the measurement of the quantization noise power in the residual echo signal $e(k)$, no background noise is added to the reference (i.e. microphone) signal, $y(k)$. These measurements, in the form of short-time averaged learning curves, are made immediately after initial startup for a period of a few seconds, before error accumulation in $\mathbf{R}^{-1}(k)$ becomes significant. Thus, precision may be studied independently from stability.

In Fig. 3.5, the short-time power of the residual echo (or error) signal $e(k)$ at the output of the FAP-RLS is plotted as a function of time for different values of b. The algorithm parameters have been set to: $\delta = \sigma_x^2 = 0.1$, $\mu = 1.0$ and $M = 4$. It can be observed that the quantization errors have no effects on the initial convergence speed of the algorithm. In the steady-state regime, a decrease in the residual error level of roughly 7 dB to 9 dB per additional bit is observed. With $b = 16$ bits, the residual error level would remain below that of the background noise usually present in applications.

5. FIXED-POINT WOA IMPLEMENTATION

Since the WOA technique makes use of the DFT, we first investigate and compare the effects of quantization on a direct computation of the DFT and on the FFT algorithm. Quantization effects in the analysis steps and synthesis steps of the WOA method are then studied separately. Since all the operations involved are stable, only the precision aspect is considered.

5.1 DFT OR FFT?

The K-point DFT of a sequence $x(k)$, $k = 0, .., K-1$ is defined by $X(n) = \sum_{k=0}^{K-1} x(k) W_K^{nk}, n = 0, ..., K-1$, where $W_K = e^{-j2\pi/K}$. In practice, instead of using the above definition (i.e. direct method), the DFT is evaluated via the FFT algorithm. However, for small values of K, the computational complexity of the FFT does not largely exceed that of the direct method. Thus, a comparative evaluation of quantization effects on both approaches is justified.

To evaluate the signal-to-quantization noise ratio (SQNR), we assume a statistical model of the quantization error as described previously. Quantization of a complex number introduces an error $q = q_r + jq_i$ with variance $2\Sigma^2$.

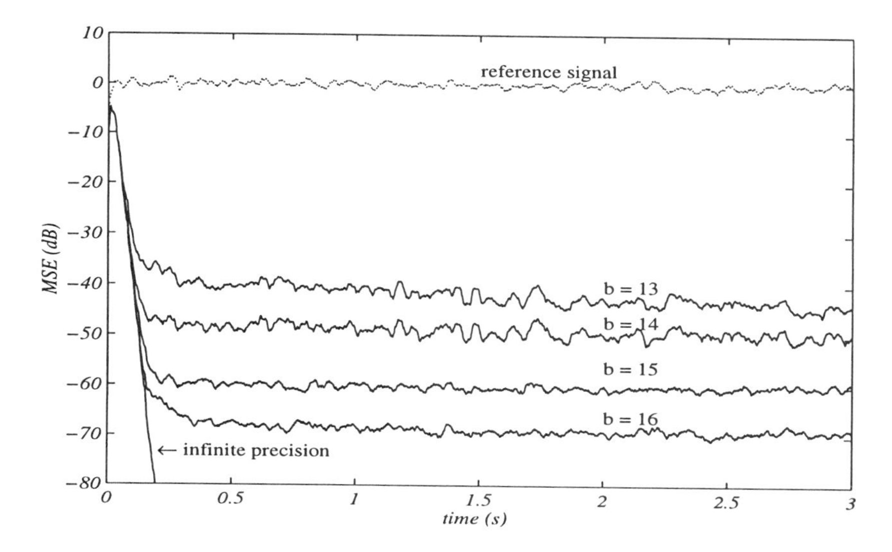

Figure 3.5 Short-time power of error signal $e(k)$ versus time in FAP-RLS for different precision b in bits.

For simplicity, the signal $x(k)$ (either real or complex) is modeled as a white noise sequence with variance σ_x^2. We assume that an error q_x of variance Σ_x^2 has been introduced in the quantization of $x(k)$. We also assume that the fixed-point DSP under consideration has at least $\log_2 K$ guard bits beyond the MSB in its accumulators, so that scaling is only necessary when the final result is being transferred to memory. To avoid overflow during the transfer, the result of the computation is in effect scaled down (shifted towards LSB) by a factor $1/K$ (since $|X(n)| \leq \sum_{k=0}^{K-1} |x(k)| < K$).

For the direct method, a straightforward calculation of the SQNR yields the following result:

$$SQNR \simeq \sigma_x^2/(2K\Sigma^2 + \Sigma_x^2) \tag{3.7}$$

For the FFT, the scaling factor $1/K$ is distributed equally over the different stages. A more elaborate calculation, involving a recursion over the butterfly index, then yields

$$SQNR \simeq \sigma_x^2/(4K\Sigma^2 + \Sigma_x^2) \tag{3.8}$$

According to (3.7)-(3.8), the direct method presents a small advantage of at most 3 dB with respect to the FFT. As an example, for a real process $x(k)$ with $\sigma_x^2 = 0.1$, $\Sigma_x = \Sigma$, $K = 16$ and $b = 16$ bits, the computed SQNR is about 76 dB for the direct method and 73 dB for the FFT. These figures perfectly match the values measured experimentally with the TMS320C54 simulator.

The small loss in SQNR with FFT is compensated by its faster operation: for $K = 16$, the FFT is 1.6 times faster than the direct method for real data and 3.4 times for complex data. Based on these observations, FFT remains advantageous in our application.

5.2 ANALYSIS BANK

We next study quantization effects in the analysis steps of the WOA implementation. The signal $x(k)$ is modeled as a real, white noise sequence with variance σ_x^2. We assume that quantization errors $q_x(k)$ and $q_h(k)$ are present in $x(k)$ and the low-pass filter $h(k)$ (also real), with variance $\Sigma_h^2 = \Sigma_x^2 = \Sigma^2$. Considering the introduction of quantization errors at the various steps in the analysis portion of the WOA method, the propagation of these errors in subsequent steps, and using results obtained in the above study of the FFT algorithm, one may show that the SQNR in each branch $x_i(m)$ at the output of the analysis bank is approximately given by

$$SQNR \simeq \|\mathbf{h}\|^2 \sigma_x^2 / [(L + 4K^2)\Sigma^2] \quad (3.9)$$

where $\mathbf{h} = [h(0), h(1), ..., h(L-1)]^t$ is a vector containing the coefficients of the analysis prototype filter. The above formula is valid provided $\|\mathbf{h}\|^2/L \ll 1$ and $\sum_{s=0}^{n_b-1} h(t+sK)^2 \simeq n_b \|\mathbf{h}\|^2/L$. These two conditions are usually satisfied in practice.

As an example, consider a real white noise process $x(k)$, with variance $\sigma_x^2 = 0.1$, going through the analysis bank in Design Example 2 in [13], for which $K = 16$, $D = 12$ and $L = 128$. In this case, the above formula yields an SQNR of about 60 dB. This result is in perfect agreement with fixed-point experiments on the TMS320C54 simulator. Such level of quantization noise is well below the background noise level in typical subband AEC applications. Accordingly, we conclude that quantization errors in the analysis banks do not represent a significant impairment.

5.3 SYNTHESIS BANK

We assume that the subband signals at the input of the synthesis bank, say $x_i(m)$ ($i = 0, ..., K-1$), result from passing a real signal $x(k)$ with variance σ_x^2 through the analysis bank, implemented with the WOA method. The analysis of the previous section actually shows that the variance of $x_i(m)$ is $\sigma_{x_i}^2 \simeq \|\mathbf{h}\|^2 \sigma_x^2 / K^2$ and that the variance of the quantization error in $x_i(m)$ is $\Sigma_{x_i}^2 = (4 + L/K^2)\Sigma^2$. Going through a similar analysis, i.e. following the progression of the signal and quantization error variances through the various steps of the synthesis portion of the WOA method, one may show that the SQNR at the output of the synthesis bank, denoted $\hat{x}(k)$, is approximately given by

$$SQNR \simeq K\sigma_x^2/[D\|\mathbf{h}\|^2(L+6K^2)\Sigma^2] \tag{3.10}$$

For the same example as in Section 4.2, (3.10) gives an SQNR of about 60 dB, in good agreement with fixed-point experiments. For this example, we find $\Sigma_{x_i}^2 \simeq 4.5\Sigma^2$ after the analysis and $\Sigma_{\hat{x}}^2 \simeq 1000\Sigma^2$ after the synthesis. Thus, quantization errors introduced in the synthesis bank are much more important than those introduced in the analysis bank. We also find that the reconstruction error at the output of the synthesis bank is about -40 dB, as a result of non-ideal selective property of the prototype filter $h(k)$. Our main conclusion here is that in a 16-bit fixed-point WOA implementation of the uniform DFT filter banks, with practical values of prototype filter length L, quantization errors are usually small compared to the reconstruction error and may be neglected.

6. EVALUATION OF COMPLETE ALGORITHM

We finally investigate the behavior of the complete subband FAP-RLS algorithm when implemented in fixed-point.

The parameters of the filter banks are chosen as follows: $K = 16$, $D = 12$, $L = 128$, $h(k)$ as in Design Example 2 in [13]. For the subband FAP-RLS algorithms, we use $N = 1000/12 = 83$ and $M = 4$ in all subbands (the values of μ and δ in each of the subbands may differ slightly). All the algorithms are implemented in 16-bit fixed-point using the TMS320C54 simulator. The 32/16-bit technique is used for the computation of $\mathbf{R}^{-1}(k)$. For the evaluation, a CSS sequence sampled at $F_s = 8$ kHz is used as the loudspeaker $x(k)$. The microphone signal $y(k)$ is obtained by convolving $x(k)$ with a room response, whose polarity is reversed every 25 seconds to simulate time-varying conditions; background noise at -30 dB is added to the result.

The short-term power of the fullband residual echo at the output of the synthesis bank is shown in Fig. 3.6. Note that reinitialization has not been used here, although it may be included in practice for safer operation.

7. CONCLUSION

In this chapter, we investigated quantization effects in the fixed-point implementation of a subband AEC system using the FAP-RLS algorithm in an oversampled, uniform DFT subband structure. The main sources of errors in both the FAP-RLS and the DFT filter banks were identified and characterized. Simple and effective solutions for stable operation of the subband AEC system were proposed and their validity confirmed by computer experiments in fixed-point environments.

As expected, the most critical aspect for stable and reliable operation of the subband FAP-RLS algorithm in fixed-point is the choice of a proper scheme

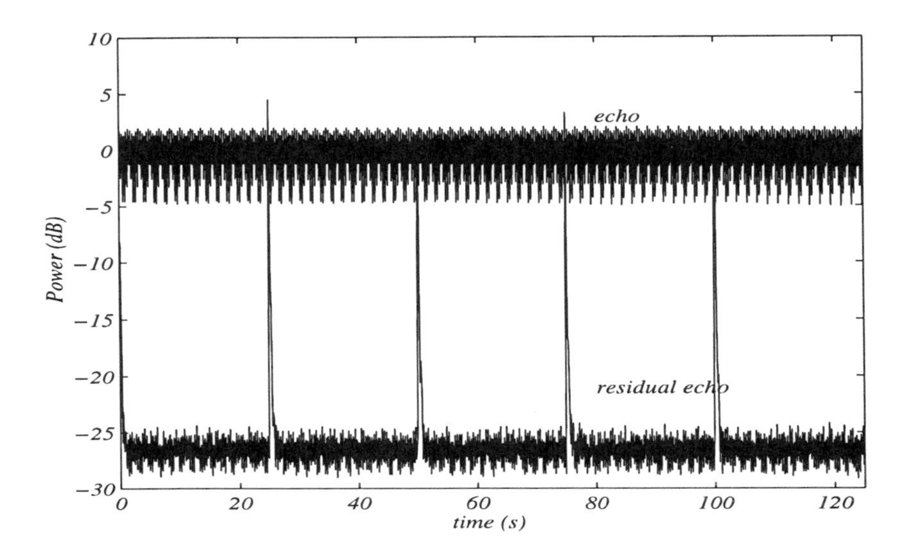

Figure 3.6 Short-term power of residual echo in fixed-point implementation of subband FAP-RLS.

for the implementation of the recursive update of the inverse data covariance matrix $\mathbf{R}^{-1}(k)$. The situation is aggravated by the necessity to scale down the entries of this matrix when the regularization parameter in FAP is small. The solution that we put forward involves the use of a mixed-precision, i.e. 16/32-bit approach in the implementation of the update for $\mathbf{R}^{-1}(k)$ along with a simple reinitialization scheme during periods of low signal power. A simple scheme to avoid error accumulation in the correlation vector was also proposed. With respect to precision, results indicate that the quantization errors introduced by the FAP-RLS and the uniform DFT filter banks do not significantly affect the performance of the algorithm in a 16-bit environment.

With the proposed implementation strategy, the FAP-RLS based subband AEC system described here provides a low-cost, efficient and reliable solution to the control of acoustic echoes in commercial applications.

References

[1] J. M. Cioffi and T. Kailath, "Windowed fast transversal filters adaptive algorithms with normalization," *IEEE Trans. Acoust., Speech, Signal Processing*, vol. 33, pp. 607-625, June 1985.

[2] R. E. Crochiere and L. R. Rabiner, *Multirate Digital Signal Processing*. Prentice Hall, 1983.

[3] S. C. Douglas, "Efficient approximate implementations of the fast affine projection algorithm using orthogonal transforms," in *Proc. IEEE ICASSP*, Atlanta, U.S.A., 1996, pp.

1656-1659.

[4] B. Farhang-Boroujeny and Z. Wang, "Adaptive filtering in subbands: design issues and experimental results for acoustic echo cancellation," *Signal Processing*, vol. 61, pp. 213-223, 1997.

[5] S. L. Gay and S. Tavathia, "The fast affine projection algorithm," in *Proc. IEEE ICASSP*, Detroit, U.S.A., 1995, pp. 3023-3026.

[6] S. L. Gay, "Dynamically regularized fast RLS with application to echo cancellation," in *Proc. IEEE ICASSP*, Atlanta, U.S.A., 1996, pp. 957-960.

[7] A. Gilloire and M. Vetterli, "Adaptive filtering in subbands with critical sampling: analysis, experiments, and application to acoustic echo cancellation," *IEEE Trans. Signal Processing*, vol. 40, pp. 1862-1875, Aug. 1992.

[8] A. Gilloire, "Recent advances in adaptive filtering algorithms for acoustic echo cancellation," in *Proc. IWAENC*, Roros, Norway, 1995, pp. 115-134.

[9] *Acoustic Echo Controllers.* ITU-T Recommendation G.167, Feb. 1994.

[10] Y. Kaneda, M. Tanaka, and J. Kojima, "An adaptive algorithm with fast convergence for multi-input sound control," in *Proc. ACTIVE*, Newport Beach (CA), U.S.A., 1995, pp. 993-1004.

[11] W. Kellermann, "Analysis and design of multirate systems for cancellation of acoustical echoes," in *Proc. IEEE ICASSP*, New York, U.S.A., 1988, pp. 2570-2573.

[12] Q. G. Liu, B. Champagne, and K. C. Ho, "On the use of a modified FAP algorithm in subbands for acoustic echo cancellation," in *Proc. 7th IEEE DSP Workshop*, Loen, Norway, 1996, pp. 354-357.

[13] Q. G. Liu, B. Champagne, and K. C. Ho, "Simple design of oversampled uniform DFT filter banks with applications to subband AEC," submitted to *Signal Processing*, Nov. 1998. (See also: Q. G. Liu *et al.*, "Simple design of filter banks...," in *Proc. IWAENC-97*, pp. 132-135.)

[14] K. Ozeki and T. Umeda, "An adaptive filtering algorithm using an orthogonal projection to an affine subspace and its properties," *Elec. and Comm. in Japan,* vol. 67-A, pp. 126-132, Feb. 1984.

[15] M. Tanaka, Y. Kaneda, S. Makino, and J. Kojima, "Fast projection algorithm and its step size control," in *Proc. IEEE ICASSP*, Detroit, U.S.A., 1995, pp. 945-948.

[16] M. H. Verhaegen, "Round-off error propagation in four generally-applicable, recursive, least-squares estimation schemes," *Automatica*, vol.25, pp. 437-444, 1989.

Chapter 4

REAL-TIME IMPLEMENTATION OF THE EXACT BLOCK NLMS ALGORITHM FOR ACOUSTIC ECHO CONTROL IN HANDS-FREE TELEPHONE SYSTEMS

Bernhard H. Nitsch

Fachgebiet Theorie der Signale, Darmstadt University of Technology

nitsch@nesi.tu-darmstadt.de

Abstract Adaptive filter algorithms that employ a block processing approach converge slower for colored excitation signals like speech than their sample by sample counterparts. Furthermore, the block processing introduces a signal delay which increases with rising block length. If the error signal of a block adaptive filter is corrected in such a way that the corrected error signal equals the error signal of the sample-by-sample version of the adaptive filter, then both algorithms are mathematically identical. This error correction method is applied to many algorithms such as the NLMS, the affine projection, and the fast Newton algorithm. In this contribution, the exact block NLMS algorithm is examined in detail. To reduce the signal delay and the numerical complexity, the error signal correction method is combined with a partitioning of the filter and with the overlap save implementation method. The resulting algorithm has a low numerical complexity, a moderate signal delay, and the same convergence behavior as the NLMS algorithm. Finally, real-time implementation aspects of the algorithm are discussed.

Keywords: Frequency Domain Adaptive Filter, NLMS Algorithm, Acoustic Echo Canceler, Real-Time Implementation

1. INTRODUCTION

Nowadays, hands-free telephony for mobile communications and for teleconferencing systems are becoming increasingly important in order to satisfy users' concern for both safety and convenience. Echoes arising from the acoustic coupling between the loudspeaker and microphone on the near end of a connection is disturbing to users on the far end. Echo control must be used to insert sufficient echo return loss for comfortable conversations. Echo control techniques typically employ adaptive filters to mimic the impulse response of the acoustic coupling to produce an echo estimate which is then subtracted from the microphone signal. The impulse response can be 300 to 600 ms long in normal sized rooms. This requires several hundreds of coefficients for the impulse response of the echo canceler.

Two challenging aspects of such long filters include the large computational complexity and the ability of the filter to track changes in the acoustic coupling. Frequency domain based adaptive filters are able to reduce the numerical complexity by using the overlap-and-save implementation method. By this method, the error signal and the correction term of the filter impulse response are calculated via the frequency domain. The overlap-and-save implementation method is based on a block processing approach which introduces a signal delay. The signal delay can be reduced by a partitioning of the filter into sub-filters of smaller length. Due to the block processing approach, the frequency domain adaptive filter converges slower than the sample-by-sample counterpart in the time domain. To improve the convergence property, an error signal correction may be employed which corrects the error signal in such a way that it fits with the error signal of the sample-by-sample version of the algorithm.

The outline of this chapter is as follows: In Section 2, the general structure of adaptive filters based on block processing is introduced. The error signal correction is explained in Section 3 and in Section 4 the partitioning of the filter. In Section 5, the PEFBNLMS algorithm is presented which combines the error signal correction, the partitioning of the filter, and the overlap-and-save implementation. Performance and implementation of the PEFBNLMS algorithm on a digital signal processor for acoustic echo control in a hands-free telephone system are elucidated in Sections 6 and 7. We conclude this chapter with a summary.

2. BLOCK PROCESSING

In this contribution, we use the following notation: Underlined lower-case characters $\underline{x}_M$ are column vectors in the time domain, whereas underlined capital letters $\underline{X}_M$ denote column vectors in the frequency domain. The index M is the length of the vectors. For vectors in the time domain, the most recent sample is the last coefficient of the vector. Bold upper-case characters $\boldsymbol{X}_{Y,X}$ are matrices which consists of Y rows and X columns. The operator diag$\{\ldots\}$ transforms a vector to a diagonal matrix whose main diagonal is formed by the elements of the vector. Finally, the superscript 'T' describes the transpose operation, the symbol '$\otimes$' an elementwise multiplication of two vectors, and the symbol '$*$' the conjugate operation of a complex value.

Adaptive filters that are based on a block processing structure collect B input signal samples before B output signal samples are calculated. Therefore, a signal delay is introduced [5]. The impulse response of the filter is adapted only once every B sampling periods. In the filter part of the adaptive filter, the error signal vector

$$\underline{e}_B(kB) = \underline{d}_B(kB) - \boldsymbol{X}_{C,B}^T(kB) \cdot \underline{w}_C(kB) \tag{4.1}$$

is calculated by a matrix vector multiplication of the excitation signal matrix

$$\boldsymbol{X}_{C,B}(kB) = \left[\underline{x}_C(kB - B + 1), \underline{x}_C(kB - B + 2), \ldots, \underline{x}_C(kB)\right] \tag{4.2}$$

and the impulse response $\underline{w}_C(kB)$ of the filter. The vector $\underline{d}_B(kB)$ is the desired signal vector, B the block length, and C the filter length. In the adaptation part, the impulse response of the filter is updated in the following way:

$$\underline{w}_C(kB + B) = \underline{w}_C(kB) + \alpha(kB) \cdot \boldsymbol{X}_{C,B}(kB) \cdot \underline{e}_B(kB), \tag{4.3}$$

where $\alpha(kB)$ is a step-factor which controls the convergence behavior of the filter.

3. THE EXACT BLOCK NLMS ALGORITHM

Since the adaptive filter described in Section 2 adapts the impulse response of the filter only once every B sampling periods the convergence speed is smaller compared to the sample-by-sample version of the algorithm for colored excitation signals like speech. To improve the convergence behavior, we can correct the error signal in such a way that the error signal is equal to the error signal one would receive using the sample-by-sample NLMS algorithm [1, 5]. To correct the error signal, we calculate B consecutive error signal samples of the NLMS algorithm using the impulse response $\underline{w}_C(kB)$ of the filter:

$$
\begin{aligned}
e^{(NLMS)}(kB-B+1) = {} & d(kB-B+1) \\
& - \underline{x}_C^T(kB-B+1) \cdot \underline{w}_C(kB) \\
e^{(NLMS)}(kB-B+2) = {} & d(kB-B+2) \\
& - \underline{x}_C^T(kB-B+2) \cdot \underline{w}_C(kB) \\
& - \frac{\alpha(kB-B+1)}{\|\underline{x}_C(kB-B+1)\|^2} \cdot e^{(NLMS)}(kB-B+1) \\
& \cdot \underline{x}_C^T(kB-B+2) \cdot \underline{x}_C(kB-B+1) \\
& \ldots \\
e^{(NLMS)}(kB) = {} & d(kB) \\
& - \underline{x}_C^T(kB) \cdot \underline{w}_C(kB) \\
& - \frac{\alpha(kB-B+1)}{\|\underline{x}_C(kB-B+1)\|^2} \cdot e^{(NLMS)}(kB-B+1) \\
& \cdot \underline{x}_C^T(kB) \cdot \underline{x}_C(kB-B+1) \\
& - \ldots \\
& - \frac{\alpha(kB-1)}{\|\underline{x}_C(kB-1)\|^2} \cdot e^{(NLMS)}(kB-1) \\
& \cdot \underline{x}_C^T(kB) \cdot \underline{x}_C(kB-1).
\end{aligned}
\tag{4.4}
$$

The first two lines of each equation of the system (4.4) correspond to (4.1). The other terms are correction terms.

In a next step, we rewrite the system of equations (4.4) in a matrix-vector-form and normalize the error signal vector

$$\underline{\breve{e}}_B^{(NLMS)}(kB) = \operatorname{diag}\left\{\underline{\breve{\alpha}}_B(kB)\right\} \cdot \underline{e}_B^{(NLMS)}(kB), \tag{4.5}$$

with the normalized step-factor vector

$$\underline{\breve{\alpha}}_B(kB) = \left[\ \frac{\alpha(kB-B+1)}{\|\underline{x}_C(kB-B+1)\|^2},\ \ \ldots,\ \ \frac{\alpha(kB)}{\|\underline{x}_C(kB)\|^2}\ \right]^T. \tag{4.6}$$

Finally, we receive

$$\underline{\breve{e}}_B^{(NLMS)}(kB) = \left[\operatorname{diag}^{-1}\left\{\underline{\breve{\alpha}}_B(kB)\right\} + \boldsymbol{K}_{B,B}(kB)\right]^{-1} \cdot \underline{e}_B(kB) \tag{4.7}$$

for the corrected and normalized error signal vector, where

$$\boldsymbol{K}_{B,B}(kB) = \begin{bmatrix} 0 & 0 & \cdots & 0 \\ \hat{r}^{(xx)}(kB-B+2,-1) & 0 & \cdots & 0 \\ \hat{r}^{(xx)}(kB-B+3,-2) & \hat{r}^{(xx)}(kB-B+3,-1) & \cdots & 0 \\ \vdots & \vdots & \ddots & \vdots \\ \hat{r}^{(xx)}(kB,-B+1) & \hat{r}^{(xx)}(kB,-B+2) & \cdots & 0 \end{bmatrix} \quad (4.8)$$

is the error signal correction matrix which is a lower triangular matrix and consists of estimates

$$\hat{r}^{(xx)}(l,m) = \underline{x}_C^T(l) \cdot \underline{x}_C(l+m) \quad (4.9)$$

of auto-correlation coefficients of the excitation signal.

If the corrected and normalized error signal vector $\breve{\underline{e}}_B^{(NLMS)}(kB)$ is used to adapt the impulse response of the filter according to (4.3), the block adaptive filter is mathematically equivalent to the NLMS algorithm and both algorithms have the same convergence properties.

4. REDUCTION OF THE SIGNAL DELAY

If we implement the exact block NLMS algorithm with reduced numerical complexity using the overlap and save implementation method [2, 5] we may perform a partitioning of the filter impulse response as illustrated in Fig. 4.1 to reduce the signal delay [3, 5]. The signal delays of the block adaptive filters are modeled in Fig. 4.1 by the gray labeled delay elements z^{-B}. The overlap-and-save method works most efficiently for a block length B equal to the filter length C of the whole filter, which would introduce an enormous signal delay. If we partition the filter into N sub-filters

$$\underline{w}_L^{(p)}(kB) = \left[\ w^{(p \cdot L+L-1)}(kB), \quad \ldots, \quad w^{(p \cdot L)}(kB)\ \right]^T \quad (4.10)$$

of smaller filter length L, we are able to use a smaller block length B for the frequency domain implementation, which reduces the signal delay.

By this method, the estimation signal vector is calculated by the superposition of the estimation signal vectors of the sub-filters

$$\underline{e}_B(kB) = \underline{d}_B(kB) - \sum_{p=0}^{N-1} \boldsymbol{X}_{L,B}^T(kB - p \cdot L) \cdot \underline{w}_L^{(p)}(kB) \quad (4.11)$$

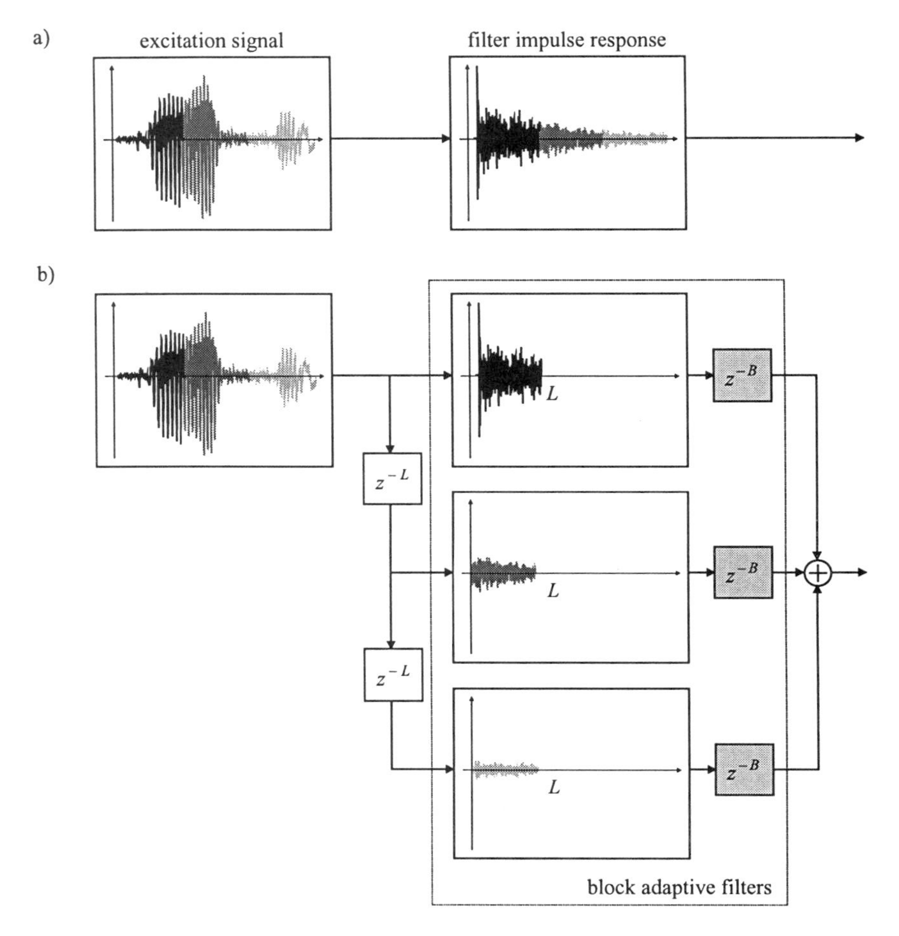

Figure 4.1 Reduction of the signal delay: The filter is divided into N sub-filters. Since the length L of the sub-filters is smaller than the length C of the whole filter, a smaller block length B can be chosen for the frequency-domain implementation, which reduces the signal delay. The signal delays of the block adaptive filters are modeled by the gray labeled delay elements z^{-B}. a) FIR filter structure. b) Equivalent partitioned FIR filter structure.

and the sub-filters are adapted in the following way:

$$\begin{bmatrix} \underline{w}_L^{(N-1)}(kB+B) \\ \vdots \\ \underline{w}_L^{(0)}(kB+B) \end{bmatrix} = \begin{bmatrix} \underline{w}_L^{(N-1)}(kB) \\ \vdots \\ \underline{w}_L^{(0)}(kB) \end{bmatrix} + \begin{bmatrix} \boldsymbol{X}_{L,B}(kB-(N-1)L) \\ \vdots \\ \boldsymbol{X}_{L,B}(kB) \end{bmatrix} \cdot \underline{\breve{e}}_B^{(NLMS)}(kB). \quad (4.12)$$

5. THE PEFBNLMS ALGORITHM

We now combine the error signal correction and the partitioning of the filter impulse response with the overlap-and-save implementation method [2, 5], which calculates the estimation signal vectors and the directions in which the impulse responses of the sub-filters should be updated via the frequency domain. The method is based on a circular convolution and correlation operation.

In the filter part of the algorithm, a circular convolution operation is used to calculate the estimation signal vector:

$$\underline{e}_B(kB) = \underline{d}_B(kB) - \boldsymbol{P}_{B,M} \cdot \boldsymbol{F}^{-1}_{M,M} \cdot \sum_{p=0}^{N-1} \underline{X}_M(kB - pL) \otimes \underline{\tilde{W}}_M^{(p)}(kB), \quad (4.13)$$

where $\boldsymbol{F}^{-1}_{M,M}$ is the inverse Fourier matrix whose coefficients are equal to

$$(\boldsymbol{F}^{-1}_{M,M})_{(y,x)} = 1/M \cdot \exp{(j2\pi/M \cdot x \cdot y)}. \quad (4.14)$$

and

$$\underline{X}_M(kB) = \boldsymbol{F}_{M,M} \cdot \underline{x}_M(kB), \quad (4.15)$$

the spectrum of the excitation signal which is calculated by a Discrete Fourier Transform. The coefficients of the Fourier matrix $\boldsymbol{F}_{M,M}$ are given by

$$(\boldsymbol{F}_{M,M})_{(y,x)} = \exp{(-j2\pi/M \cdot x \cdot y)} \quad (4.16)$$

and the vectors

$$\underline{\tilde{W}}_M^{(p)}(kB) = \boldsymbol{F}_{M,M} \cdot \begin{bmatrix} \boldsymbol{J}_{L,L} \cdot \underline{w}_L^{(p)}(kB) \\ \underline{0}_{M-L} \end{bmatrix} \quad (4.17)$$

are defined as transfer functions of the extended impulse responses of the sub-filters whose coefficients are mirrored using the mirror matrix

$$\boldsymbol{J}_{L,L} = \begin{bmatrix} 0 & \cdots & 0 & 1 \\ 0 & \cdots & 1 & 0 \\ \vdots & \ddots & \vdots & \vdots \\ 1 & \cdots & 0 & 0 \end{bmatrix}. \quad (4.18)$$

Due to the circular nature of the frequency domain implementation, only B coefficients of the resulting vector of the transformation into the time domain of (4.13) are correct. The others contain wraparound errors [7]. Using the projection matrix

$$\boldsymbol{P}_{B,M} = [\ \boldsymbol{0}_{B,M-B}, \quad \boldsymbol{I}_{B,B}\ ] \quad (4.19)$$

on the filter part, only valid coefficients are used to calculate the error signal vector.

In the adaptation part, the correction terms of the impulse responses of the sub-filters are calculated using a circular correlation operation:

$$\begin{bmatrix} \underline{w}_L^{(N-1)}(kB+B) \\ \vdots \\ \underline{w}_L^{(0)}(kB+B) \end{bmatrix} = \begin{bmatrix} \underline{w}_L^{(N-1)}(kB) \\ \vdots \\ \underline{w}_L^{(0)}(kB) \end{bmatrix} + \begin{bmatrix} \boldsymbol{Q}_{L,M} \cdot \boldsymbol{F}_{M,M}^{-1} \cdot \left[\underline{X}_M^*(kB-(N-1)L) \otimes \underline{\tilde{E}}_M^{(NLMS)}(kB) \right] \\ \vdots \\ \boldsymbol{Q}_{L,M} \cdot \boldsymbol{F}_{M,M}^{-1} \cdot \left[\underline{X}_M^*(kB) \otimes \underline{\tilde{E}}_M^{(NLMS)}(kB) \right] \end{bmatrix}, \quad (4.20)$$

where

$$\underline{\tilde{E}}_M^{(NLMS)}(kB) = \boldsymbol{F}_{M,M} \cdot \begin{bmatrix} \underline{0}_{M-B} \\ \underline{\breve{e}}_B^{(NLMS)}(kB) \end{bmatrix} \quad (4.21)$$

is the spectrum of the corrected and normalized error signal, which has been extended, and

$$\boldsymbol{Q}_{L,M} = \begin{bmatrix} \boldsymbol{J}_{L,L}, & \boldsymbol{0}_{L,M-L} \end{bmatrix} \quad (4.22)$$

the projection matrix of the adaptation part, which prevents errors due to the circular nature of the correlation.

This algorithm is called the partitioned exact frequency-domain block NLMS (PEFBNLMS) algorithm. For a better understanding, a block diagram of the algorithm is presented in Figs. 4.2 and 4.3. The boxes marked with 'S/P' or 'P/S' are serial-to-parallel converters and parallel-to-serial converters, respectively. The elements described with 'FFT' or 'IFFT' perform a fast Fourier transform or an Inverse fast Fourier transform and the box marked with 'overlap' overlaps consecutive input signal vectors due to the overlap and save sectioning method. The delay elements z_B^{-1} and z_B^{-d}, $d = L/B$ delay the input vector for one or d iterations of the algorithms. For a numerically efficient implementation, we assume that the sub-filter length L is a multiple of the block length B so that the excitation signal spectrum, which is calculated for the first sub-filter, can be reused for sub-filters of higher order p in following iterations. The excitation signal spectrum is therefore calculated only for the first sub-filter.

6. PERFORMANCE

To examine the numerical complexity and the memory requirement of the PEFBNLMS algorithm, we may count the number of additions C_a, multiplication C_m, and divisions C_d needed for one iteration of the algorithm in B sampling

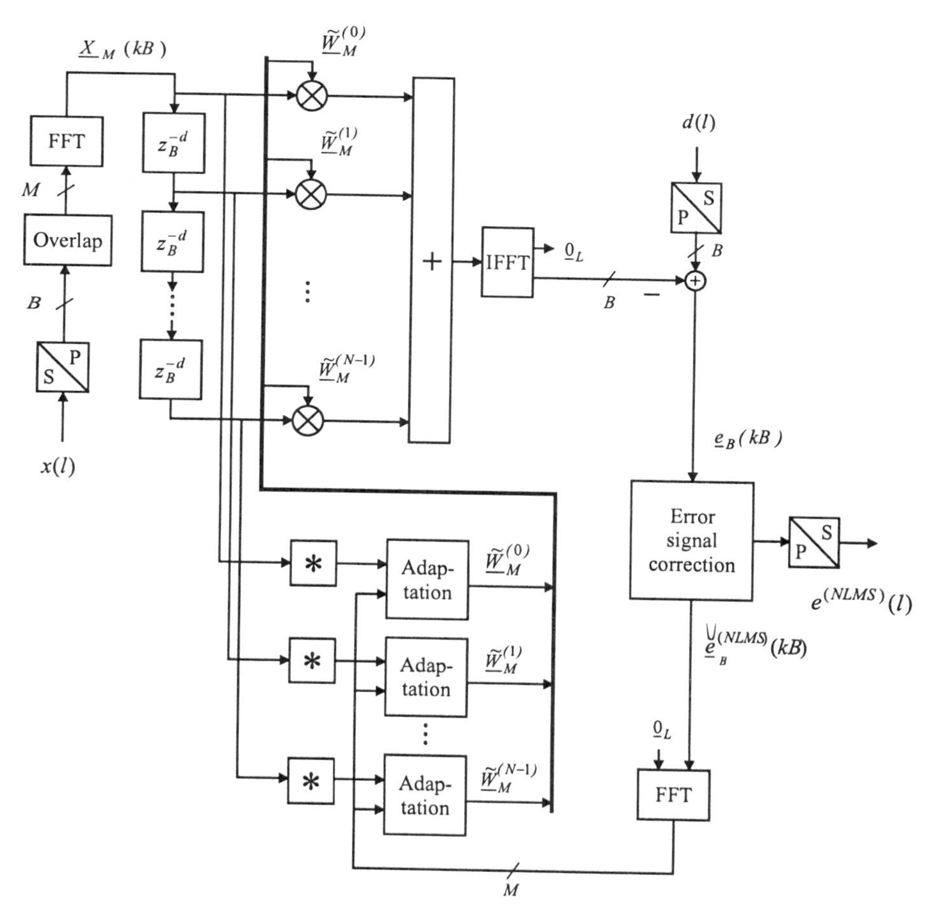

Figure 4.2 Block diagram of the PEFBNLMS algorithm: The PEFBNLMS algorithm uses the overlap-and-save implementation method to reduce the numerical complexity and a partitioning of the filter to decrease the signal delay introduced by the block processing approach. The convergence behavior is improved by an error signal correction.

periods and the required number of memory words $\mathcal{M}$. If we normalize the numerical complexity by the block length B and the numerical complexity of the NLMS algorithm we receive the relative cost of the PEFBNLMS algorithm compared to the time domain NLMS algorithm [6].

Figure 4.4 shows the relative numerical complexity and the relative memory requirement of the PEFBNLMS algorithm for filter lengths C between 1000 and 4000 coefficients and for block lengths B of 32, 64, 128, and 256 samples. For each point in Fig. 4.4, the algorithm parameters B, L, M, and N are optimized for minimal numerical complexity. Therefore, we receive a continuous curve for the relative numerical complexity and a noncontinuous curve for the relative

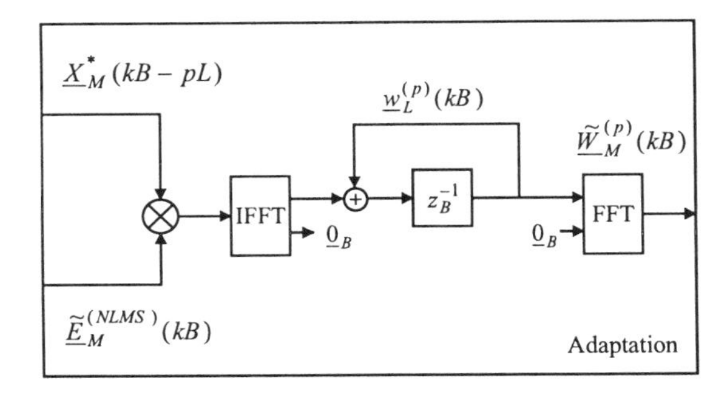

Figure 4.3 Block diagram of the adaptation of the sub-filters: The directions in which the impulse responses of the sub-filters are corrected are calculated with a circular correlation operation via the frequency-domain with reduced numerical complexity.

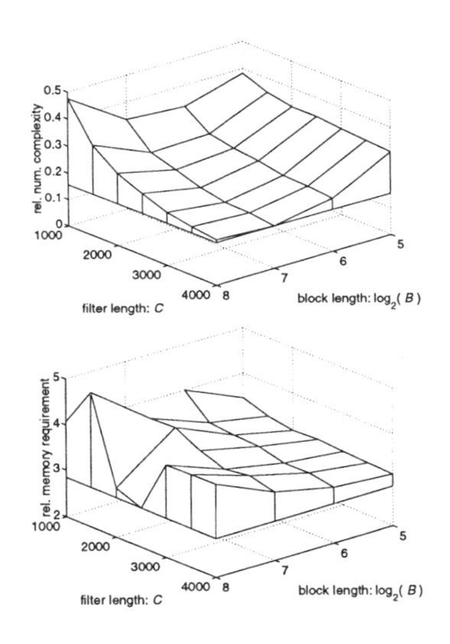

Figure 4.4 Complexity of the PEFBNLMS algorithm compared to the time-domain NLMS algorithm. Top: Relative numerical complexity. Bottom: Relative memory requirement.

memory requirement. In Table 4.1, some of the optimal parameter sets are specified.

The PEFBNLMS algorithm reduces the numerical complexity by a factor of between two and six, but requires more memory. With increasing filter length C, the relative numerical complexity decreases because fast Fourier transforms with increased length can be used which work more efficiently. A local minimum exists depending on the block length. Since the numerical complexity of the error signal correction part increases with rising block length B, compare with (4.7), and the complexity of the filter and adaptation part

Table 4.1 Relative numerical complexity $\mathcal{C}_{rel}$ and relative memory requirement $\mathcal{M}_{rel}$ of the PEFBNLMS algorithm compared to the time-domain NLMS algorithm.

C	B	L	N	$\mathcal{C}_a$	$\mathcal{C}_m$	$\mathcal{C}_d$	$\mathcal{M}$	$\mathcal{C}_{rel}$	$\mathcal{M}_{rel}$
2000	32	96	21	59712	24854	32	12768	0.33	3.19
2000	64	192	11	85416	37542	64	13856	0.24	3.46
2000	128	384	6	152628	76358	128	16032	0.22	4.01
2000	256	256	8	350332	209542	256	12192	0.27	3.05
3000	32	96	32	87508	36118	32	18992	0.32	3.16
3000	64	192	16	114596	49062	64	19696	0.21	3.28
3000	128	384	8	179236	86598	128	21104	0.17	3.52
3000	256	768	4	384596	219782	256	23920	0.20	3.99
4000	32	96	42	112868	46358	32	24832	0.31	3.10
4000	64	192	21	143776	60582	64	25536	0.20	3.19
4000	128	384	11	218648	101958	128	27712	0.16	3.46
4000	256	768	6	442948	242310	256	32064	0.17	4.01

decreases due to the overlap-and-save implementation method, there exists an optimal point where the complexity is minimal.

7. REAL-TIME IMPLEMENTATION

For a real-time implementation of the PEFBNLMS algorithm, we make some modifications that save computational power and memory. Firstly, the estimates of the auto-correlation coefficients are calculated recursively:

$$\hat{r}^{(xx)}(l,m) = \hat{r}^{(xx)}(l-1,m) + x(l)\cdot x(l+m) - x(l-C)\cdot x(l-C+m). \quad (4.23)$$

Since for low signal delays the block length B is normally chosen much smaller than the filter length C, the recursive calculation saves a lot of operations compared to a direct calculation according to (4.9). If the PEFBNLMS algorithm is implemented on a floating point processor, the estimates of the auto-correlation coefficients should be reinitialized periodically to prevent a divergence of the algorithm due the roundoff errors by the additions in (4.23). A computationally efficient method reinitializes the auto-correlation function every C sampling periods. By this method, the second summand of (4.23) can be used to calculate the exact auto-correlation coefficients for the reinitialization with very little additional computational power.

Secondly, we can combine the recursive calculation of the auto-correlation function with the matrix inversion needed for the error signal correction, compare with (4.7). If the error signal correction is performed recursively, only two times B coefficients of the auto-correlation function must be stored: the auto-correlation coefficients used for the error signal correction and the coefficients for the reinitialization procedure. By this method, the auto-correlation function $\hat{r}^{(xx)}(kB - B + 1, m)$ for the time instant $kB - B + 1$ is calculated and then the error signal sample $e(kB + B + 1)$ is corrected by solving the equation described by the first row of the matrix $[\mathrm{diag}^{-1}\left\{\underline{\breve{\alpha}}_B(kB)\right\} + \boldsymbol{K}_{B,B}(kB)]$ and the vector $\underline{\breve{e}}_B^{(NLMS)}(kB)$ of the following matrix vector product:

$$\underline{e}_B(kB) = \left[\mathrm{diag}^{-1}\left\{\underline{\breve{\alpha}}_B(kB)\right\} + \boldsymbol{K}_{B,B}(kB)\right] \cdot \underline{\breve{e}}_B^{(NLMS)}(kB). \qquad (4.24)$$

Then the auto-correlation function $\hat{r}^{(xx)}(kB - B + 2, m)$ for the next time instant is calculated and the next error signal sample $e(kB + B + 2)$ is corrected by solving the equation described by the second row of the matrix $[\mathrm{diag}^{-1}\left\{\underline{\breve{\alpha}}_B(kB)\right\} + \boldsymbol{K}_{B,B}(kB)]$ and the vector $\underline{\breve{e}}_B^{(NLMS)}(kB)$. These operations are continued until the all coefficients of the vector $\underline{\breve{e}}_B^{(NLMS)}(kB)$ are calculated.

The PEFBNLMS algorithm was implemented on the floating point digital signal processor (DSP) TMS320C44 of Texas Instruments for a hands-free telephone system. The DSP provides a computational power of 50 MIPS and the AD and DA converters work with a sampling frequency of 8 kHz. Table 4.2 illustrates how far the filter length C_{max} could be increased by the PEFBNLMS algorithm compared to the NLMS algorithm for the real-time implementation and the respective algorithm parameters. The signal delay mentioned in Table 4.2 only consists of the delay produced by the structure of the algorithm, the delay introduced the the AD and DA converters is not included.

The filter length can be increased by a factor between 1.2 and 4 by the PEFBNLMS algorithm. Since the instruction set of the DSP supports simple filter operations with efficient instructions and the fast Fourier transform is less well supported, the relative numerical complexity of the real-time implementation is larger than the complexity calculated theoretically in Section 6.

Figure 4.5 shows two typical convergence curves of the algorithm. To guarantee reproducible measurements, another DSP system was used to simulate the loudspeaker-room-microphone (LRM) system. The FIR filter that simulates the LRM system had the same length as the compensator. Speech was used as the excitation signal and white noise to simulate the background noise in the local room. The signal-to-noise ratio during speech activity periods was approximately equal to 30 dB. To simulate a time variant LRM system, the impulse response of the LRM system was changed after 11 s so that the tracking capability could be examined. A step-factor control [4] locks the adaptation during

Table 4.2 Maximum reachable filter length C_{max} in the real-time implementation.

algorithm	B_{opt}	L_{opt}	M_{opt}	N_{opt}	C_{max}	signal delay [ms]	filter length [ms]
NLMS alg.	-	-	-	-	940	0.25	117.5
PEFBNLMS alg.	32	96	128	12	1152	8	144
	64	192	256	11	2112	16	264
	128	384	512	8	3072	32	384
	256	768	1024	5	3840	64	480

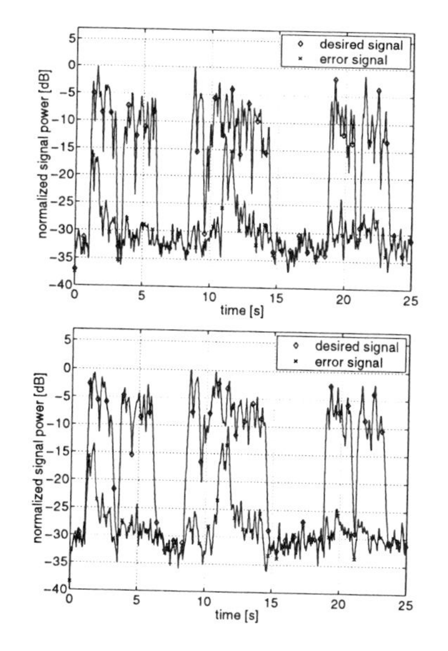

Figure 4.5 Typical convergence curves of the PEFBNLMS algorithm. Top: $B_1 = 64, L_1 = 448$, $M_1 = 512$, $N_1 = 2$, $\alpha_1 = 1$. Bottom: $B_2 = 128$, $L_2 = 896$, $M_2 = 1024$, $N_2 = 1$, $\alpha_2 = 1$.

speech pauses by choosing a step-factor $\alpha = 0$. Because the PEFBNLMS algorithm is mathematically identical to the NLMS algorithm, the convergence behavior does not depend on the block length B.

8. CONCLUSIONS

In this contribution the PEFBNLMS algorithm was presented. The algorithm is based on a block processing approach which introduces a signal delay and may decrease the convergence speed of the filter. By a correction of the error signal in such a way that this error signal is equal to the error signal of the NLMS algorithm, the convergence behavior was improved. The PEFBNLMS algorithm uses a partitioning of the filter to reduce the signal delay and the overlap-and-save implementation method to decrease the numerical complexity. The resulting algorithm needs two to six times less operations as compared to the time-domain NLMS algorithm and has a moderate signal delay. The convergence behavior of the PEFBNLMS algorithm is equal to the NLMS algorithm because both algorithms are mathematically equivalent.

References

[1] J. Benesty and P. Duhamel, "A fast exact least mean square adaptive algorithm," *IEEE Trans. Signal Processing*, vol. 40, pp. 2904-2920, Dec. 1992.

[2] G. A. Clark, S. R. Parker, and S. K. Mitra, "A unified approach to time- and frequency-domain realizations of FIR adaptive digital filters," *IEEE Trans. Acoust., Speech, Signal Processing*, vol. 31, pp. 1073-1083, Oct. 1983.

[3] G. P. M. Egelmeers, *Real time realization concepts of large adaptive filters*. Ph.D. dissertation, Technical University Eindhoven, Netherlands, 1995.

[4] B. H. Nitsch, "Implementation of a block adaptive filter working in the frequency domain combined with a robust adaptation control," in *Proc. EUSIPCO*,1998, pp. 1225-1229.

[5] B. H. Nitsch, "The partitioned exact frequency domain block NLMS algorithm, a mathematical exact version of the NLMS algorithm working in the frequency domain," *AEÜ International Journal of Electronics and Communictions*, vol. 52, pp. 293-301, Oct. 1998.

[6] B. H. Nitsch, *Adaptive Filter im Frequenzbereich für Freisprecheinrichtungen*. Ph.D. dissertation, Fachgebiet Theorie der Signale, Darmstadt University of Technology, Germany, 2000.

[7] L. Pelkowitz, "Frequency domain analysis of wraparound error in fast convolution algorithms," *IEEE Trans. Acoust., Speech, Signal Processing*, vol. 29, pp. 413-422, June 1981.

Chapter 5

DOUBLE-TALK DETECTION SCHEMES FOR ACOUSTIC ECHO CANCELLATION

Tomas Gänsler
Bell Laboratories, Lucent Technologies
gaensler@bell-labs.com

Jacob Benesty
Bell Laboratories, Lucent Technologies
jbenesty@bell-labs.com

Steven L. Gay
Bell Laboratories, Lucent Technologies
slg@bell-labs.com

Abstract Double-talk detectors (DTDs) are vital to the operation and performance of acoustic echo cancelers. In this chapter, we enlighten important aspects needed to be considered when choosing and designing a DTD. The generic double-talk detector scheme along with fundamental means of performance evaluation are discussed and a number of double-talk detectors suitable for acoustic echo cancelers are presented.

Keywords: Double-Talk Detector, Acoustic Echo Canceler, Adaptive Algorithms, LMS, NLMS, Coherence, Cross-Correlation

1. INTRODUCTION

The design of a good double-talk detector is much more of an art than the design of the adaptive filter itself. [1][1]

Ideally, acoustic echo cancelers (AECs) remove undesired echoes that result from coupling between the loudspeaker and the microphone used in full-duplex hands-free telecommunication systems. Figure 5.1 shows a basic AEC diagram.

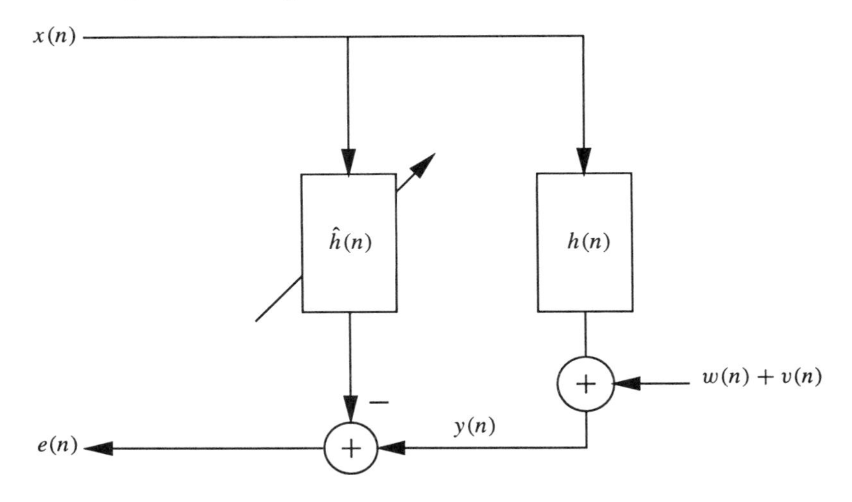

Figure 5.1 Block diagram of a basic AEC setup.

The far-end speech signal $x(n)$ goes through the echo path represented by a filter $h(n)$, then it is picked up by the microphone together with the near-end talker signal $v(n)$ and ambient noise $w(n)$. The microphone signal is denoted $y(n)$. Most often the echo path is modeled by an adaptive FIR filter, $\hat{h}(n)$, which subtracts a replica of the echo from the return channel and thereby achieves cancellation. This may look like a simple straightforward system identification task for the adaptive filter; however, in most conversation there are so called *double-talk* situations that make the identification much more problematic than what it might appear at a first glance. Double-talk occurs when the two talkers on both sides speak simultaneously, i.e. $x(n) \neq 0$ and $v(n) \neq 0$ (the situation with near-end talk only, $x(n) = 0$ and $v(n) \neq 0$, can be regarded as an "easy-to-detect" double-talk case). In the double-talk situation, the near-end speech acts as a large level uncorrelated noise to the adaptive algorithm. The disturbing near-end speech may cause the adaptive filter to diverge. Hence, annoying audible echo will pass through to the far-end. The common way to alleviate this problem is to slow down or completely halt the filter adaptation when presence of near-end speech is detected. This is the very important role of the so called double-talk detector (DTD). The basic double-talk detection scheme starts with computing a detection statistic, ξ, and comparing it with a preset threshold, T. Different methods, i.e. DTDs, have been proposed to form the detection statistic. Important issues that have to be addressed are:

(i) What basic knowledge is needed in order to devise a sufficient DTD solution?

(ii) What characterize "good" double-talk detectors?

Primarily, we must know under what circumstances double-talk disturbs the adaptive filter. The performance of AECs are most often evaluated through their mean square error performance (MSE) or preferably, through their misalignment $\varepsilon = \|\mathbf{h} - \hat{\mathbf{h}}\|_2 / \|\mathbf{h}\|_2$ in different situations. Question (i) stated above can be answered by the misalignment formula:

$$E\{\varepsilon^2\} \approx \frac{\mu_0}{2\mathrm{SNR}_{\mathrm{en}}}, \tag{5.1}$$

where $E\{\cdot\}$ denotes mathematical expectation and μ_0 is a constant parameter of the adaptive algorithm. That is, the level of convergence (misalignment performance) is completely governed by *echo to near-end speech and noise* power ratio, $\mathrm{SNR}_{\mathrm{en}}$. If we have equally strong talkers, we find that when the echo path has a high attenuation the adaptive filter will be very sensitive to double-talk. On the contrary, a low attenuation or amplification means lower sensitivity. This is important to remember when setting the parameters of the DTD. Equation (5.1) is easily found by assuming the following: Far- and near-end signals are uncorrelated stochastic processes and the adaptive algorithm is NLMS, see Chapter 1 in this book and LMS properties in [3]. Furthermore, if we suppose that the signals are white noise, the misalignment is exactly the inverse of the *echo return loss enhancement* (ERLE) which reflects the echo attenuation provided by the AEC. Even though these assumptions describe an oversimplified model of the AEC situation, it gives the insight needed to what governs divergence.

As far as (ii) is concerned, we can partially characterize a DTD through the use of general detection theory [4, 5, 6]. By means of detection probability and false alarm rates we can objectively evaluate and compare the performance of different DTDs. Moreover, the theory justifies a method to select the threshold T which has been missing in this field of DTD design. One must, however, also accompany these performance measures with a joint evaluation of the DTD and the echo canceler.

A large number of DTD schemes have been proposed since the dawn of echo cancellation [7]. The Geigel algorithm [8] has proven successful in line echo cancelers; however, it does not always provide reliable performance when used in the acoustic situation. This is because it assumes a minimum echo path attenuation which may not be valid in the acoustic case. Other methods based on cross-correlation and coherence [9, 10, 4, 5] have been studied which appear to be more appropriate for AEC applications. Spectral comparing methods [11] and two-microphone solutions have also been proposed [12]. A DTD based on multi statistic testing in combination with modeling of the echo path by two filters is proposed in [13]. The objective of this chapter is to summarize some of the DTD proposals and present evaluation methods. Many results in this chapter are derived from the papers [5, 6].

The chapter is organized as follows: Section 2 introduces the AEC notations and describes the general DTD scheme. A number of double-talk detection algorithms that have been proposed and used in acoustic echo cancelers are presented in Section 3. Section 4 gives a discussion of the abilities of different DTD schemes and summarizes the important aspects that need to be considered for a successful double-talk detector implementation.

2. BASICS OF AEC AND DTD

In this section, we give the basics of an AEC combined with a DTD. We first formulate the AEC problem.

2.1 AEC NOTATIONS

The AEC setup in Fig. 5.1 is described in mathematical terms as:

$$y(n) = \mathbf{h}^T \mathbf{x}(n) + v(n) + w(n), \tag{5.2}$$

where

$$\mathbf{h} = \begin{bmatrix} h_0 & h_1 & \cdots & h_{N-1} \end{bmatrix}^T,$$
$$\mathbf{x}(n) = \begin{bmatrix} x(n) & x(n-1) & \cdots & x(n-N+1) \end{bmatrix}^T,$$

and N is the length of the echo path response, $\mathbf{h}$. The error signal is defined as

$$\begin{aligned} e(n) &= y(n) - [\ \hat{\mathbf{h}}^T \quad \mathbf{0}^T\ ]\mathbf{x}(n) \\ &= \Delta\mathbf{h}^T \mathbf{x}(n) + v(n) + w(n), \end{aligned} \tag{5.3}$$

where

$$\hat{\mathbf{h}} = \begin{bmatrix} \hat{h}_0 & \hat{h}_1 & \cdots & \hat{h}_{L-1} \end{bmatrix}^T \tag{5.4}$$

is the adaptive filter coefficient vector of length L (generally less than N), and

$$\Delta\mathbf{h} = \mathbf{h} - \begin{bmatrix} \hat{\mathbf{h}} \\ \mathbf{0} \end{bmatrix}. \tag{5.5}$$

2.2 THE GENERIC DTD

All types of double-talk detectors basically operate in the same manner. Thus, the general procedure for handling double-talk is described by the following:

1. A detection statistic ξ is formed using available signals, e.g. x, y, e, etc., and the estimated filter coefficients $\hat{h}$.

2. The detection statistic ξ is compared to a preset threshold T, and double-talk is declared if $\xi < T$.

3. Once double-talk is declared, the detection is held for a minimum period of time T_{hold}. While the detection is held, the filter adaptation is disabled.

4. If $\xi \geq T$ consecutively over a time T_{hold}, the filter resumes adaptation, while the comparison of ξ to T continues until $\xi < T$ again.

The hold time T_{hold} in Step 3 and Step 4 is necessary to suppress detection dropouts due to the noisy behavior of the detection statistic. Although there are some possible variations, most of the DTD algorithms keep this basic form and only differ in how to form the detection statistic.

An "optimum" decision variable ξ for double-talk detection will behave as follows:

(i) if $v(n) = 0$ (double-talk is not present), $\xi \geq T$.

(ii) if $v(n) \neq 0$ (double-talk is present), $\xi < T$.

(iii) ξ is insensitive to echo path variations when $v(n) = 0$.

The threshold T must be a constant, independent of the data. Moreover, it is desirable that the decisions are made without introducing any delay (or minimize the introduced delay) in the updating of the adaptive filter. The delayed decisions will otherwise affect the AEC algorithm negatively.

2.3 A SUGGESTION TO PERFORMANCE EVALUATION OF DTDS

The role of the threshold T is essential to the performance of the double-talk detector. To select the value of T and to compare different DTDs objectively one could view the DTD as a classical binary detection problem. By doing so, it is possible to rely on a well established detection theory. This approach to characterize DTDs was proposed in [4, 6].

The general characteristics of a binary detection scheme are:

- *Probability of False Alarm* (P_{f}): Probability of declaring detection when a target, in our case double-talk, is not present.

- *Probability of Detection* (P_{d}): Probability of successful detection when a target is present.

- *Probability of Miss* ($P_{\text{m}} = 1 - P_{\text{d}}$): Probability of detection failure when a target is present.

A well designed DTD maximizes P_{d} while minimizing P_{f} even in a low SNR. In general, higher P_{d} is achieved at the cost of higher P_{f}. There should be a tradeoff in performance depending on the penalty or cost function of a false alarm.

One common approach to characterize different detection methods is to represent the detection characteristic $P_{\rm d}$ as a function of SNR under a given constraint on the false alarm probability $P_{\rm f}$. This is known as a receiver operating characteristic (ROC). The $P_{\rm f}$ constraint can be interpreted as the maximum tolerable false alarm rate.

Evaluation of a DTD is carried out by estimating the performance parameters, $P_{\rm d}$ ($P_{\rm m}$) and $P_{\rm f}$. A principle for this technique can be found in [6]. Though in the end, one should accompany these performance measures with a joint evaluation of the DTD and the AEC. This is due to the fact that the response time of the DTD can seriously affect the performance of the AEC and this is in general not shown in the ROC curve.

3. DOUBLE-TALK DETECTION ALGORITHMS

In this section, we explain different DTD algorithms that can be useful for AEC. We start with the Geigel algorithm since it was the very first DTD.

3.1 GEIGEL ALGORITHM

A very simple algorithm due to A. A. Geigel [8] is to declare the presence of near-end speech whenever

$$\xi^{(\rm g)} = \frac{\max\{\,|x(n)|, \ldots, |x(n-L_{\rm g}+1)|\,\}}{|y(n)|} < T, \tag{5.6}$$

where $L_{\rm g}$ and T are suitably chosen constants. This detection scheme is based on a waveform level comparison between the microphone signal $y(n)$ and the far-end speech $x(n)$ assuming the near-end speech $v(n)$ in the microphone signal will be typically stronger than the echo $\mathbf{h}^T\mathbf{x}$. The maximum or l_∞ norm of the $L_{\rm g}$ most recent samples of $x(n)$ is taken for the comparison because of the undetermined delay in the echo path. The threshold T is to compensate for the energy level of the echo path response h, and is often set to 2 for line echo cancelers because the hybrid loss is typically about 6 dB or more. For an AEC, however, it is not easy to set a universal threshold to work reliably in all the various situations because the loss through the acoustic echo path can vary greatly depending on many factors. For $L_{\rm g}$, one easy choice is to set it the same as the adaptive filter length L since we can assume that the echo path is covered by this length. This detector also has the advantage that it can be implemented very efficiently by block updating the numerator of (5.6).

3.2 CROSS-CORRELATION METHOD

In [9] the cross-correlation coefficient vector between $\mathbf{x}$ and e was proposed as a means for double-talk detection. A similar idea using the cross-correlation coefficient vector between $\mathbf{x}$ and y has proven more robust and reliable [10, 6].

This section will therefore focus on the cross-correlation coefficient vector between $\mathbf{x}$ and y which is defined as

$$\begin{aligned}\mathbf{c}_{xy}^{(1)} &= \frac{E\{\mathbf{x}(n)y(n)\}}{\sqrt{E\{x^2(n)\}E\{y^2(n)\}}} \\ &= \frac{\mathbf{r}_{xy}}{\sigma_x \sigma_y} \\ &= \begin{bmatrix} c_{xy,0}^{(1)} & c_{xy,1}^{(1)} & \cdots & c_{xy,L-1}^{(1)} \end{bmatrix}^T,\end{aligned} \tag{5.7}$$

where $c_{xy,i}^{(1)}$ is the cross-correlation coefficient between $x(n-i)$ and $y(n)$.

The idea here is to compare

$$\begin{aligned}\xi^{(1)} &= \|\mathbf{c}_{xy}^{(1)}\|_\infty \\ &= \max_i |c_{xy,i}^{(1)}|, \quad i = 0, 1, ..., L-1\end{aligned} \tag{5.8}$$

to a threshold level T. The decision rule will be very simple: if $\xi^{(1)} \geq T$, then double-talk is not present; if $\xi^{(1)} < T$, then double-talk is present.

Although the l_∞ norm used in (5.7) is perhaps the most natural, other scalar metrics, e.g., l_1, l_2, could alternatively be used to assess the cross-correlation coefficient vectors. However, there is a fundamental problem here which is not linked to the type of metric used. The problem is that these cross-correlation coefficient vectors are not well normalized. Indeed, we can only say in general that $\xi^{(1)} \leq 1$. If $v(n) = 0$, that does not imply that $\xi^{(1)} = 1$ or any other known value. We do not know the value of $\xi^{(1)}$ in general. The amount of correlation will depend a great deal on the statistics of the signals and of the echo path. As a result, the best value of T will vary a lot from one experiment to another. So there is no natural threshold level associated with the variable $\xi^{(1)}$ when $v(n) = 0$.

Next section presents a decision variable that exhibits this behavior. This is achieved by normalizing the cross-correlation vector between $\mathbf{x}$ and y.

3.3 NORMALIZED CROSS-CORRELATION METHOD

There is a simple way to normalize the cross-correlation vector between a vector $\mathbf{x}$ and a scalar y in order to have a natural threshold level for ξ when $v(n) = 0$.

Suppose that $v(n) = 0$. In this case:

$$\sigma_y^2 = \mathbf{h}^T \mathbf{R}_{xx} \mathbf{h}, \tag{5.9}$$

where $\mathbf{R}_{xx} = E\{\mathbf{x}(n)\mathbf{x}^T(n)\}$. Since $y(n) = \mathbf{h}^T \mathbf{x}(n)$, we have

$$\mathbf{r}_{xy} = \mathbf{R}_{xx} \mathbf{h}, \tag{5.10}$$

and (5.9) can be re-written as

$$\sigma_y^2 = \mathbf{r}_{xy}^T \mathbf{R}_{xx}^{-1} \mathbf{r}_{xy}. \tag{5.11}$$

In general for $v(n) \neq 0$ we have,

$$\sigma_y^2 = \mathbf{r}_{xy}^T \mathbf{R}_{xx}^{-1} \mathbf{r}_{xy} + \sigma_v^2. \tag{5.12}$$

If we divide (5.11) by (5.12) and take the square root, we obtain the decision variable [5, 14]

$$\begin{aligned} \xi^{(2)} &= \sqrt{\mathbf{r}_{xy}^T (\sigma_y^2 \mathbf{R}_{xx})^{-1} \mathbf{r}_{xy}} \\ &= \|\mathbf{c}_{xy}^{(2)}\|_2, \end{aligned} \tag{5.13}$$

where

$$\mathbf{c}_{xy}^{(2)} = (\sigma_y^2 \mathbf{R}_{xx})^{-1/2} \mathbf{r}_{xy} \tag{5.14}$$

is what we will call the normalized cross-correlation vector between $\mathbf{x}$ and y.

Substituting (5.10) and (5.12) into (5.13), we show that the decision variable is:

$$\xi^{(2)} = \frac{\sqrt{\mathbf{h}^T \mathbf{R}_{xx} \mathbf{h}}}{\sqrt{\mathbf{h}^T \mathbf{R}_{xx} \mathbf{h} + \sigma_v^2}}. \tag{5.15}$$

We easily deduce from (5.15) that for $v(n) = 0$, $\xi^{(2)} = 1$ and for $v(n) \neq 0$, $\xi^{(2)} < 1$. Note also that $\xi^{(2)}$ is not sensitive to changes of the echo path when $v = 0$. Moreover, a fast version of this algorithm can be derived by recursively updating $\mathbf{R}_{xx}^{-1} \mathbf{r}_{xy}$ using the Kalman gain $\mathbf{R}_{xx}^{-1} \mathbf{x}$ [3].

For the particular case when x is white Gaussian noise, the autocorrelation matrix is diagonal: $\mathbf{R}_{xx} = \sigma_x^2 \mathbf{I}$. Then (5.14) becomes:

$$\begin{aligned} \mathbf{c}_{xy}^{(2)} &= \frac{\mathbf{r}_{xy}}{\sigma_x \sigma_y} \\ &= \mathbf{c}_{xy}^{(1)}. \end{aligned} \tag{5.16}$$

Note that, in general, what we are doing in (5.13) is equivalent to prewhitening the signal $\mathbf{x}$, which is one of many known "generalized cross-correlation" techniques [15]. Thus, when $\mathbf{x}$ is white, no prewhitening is necessary and $\mathbf{c}_{xy}^{(2)} = \mathbf{c}_{xy}^{(1)}$. This suggests a more practical implementation, whereby matrix operations are replaced by an adaptive prewhitening filter [16].

3.4 COHERENCE METHOD

Instead of using the cross-correlation vector, a detection statistic can be formed by using the squared magnitude coherence. A DTD based on coherence

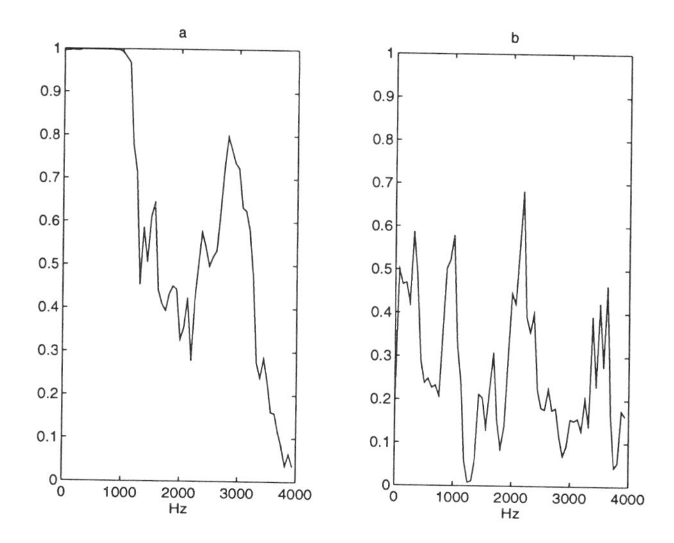

Figure 5.2 Estimated coherence using the multiple window method. (a) Far-end talker is active only. (b) Double-talk situation where the far- and near-end signals powers are equal. Echo path attenuation is 6 dB and the ambient noise power is 25 dB lower than the far-end speech.

was proposed in [4]. The idea is to estimate the coherence between $\mathbf{x}(n)$ and $\mathbf{y}(n)$. The coherence is close to one when there is no double-talk and it is close to zero in a double-talk situation. Figure 5.2 shows an example of estimated coherence between loudspeaker and microphone signals in the presence and absence of double-talk. The squared coherence is defined as,

$$\gamma_{xy}^2(k) = \frac{|S_{xy}(k)|^2}{S_{xx}(k)S_{yy}(k)}, \tag{5.17}$$

where $S_{..}(k)$ is the DFT based cross-power spectrum and k is the DFT frequency index. As decision parameter, an average over a few frequencies is used as detection statistics,

$$\xi^{(3)} = \frac{1}{I}\sum_{i=0}^{I-1}\gamma_{xy}^2(k_i), \tag{5.18}$$

where I is the number of intervals used. Typical choices of these parameters are $I = 3$ and k_0, k_1, k_2 are the intervals chosen such that their center correspond to approximately 300, 1200, and 1800 Hz respectively. This gives in practice a significantly better performance than averaging over the whole frequency range since there is a poorer speech-to-noise ratio in the upper frequencies (the average speech spectrum declines with about 6 dB/octave above 2 kHz).

Estimation of the spectra in (5.17) can be made by using the multiple window technique [17], where

$$\hat{S}_{xx}(k) = \frac{1}{P}\sum_{p=0}^{P-1} |X_p(k)|^2, \tag{5.19}$$

$$\hat{S}_{xy}(k) = \frac{1}{P}\sum_{p=0}^{P-1} X_p(k)Y_p^*(k), \tag{5.20}$$

where $X_p(k)$ is the $p:th$ eigenspectrum

$$X_p(k) = \sum_{n=0}^{L_c-1} x(L_c - 1 - n)\phi_p(n)e^{-j2\pi\frac{k}{L_c}n} \tag{5.21}$$

and $Y_p(k)$ is analogously defined. The window $\phi_p(n)$ is the $p:th$ discrete spheroidal wave function [18]. L_c is the block length of the DFT. The multiple window method has advantages such as easy tradeoff between bias and variance. Another possibility is to use the Welch spectrum estimation method [19].

Since this DTD as well as the correlation methods in Sections 3.2 and 3.3 are based on block processing of the signals, there is a tradeoff between calculation complexity and time between decisions. It is desirable to keep the time between decisions as short as possible in order to have as low detection failures as possible (both false alarm and detection miss).

3.5 NORMALIZED CROSS-CORRELATION MATRIX

Obviously, the cross-correlation and coherence methods are related in some sense. This link can be established by extending the definition of the cross-correlation method to incorporate correlation between two vectors $\mathbf{x}$ and $\mathbf{y}$ instead of only the scalar $y(n)$ [5]. Define the normalized cross-correlation matrix $\mathbf{C}_{xy}$ between two vectors $\mathbf{x}$ and $\mathbf{y}$ as follows

$$\mathbf{C}_{xy} = \mathbf{R}_{xx}^{-1/2}\mathbf{R}_{xy}\mathbf{R}_{yy}^{-1/2}, \tag{5.22}$$

where

$$\mathbf{y}(n) = \begin{bmatrix} y(n) & y(n-1) & \cdots & y(n-N+1) \end{bmatrix}^T$$

is a vector of size N. There are two interesting cases:

(i) $N = 1$, $\mathbf{C}_{xy} = \mathbf{c}_{xy}^{(2)}$ (normalized cross-correlation vector between $\mathbf{x}$ and y).

(ii) $N = L = 1$, $\mathbf{C}_{xy} = c_{xy,0}^{(1)}$ (cross-correlation coefficient between x and y).

By extension to (5.13), we then form the detection statistic

$$\xi^{(4)} = \frac{1}{\sqrt{N}} \|\mathbf{C}_{xy}\|_{\mathrm{F}} = \frac{1}{\sqrt{N}} \sqrt{\mathrm{tr}(\mathbf{C}_{xy}^T \mathbf{C}_{xy})}, \tag{5.23}$$

where the subscript "F" denotes the Frobenius norm. We note that for case (i), $\xi^{(4)} = \xi^{(2)}$ as before. Again, we can interpret this formulation as a "generalized cross-correlation", where now both $\mathbf{x}$ and $\mathbf{y}$ are prewhitened, which is also known as the "smoothed coherence transform" (SCOT) [15].

The link between the normalized cross-correlation matrix and the coherence can now be established as follows: Suppose that $N = L \to \infty$. In this case, a Toeplitz matrix is asymptotically equivalent to a circulant matrix if its elements are absolutely summable [20], which is the case for the intended application. Hence we can decompose $\mathbf{R}_{ab}$ as

$$\mathbf{R}_{ab} = \mathbf{F}^{-1}\mathbf{S}_{ab}\mathbf{F}, \tag{5.24}$$

where $\mathbf{F}$ is the discrete Fourier transform (DFT) matrix and

$$\mathbf{S}_{ab} = \mathrm{diag}\{S_{ab}(0), S_{ab}(1), \cdots, S_{ab}(L-1)\} \tag{5.25}$$

is a diagonal matrix formed by the first column of $\mathbf{FR}_{ab}$, and

$$\begin{aligned} S_{ab}(k) &= \sum_{m=-\infty}^{+\infty} E\{a(n)b(n-m)\}\mathrm{e}^{-i2\pi km/L} \\ &= \sum_{m=-\infty}^{+\infty} R_{ab}(m)\mathrm{e}^{-i2\pi km/L} \end{aligned} \tag{5.26}$$

is the DFT cross-power spectrum. Now:

$$\begin{aligned} \mathrm{tr}(\mathbf{C}_{xy}^T\mathbf{C}_{xy}) &= \mathrm{tr}(\mathbf{R}_{yy}^{-1/2}\mathbf{R}_{yx}\mathbf{R}_{xx}^{-1}\mathbf{R}_{xy}\mathbf{R}_{yy}^{-1/2}) \\ &= \mathrm{tr}(\mathbf{R}_{yx}\mathbf{R}_{xx}^{-1}\mathbf{R}_{xy}\mathbf{R}_{yy}^{-1}) \end{aligned} \tag{5.27}$$

since $\mathrm{tr}(\mathbf{AB}) = \mathrm{tr}(\mathbf{BA})$. Using (5.24), we easily find that

$$\begin{aligned} \mathrm{tr}(\mathbf{C}_{xy}^T\mathbf{C}_{xy}) &= \mathrm{tr}(\mathbf{S}_{yx}\mathbf{S}_{xx}^{-1}\mathbf{S}_{xy}\mathbf{S}_{yy}^{-1}) \\ &= \sum_{k=0}^{L-1} |\gamma_{xy}(k)|^2, \end{aligned} \tag{5.28}$$

where

$$\gamma_{xy}(k) = \frac{S_{xy}(k)}{\sqrt{S_{xx}(k)S_{yy}(k)}} \tag{5.29}$$

is the discrete coherence function. Thus, asymptotically we have

$$\begin{aligned}
\xi^{(4)} &\approx \sqrt{\frac{1}{L}\sum_{k=0}^{L-1}|\gamma_{xy}(k)|^2} \\
&= \sqrt{\frac{1}{L}\sum_{k=0}^{L-1}\frac{|H(k)|^2}{|H(k)|^2+\kappa(k)}},
\end{aligned} \tag{5.30}$$

where $H(k)$ is the transfer function of h and

$$\kappa(k) = \frac{S_{vv}(k)}{S_{xx}(k)} \geq 0 \tag{5.31}$$

is the near-end talker to far-end talker spectral ratio at frequency k. Except for an unrestricted frequency range, this form is identical to the coherence-based double-talk detector presented in Section 3.4. We find that this idea is very appropriate since when $v(n) = 0$, the two signals x and y are completely coherent and then $|\gamma_{xy}(k)| = 1, \forall k$, and $\xi^{(4)} \approx 1$; when $v \neq 0$, $|\gamma_{xy}(k)| < 1, \forall k$, and $\xi^{(4)} < 1$.

3.6 TWO-PATH MODEL

An interesting approach to double-talk handling was proposed in [13]. This method was introduced for network echo cancellation. However, it has proven far more useful for the AEC application. In this method two filters model the echo path, one background filter which is adaptive as in a conventional AEC solution and one foreground filter which is not adaptive. The foreground filter cancels the echo. Whenever the background filter performs better than the foreground, its coefficients are copied to the foreground. Coefficients are copied only when a set of conditions are met, which should be compared to the single statistic decision declaring "no double-talk" in a traditional DTD presented in the previous sections.

The basic set of conditions found in [13] are given by (5.32)-(5.34). Copying is *inhibited*, equivalent to double-talk is detected, if any of (5.32)-(5.34) is fulfilled,

$$\xi^{(e)} = \frac{P(e_\text{f})}{P(e_\text{b})} < T_\text{e} \tag{5.32}$$

$$\xi^{(y)} = \frac{P(y)}{P(e_\text{b})} < T_y \tag{5.33}$$

$$\xi^{(x)} = \frac{P(x)}{P(y)} < 1 \tag{5.34}$$

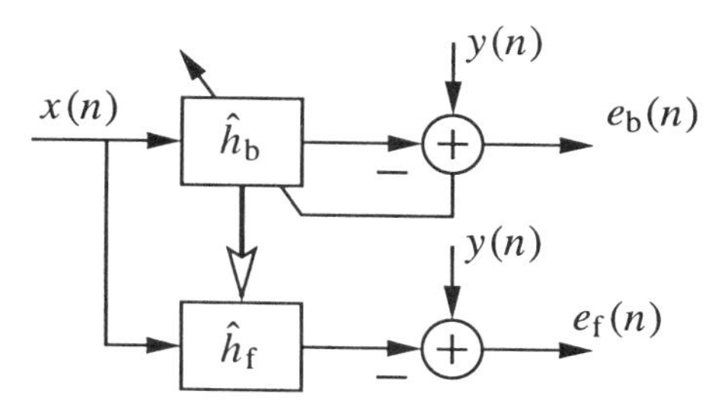

Figure 5.3 Two-path adaptive filtering. The adaptive filter estimates the impulse response $\hat{h}_b$. If, according to some criteria, $\hat{h}_b$ is determined to be a better estimate than the earlier estimate $\hat{h}_f$, the coefficients in $\hat{h}_b$ are copied to $\hat{h}_f$. The latter is then used to calculate the residual echo signal $e_f(n)$.

where $P(a)$ is the short time smoothed absolute magnitude of a signal $a(n)$,

$$P(a) = \sum_{l=0}^{L_{\mathrm{TP}}-1} |a(n-l)|. \tag{5.35}$$

The last condition (5.34) is basically the same as in the Geigel DTD with a unity threshold, i.e., the echo path is assumed not to attenuate the far-end speech. A hangover time T_{hold} is also imposed when (5.34) is fulfilled. If all three conditions are not satisfied over D consecutive decisions, copying of background coefficients is resumed.

A great advantage with this algorithm is that it is not sensitive to echo path changes since the background filter is allowed to track changes freely and as soon as it performs better than the foreground it is copied over. The penalty is of course the slower convergence caused by the delay introduced by copying the coefficients.

3.7 DTD COMBINATIONS WITH ROBUST STATISTICS

All practical double-talk detectors have a probability of miss, i.e. $P_{\mathrm{m}} \neq 0$. Requiring the probability of miss to be smaller will undoubtedly increase the probability of false alarms hence slowing down the convergence rate. As a consequence, no matter what DTD is used, undetected near-end speech will perturb the adaptive algorithm from time to time. Figure 5.4 shows the remaining undetected near-end speech (double-talk) after double-talk detection with a Geigel detector with $T = 2$. The impact of this perturbation is governed by the echo to near-end speech ratio as described in Section 1.

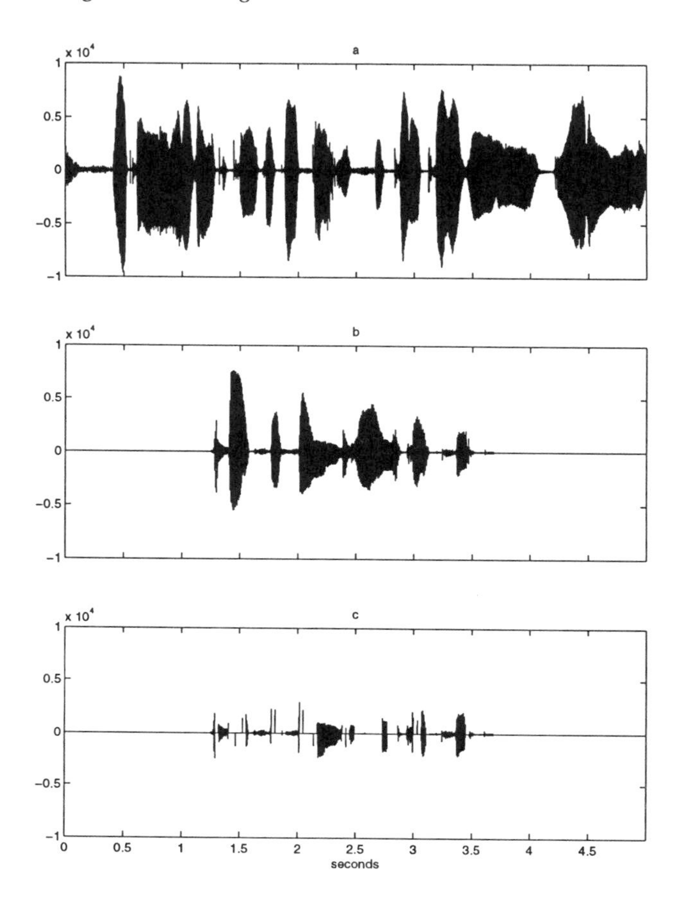

Figure 5.4 (a) Far-end speech. (b) Near-end speech, i.e. double-talk. (c) Near-end speech gated with the decision of the DTD. These are the disturbances that actually enters the adaptive algorithm. Average far- to near-end ratio: 6 dB (1.125-3.625 s).

In practice, what has been done in the past is, first the DTD is designed to be "as good as" one can afford and then, the adaptive algorithm is slowed down so that it copes with the detection errors made by the DTD. This is natural to do since if the adaptive algorithm is very fast, it can react faster to situation changes (e.g. double-talk) than the DTD and thus can diverge. However, this approach severely penalizes the convergence rate of the AEC when the situation is good, i.e. far-end but no near-end talk is present.

In the light of these facts, it may be fruitful to look at adaptive algorithms that can handle at least a small amount of double-talk without diverging. This approach has been studied and proven very successful in the line echo canceler case [21], where the combination of outlier resistant adaptive algorithms and a

Geigel DTD were studied. For the acoustic case one could use any appropriate DTD and combine it with a robust adaptive algorithm.

The approach can be exemplified by a robust version of the NLMS algorithm:

$$\begin{aligned}\hat{\mathbf{h}}(n) &= \hat{\mathbf{h}}(n-1)\\ &+ \frac{\mu_0 \mathbf{x}(n)}{\mathbf{x}^T(n)\mathbf{x}(n)+\delta}\psi\left[\frac{|e(n)|}{s(n-1)}\right]\text{sign}\{e(n)\}s(n-1). \quad (5.36)\end{aligned}$$

As for any AEC/DTD, adaptation is inhibited by setting the step-size parameter to zero when double-talk is detected. The scaled non-linearity $\psi(\cdot)$ in (5.35) can be chosen to be the limiter [22],

$$\psi(\frac{|e(n)|}{s(n)}) = \min\{\frac{|e(n)|}{s(n)}, k_0\}, \qquad (5.37)$$

where $s(n)$ is an adaptive scale factor. Making the scale factor adaptive and supervised by the DTD is the key to the success of this approach. The scale factor should reflect the background noise level at the near-end, be robust to short burst disturbances (double-talk) and track long term changes of the residual error (echo path changes). To fulfill these requirements one can choose the scale factor estimate as

$$s(n) = \lambda s(n-1) + \frac{1-\lambda}{\beta}s(n-1)\psi(\frac{|e(n)|}{s(n-1)}), \qquad (5.38)$$

where $s_{-1} = \sigma_x$. Adaptation of $s(n)$ is performed as long as the DTD has not detected double-talk. Justification and details of the above derivations can be found in [21].

4. DISCUSSION

In this chapter we have presented double-talk detection algorithms suitable for acoustic echo cancellation. Because of the often unknown attenuation and the continuously time-varying nature of acoustic echo paths, devising an appropriate DTD is more challenging than in the line echo canceler case. There are basically two types of double-talk detectors. First, those which form their test statistics from estimated level or power of far-end, near-end including echo, or residual echo signals. Secondly, detectors that make their decisions from cross-correlation or coherence estimates of the same involved signals. In this group we also find detectors utilizing the estimate of the echo path since these estimates are derived through cross-correlation as well. Double-talk detectors based on cross-correlation techniques exhibit desirable properties needed for the acoustic case. Mainly, they have very low sensitivity to the attenuation of the echo path. However, a problem that has to be considered and designed for, is the longer response time which may result. This is due to the fact that good (low variance) test statistics need to be based on more data.

The Holy Grail of double-talk detectors should be insensitive to echo path variations, have equal performance whether the echo canceler is converged or not, have quick response time and be sensitive to low near-end speech levels. Moreover, the DTD must not slow down convergence rate of the AEC which can result either from erroneous decisions or introduction of delays. Some of these properties can be characterized by probability of detection and false alarm. What may be fruitful is to develop this concept further in order to introduce a theoretically founded design principle for the DTD that accounts for all important aspects.

Notes

1. Originally from [1]. This quotation was borrowed from [2].

References

[1] S. B. Weinstein, "Echo cancellation in the telephone network," *IEEE Commun. Soc. Mag.*, pp. 9-15, 1977.

[2] C. R. Johnson, "On the interaction of adaptive filtering, identification, and control," *IEEE Signal Proc. Mag.*, vol. 12, pp. 22-37, March 1995.

[3] S. Haykin, *Adaptive Filter Theory*. New Jersey: Prentice-Hall, Inc, 1996.

[4] T. Gänsler, M. Hansson, C.-J. Ivarsson, and G. Salomonsson, "A double-talk detector based on coherence," *IEEE Trans. Commun.*, vol. 44, pp. 1421-1427, Nov. 1996.

[5] J. Benesty, D. R. Morgan, and J. H. Cho, "An new class of doubletalk detectors based on cross-correlation," *IEEE Trans. Speech Audio Processing*. To appear.

[6] J. H. Cho, D. R. Morgan, and J. Benesty, "An objective technique for evaluating doubletalk detectors in acoustic cancelers," *IEEE Trans. Speech Audio Processing*, vol. 7, pp. 718-724, Nov. 1999.

[7] M. M. Sondhi, "An adaptive echo canceler," *Bell Syst. Techn. J.*, vol. XLVI, pp. 497-510, Mar. 1967.

[8] D. L. Duttweiler, "A twelve-channel digital echo canceler," *IEEE Trans. Commun.*, vol. 26, pp. 647-653, May 1978.

[9] H. Ye and B. X. Wu, "A new double-talk detection algorithm based on the orthogonality theorem," *IEEE Trans. Commun.*, vol. 39, pp. 1542-1545, Nov. 1991.

[10] R. D. Wesel, "Cross-correlation vectors and double-talk control for echo cancellation," Unpublished work, 1994.

[11] J. Prado and E. Moulines, "Frequency-domain adaptive filtering with applications to acoustic echo cancellation," *Ann. Télécomun.*, vol. 49, pp. 414-428, 1994.

[12] S. M. Kuo and Z. Pan, "An acoustic echo canceller adaptable during double-talk periods using two microphones," *Acoustics Letters*, vol. 15, pp. 175-179, 1992.

[13] K. Ochiai, T. Araseki, and T. Ogihara, "Echo canceler with two echo path models," *IEEE Trans. Commun.*, vol. COM-25, pp. 589-595, June 1977.

[14] J. Benesty, D. R. Morgan, and J. H. Cho, "A family of doubletalk detectors based on cross-correlation," in *Proc. IWAENC*, Sep. 1999, pp. 108-111.

[15] C. H. Knapp and C. G. Carter, "The generalised correlation method for estimation of time delay," *IEEE Trans. Acoust., Speech and Signal Processing*, vol. 24, pp. 320-327, Aug. 1976.

[16] J. R. Zeidler, "Performance analysis of LMS adaptive prediction filters," *Proc. of the IEEE*, vol. 78, pp. 1781-1806, Dec. 1990.

[17] D. J. Thomson, "Spectrum estimation and harmonic analysis," *Proc. of the IEEE*, vol. 70, pp. 1055-1096, Sep. 1982.

[18] D. Slepian, "Prolate spheroidal wave funcions, Fourier analysis, and uncertainty-V," *Bell Syst. Tech. J.*, vol. 40, pp. 1371-1429, 1978.

[19] P. D. Welch, "The use of fast Fourier transform for the estimation of power spectra: a method based on time averaging over short, modified periodograms," *IEEE Trans. Audio Electroacoustics*, vol. AU-15, pp. 70-73, June 1967.

[20] R. M. Gray, "On the asymptotic eigenvalue distribution of Toeplitz matrices," *IEEE Trans. Inform. Theory*, vol. IT-18, pp. 725-730, Nov. 1972.

[21] T. Gänsler, S. L. Gay, M. M. Sondhi, and J. Benesty, "Double-talk robust fast converging algorithms for network echo cancellation," *IEEE Trans. Speech Audio Processing*. To appear.

[22] P. J. Huber, *Robust Statistics*. pages 68-71, 135-138. New York: Wiley, 1981.

II
MULTI-CHANNEL ACOUSTIC ECHO CANCELLATION

Chapter 6

MULTI-CHANNEL SOUND, ACOUSTIC ECHO CANCELLATION, AND MULTI-CHANNEL TIME-DOMAIN ADAPTIVE FILTERING

Jacob Benesty
Bell Laboratories, Lucent Technologies
jbenesty@bell-labs.com

Tomas Gänsler
Bell Laboratories, Lucent Technologies
gaensler@bell-labs.com

Peter Eneroth
Department of Applied Electronics
Lund University
peter.eneroth@tde.lth.se

Abstract In this chapter, we discuss why multi-channel sound is important for telecommunication and for what kind of applications. In hands-free systems, multi-channel acoustic echo cancelers are absolutely necessary for full duplex communication. We explain the fundamental difference with the mono-channel case and study the nonuniqueness problem. We also give a general framework for multi-channel time-domain adaptive algorithms, since they are the heart of any echo canceler.

Keywords: Acoustic Echo Cancellation, Multi-Channel, Adaptive Algorithms, LMS, APA, RLS, FRLS

1. INTRODUCTION

One may ask a legitimate question: Why do we need multi-channel sound for telecommunication? Let's take the following example. When we are in a room with several people talking, laughing, or just communicating with each other, thanks to our binaural auditory system we can concentrate on one particular talker (if several persons are talking at the same time), localize or identify a person who is talking, and somehow we are able to process a noisy or a reverberant speech signal in order to make it intelligible. On the other hand, with only one ear or, equivalently, if we record what happens in the room with one microphone and listen to this monophonic signal, it will likely make all of the above mentioned tasks more difficult. So multi-channel sound teleconferencing systems provide a realistic presence that actual mono-channel systems cannot offer.

One other promising application in modern communications is desktop conferencing [1], which can involve several participants over a widely distributed area. This kind of conferencing with stereo (or multi-channel) sound will likely grow rapidly in the near future, especially over the Internet. The general scenario is as follows. Several persons in different locations would like to communicate with each other and each one of them has a workstation. Each participant would like to see video of the other participants arranged in a rational way and to hear them in a way that facilitates identification and understanding. For example, the voice of a participant whose picture is located on the left of his screen, should appear to come from the left. The location of auditory images in perceptual space is controlled by interchannel level and time differences and is mediated by the binaural auditory system.

In such hands-free systems, multi-channel acoustic echo cancelers (MCAECs) are absolutely necessary for full-duplex communication [2]. Let P be the number of channels. For a teleconferencing system, the MCAEC consists of P^2 adaptive filters aiming at identifying P^2 echo paths from P loudspeakers to P microphones. Figure 6.1 illustrates the concept of stereophonic (two-channel) echo cancellation between a transmission room and a receiving room for one microphone; similar analysis will apply to the other microphone signals. The transmission room is sometimes referred to as the far-end and the receiving room as the near-end. Clearly, according to this scheme, stereophonic acoustic echo cancellation consists of direct identification of a multi-input, unknown linear system, consisting of the parallel combination of two acoustic paths (h_1, h_2) extending through the receiving room from the loudspeakers to the microphone. The stereophonic AEC tries to model this unknown system by a pair of adaptive filters ($\hat{h}_1, \hat{h}_2$).

Although very similar, multi-channel acoustic echo cancellation (MCAEC) is fundamentally different from traditional mono echo cancellation. A multi-channel echo canceler straightforwardly implemented not only would have to track changing echo paths in the receiving room *but also in the transmission room*! For example, the canceler has to reconverge if one talker stops talking and another starts talking at a different location in the transmission room. There is no adaptive algorithm that can track such a change sufficiently fast and this scheme therefore results in poor echo suppression. Thus, a generalization of the mono AEC in the multi-channel case does not result in satisfactory performance.

The theory explaining the problem of MCAEC was described in [2] and [3]. The fundamental problem is that the multiple channels may carry linearly related signals which in turn may make the normal equation to be solved by the adaptive algorithm singular. This implies that there is no unique solution to the equation but an infinite number of solutions and it can be shown that all but the true one depend on the impulse responses of the transmission room. As a result, intensive studies have been made of how to handle this properly. It is shown in [3] that the only solution to the nonuniqueness problem is to reduce the correlation between the different signals and an efficient low complexity method for this purpose is also given.

Lately, attention has been focused on the investigation of other methods that decrease the cross-correlation between the channels in order to get well behaved estimates of the echo paths [4, 5, 6, 7, 8]. The main problem is how to reduce the correlation sufficiently without affecting the stereo perception and the sound quality.

The performance of the MCAEC is more severely affected by the choice of algorithm than the monophonic counterpart. This is easily recognized since the performance of most adaptive algorithms depends on the condition number of the input signal. In the multi-channel case, the condition number is very high; as a result, algorithms such as the Least Mean Squares (LMS) or the Normalized LMS (NLMS) that does not take into account the cross-correlation among all the input signals converge very slowly to the true solution. It is therefore highly interesting to study multi-channel adaptive filtering algorithms. Straightforward extensions of single channel algorithms may not be the best choice for the MCAEC application. Standard algorithms (and their fast versions, i.e. low complexity versions) are the Recursive Least Squares (RLS), LMS, and Affine Projection Algorithms (APA). A framework for multi-channel adaptive filtering can be found in [9]. Results from this and related papers will also be discussed below.

This chapter is organized as follows: Section 2 describes the fundamental difference between mono and multi-channel acoustic echo cancellation and explains the nonuniqueness problem. In Section 3 a summary of recent research on methods for decorrelating the stereo channels is given. Section 4 gives the principle of a hybrid approach; using both mono and stereo AECs. Section 5 reviews the highlights of the improved multi-channel time-domain adaptive filter algorithms suitable for MCAEC. A discussion of new prospects and challenges in this field is given in Section 6.

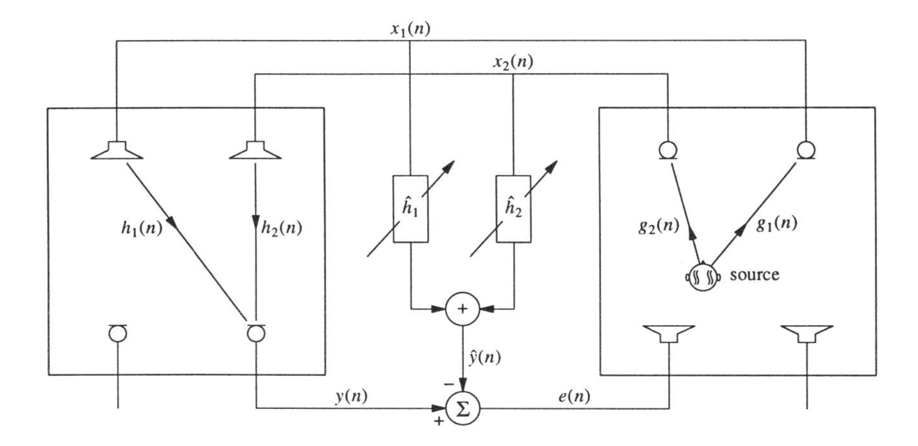

Figure 6.1 Schematic diagram of stereophonic acoustic echo cancellation.

2. MULTI-CHANNEL IDENTIFICATION AND THE NONUNIQUENESS PROBLEM

In this section we show that the normal equation for the multi-channel identification problem has not a unique solution as in the single-channel case. Indeed, since the P input signals are obtained by filtering from a common source, a problem of nonuniqueness is expected [2]. In the following discussion, we suppose that the length (L) of the impulse responses (in the transmission and receiving rooms) is equal to the length of the modeling filters. For the notations used in this section, see Fig. 6.1.

Assume that the system (transmission room) is linear and time invariant; therefore, we have the following $[P(P-1)/2]$ relations [9]:

$$\mathbf{x}_i^T(n)\mathbf{g}_j = \mathbf{x}_j^T(n)\mathbf{g}_i, \ i, j = 1, 2, ..., P; \ i \neq j, \tag{6.1}$$

where

$$\mathbf{x}_i(n) = \left[\begin{array}{cccc} x_i(n) & x_i(n-1) & \cdots & x_i(n-L+1) \end{array} \right]^T, \ i = 1, 2, ..., P$$

are vectors of signal samples at the microphone outputs in the transmission room, ^{T} denotes the transpose of a vector or a matrix, and the impulse response vectors between the source and the microphones are defined as

$$\mathbf{g}_i = \left[\begin{array}{cccc} g_{i,0} & g_{i,1} & \cdots & g_{i,L-1} \end{array} \right]^T, \ i = 1, 2, ..., P.$$

Now, let us define the recursive least squares error criterion with respect to the modeling filters:

$$J_1(n) = \sum_{l=1}^{n} \lambda^{n-l} e^2(l), \tag{6.2}$$

where λ $(0 < \lambda \leq 1)$ is a forgetting factor.

$$e(n) = y(n) - \sum_{i=1}^{P} \hat{\mathbf{h}}_i^T \mathbf{x}_i(n) \tag{6.3}$$

is the error signal at time n between the microphone output $y(n)$ in the receiving room and its estimate, where

$$\hat{\mathbf{h}}_i = \begin{bmatrix} \hat{h}_{i,0} & \hat{h}_{i,1} & \cdots & \hat{h}_{i,L-1} \end{bmatrix}^T, \; i = 1, 2, ..., P$$

are the P modeling filters.

The minimization of (6.2) leads to the normal equation:

$$\mathbf{R}(n)\hat{\mathbf{h}}(n) = \mathbf{r}(n), \tag{6.4}$$

where

$$\begin{aligned} \mathbf{R}(n) &= \sum_{l=1}^{n} \lambda^{n-l} \mathbf{x}(l)\mathbf{x}^T(l) \\ &= \begin{bmatrix} \mathbf{R}_{1,1}(n) & \mathbf{R}_{1,2}(n) & \cdots & \mathbf{R}_{1,P}(n) \\ \mathbf{R}_{2,1}(n) & \mathbf{R}_{2,2}(n) & \cdots & \mathbf{R}_{2,P}(n) \\ \vdots & \vdots & \ddots & \vdots \\ \mathbf{R}_{P,1}(n) & \mathbf{R}_{P,2}(n) & \cdots & \mathbf{R}_{P,P}(n) \end{bmatrix} \end{aligned} \tag{6.5}$$

is an estimate of the input signal covariance matrix,

$$\begin{aligned} \mathbf{r}(n) &= \sum_{l=1}^{n} \lambda^{n-l} y(l)\mathbf{x}(l) \\ &= \begin{bmatrix} \mathbf{r}_1^T(n) & \mathbf{r}_2^T(n) & \cdots & \mathbf{r}_P^T(n) \end{bmatrix}^T \end{aligned} \tag{6.6}$$

is an estimate of the cross-correlation vector between the input and output signals (in the receiving room), and

$$\hat{\mathbf{h}}(n) = \begin{bmatrix} \hat{\mathbf{h}}_1^T(n) & \hat{\mathbf{h}}_2^T(n) & \cdots & \hat{\mathbf{h}}_P^T(n) \end{bmatrix}^T,$$

$$\mathbf{x}(n) = \begin{bmatrix} \mathbf{x}_1^T(n) & \mathbf{x}_2^T(n) & \cdots & \mathbf{x}_P^T(n) \end{bmatrix}^T.$$

Consider the following vector:

$$\mathbf{u} = \begin{bmatrix} \sum_{i=2}^{P} \mathbf{g}_i^T & -\mathbf{g}_1^T & \cdots & -\mathbf{g}_1^T \end{bmatrix}^T.$$

We can verify using (6.1) that $\mathbf{R}(n)\mathbf{u} = \mathbf{0}_{PL\times 1}$, so $\mathbf{R}(n)$ is not invertible. Thus, there is no unique solution to the problem and an adaptive algorithm will drive to any one of many possible solutions, which can be very different from the "true" desired solution $\hat{\mathbf{h}} = \mathbf{h}$. These nonunique "solutions" are dependent on the impulse responses in the transmission room. This, of course, is intolerable because $\mathbf{g}_i$ can change instantaneously, for example, as one person stops talking and another starts [2, 3].

Definition: The quantity

$$\varepsilon = \|\mathbf{h} - \hat{\mathbf{h}}\| / \|\mathbf{h}\|, \tag{6.7}$$

where $\|\cdot\|$ denotes the two-norm vector, is called the misalignment and measures the mismatch between the impulse responses of the receiving room and the modeling filters. In the multi-channel case, it is possible to have good echo cancellation even when the misalignment is large. However, in such a case, the cancellation will degrade if $\mathbf{g}_i$ change. A main objective of MCAEC research is to avoid this problem.

The only one way to decrease the misalignment is to decorrelate partially (or in totality) two by two the P input signals. The correlation between two channels can be linked to ill-conditioning of the correlation matrix by means of the coherence magnitude [3]. Ill-conditioning can therefore be monitored by the coherence function which serves as a measure of achieved decorrelation. In the next section we summarize a number of approaches that have been developed recently for reducing the cross-correlation.

3. SOME DIFFERENT SOLUTIONS FOR DECORRELATION

If we have P different channels, we need to decorrelate them partially and mutually. In the following, we show how to decorrelate partially two channels. The same process should be applied for all the channels. It is well-known that the coherence magnitude between two processes is equal to 1 if and only if they are linearly related. In order to weaken this relation some non-linear or time-varying transformation of the stereo channels has to be made. Such a transformation reduces the coherence and hence the condition number of the covariance matrix, thereby improving the misalignment. The transformation has to be performed cautiously so that it is inaudible and has no effect on stereo perception.

A simple non-linear method that gives good performance uses a half-wave rectifier [3]. For this method there still may be a linear relation between the nonlinearly transformed channels e.g. if $\forall n\ x_1(n) \geq 0$ and $x_2(n) \geq 0$ or if we have $ax_1(n - \tau_1) = x_2(n - \tau_2)$ with $a > 0$. In practice however, these cases never occur because we always have zero-mean signals and $\mathbf{g}_1$, $\mathbf{g}_2$ are in practice never related by just a simple delay.

An improved version of this technique is to use positive and negative half-wave rectifiers on each channel respectively,

$$x_1'(n) = x_1(n) + \alpha \frac{x_1(n) + |x_1(n)|}{2} \tag{6.8}$$

$$x_2'(n) = x_2(n) + \alpha \frac{x_2(n) - |x_2(n)|}{2}. \tag{6.9}$$

This principle removes the linear relation in the special signal cases given above. Experiments show that stereo perception is not affected by this method even with α as large as 0.5. Also, the distortion introduced for speech is hardly audible because of the nature of the speech signal and psychoacoustic masking effects [10]. This kind of distortion is also acceptable for some music signals but may be objectionable for pure tones.

In [6] a similar approach with non-linearities is proposed. The idea is expanded so that four adaptive filters operate on different non-linearly processed signals to estimate the echo paths. These non-linearities are chosen such that the input signals of two of the adaptive filters are independent which thus represent a "perfect" decorrelation. Tap estimates are then copied to a fixed two-channel filter which performs the echo cancellation with the unprocessed signals. The advantage of the method would be that the NLMS algorithm could be used instead of more sophisticated algorithms.

Another approach that makes it possible to use the NLMS algorithm is to decorrelate the channels by means of complementary comb filtering [2, 11]. The technique is based on removing the energy in a certain frequency band of the speech signal in one channel. This means the coherence would become zero in this band and thereby resulting in fast alignment of the estimate even when using the NLMS algorithm. Energy is removed complementarily between the channels so that the stereo perception is not severely affected. The method works well for frequencies above 1 kHz but must be combined with some other decorrelation technique for the lower frequencies.

Two methods based on introducing time-varying filters in the transmission path were presented in [7, 8]. To show that this gives the right effect we can look at the condition for perfect cancellation (under the assumption that we do not have any zeros in the input signal spectrum) in the frequency domain which can be written as,

$$\tilde{H}_1(\omega)G_1(\omega)T_1(\omega, n) + \tilde{H}_2(\omega)G_2(\omega)T_2(\omega, n) = 0, \quad \forall \omega \in [-\pi, \pi]. \tag{6.10}$$

where $\tilde{H}_i(\omega)$, $i = 1, 2$ are the Fourier transforms of the filter coefficient errors $\tilde{h}_i(n) = h_i(n) - \hat{h}_i(n)$. $G_i(\omega)$ $i = 1, 2$, are the Fourier transformed transmission room echo paths and $T_i(\omega, n)$, $i = 1, 2$, are the frequency representation of the introduced time-varying filters. We see again for constant $T_i(\omega)$, $i = 1, 2$, the solution of (6.10) is nonunique. However, for a time-varying $T_i(\omega, n)$, $i = 1, 2$, (6.10) can only be satisfied if $\tilde{H}_i(\omega) = 0$, $i = 1, 2$, $\omega \in [-\pi, \pi]$, which is the idea for this method. In [7], left and right signals are filtered through two independent time-varying first order all pass filters. Stochastic time-variation is introduced by making the pole position of the filter a random walk process. The actual position is limited by the constraints of stability and inaudibility of the introduced distortion. While significant reduction in correlation can be achieved for higher frequencies with the imposed constraints, the lower frequencies are still fairly unaffected by the time-variation. In [8] a periodically varying filter is applied to either left or right channel. The signal is either delayed by one sample or passed through without delay. A transition zone between the delayed and non-delayed state is also employed in order to reduce the obvious click sounds that otherwise would be produced. These methods may also affect the stereo perception.

Although the idea of adding independent perceptually shaped noise to the channels was mentioned in [2, 3], thorough investigations of the actual benefit of the technique was not presented. Results regarding variants of this idea can be found in [4, 5]. A pre-processing unit estimating the masking threshold and adding an appropriate amount of noise was proposed in [5]. It was also noted that adding a masked noise to each channel may affect the spatialization of the sound even if the noise is inaudible at each channel separately. This effect can be controlled through correction of the masking threshold when appropriate. In [4] the improvement of misalignment was studied in the SAEC when a perceptual audio coder was added in the transmission path. Reduced correlation between the channels were shown by means of coherence analysis and improved convergence rate of the adaptive algorithm was observed. A low complexity method for achieving additional decorrelation by modifying the decoder was also proposed. The encoder cannot quantize every single frequency band optimally due to large overhead. This has the effect that there is a margin to the masking threshold which can be exploited. In the presented method the masking threshold is estimated from the MDCT (Modified Discrete Cosine) coefficients delivered by the encoder and an appropriate inaudible amount of decorrelating noise is added to the signals.

In the rest of this chapter, we suppose that the normal equation has a unique solution. So one of the previous decorrelation methods is used. But the channel input signals can still be highly correlated.

4. THE HYBRID MONO/STEREO ACOUSTIC ECHO CANCELER

This solution was proposed recently [12] and it is a good compromise between the complexity of a full-band stereo AEC and spatial realism. The un-

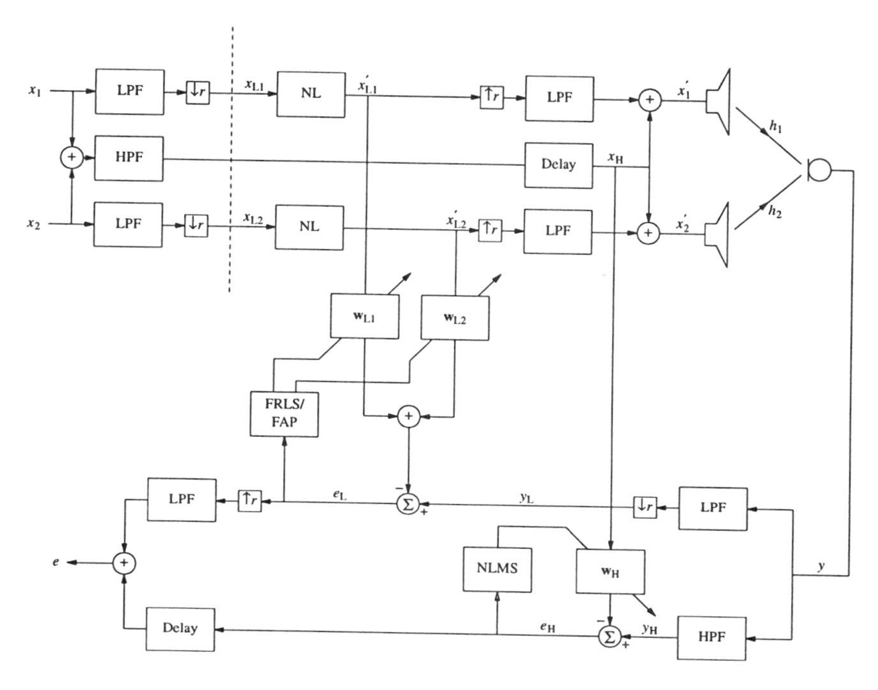

Figure 6.2 Hybrid mono/stereo acoustic echo canceler.

derstanding of the stereo effect from a psychoacoustical point of view is not easy, but many experiments show that the dominant stereophonic cues are located below about 1 kHz [13]. In many applications, this information can be exploited for efficient transmission of the microphone signals to the receiving room and also for devising an efficient AEC.

Based on the above psychoacoustical principle, Fig. 6.2 shows a way to transmit the two microphone signals to the receiving room and a set of AECs matched to these signals. First, the two signals x_1 and x_2 (left and right) are lowpass filtered in order to keep just the frequencies below f_c for realizing the stereo effect, where the crossover frequency f_c is on the order of 1 kHz. Then, the sum of the left and right channels is highpass filtered to keep the frequencies above f_c as a monophonic signal. Finally, the left (resp. right) loudspeaker signal in the receiving room is the sum of the low frequencies of the left (resp. right) channel and high frequencies of both channels. We now observe that two AECs are necessary: one mono AEC for high frequencies and one stereo AEC for low frequencies. Note that we have to put a nonlinear transformation (NL) in each low-frequency channel to help the adaptive algorithm to converge toward the "true" solution.

The above structure is much simpler than a fullband system, despite the fact that we have two different AECs. Indeed, for the stereo AEC, since the maximum frequency is f_c, we can subsample the signals by a factor $r = f_s/(2f_c)$, where f_s is the sampling rate of the system. As a result, the arithmetic complexity is divided by r^2 in comparison with a fullband implementation (the number of taps and the number of computations per second are both reduced by r). In this case, we can afford to use a rapidly converging adaptive algorithm like the two-channel FRLS [14]. On the other hand, the simple NLMS algorithm can be used to update the filter coefficients in the high frequency band.

Thus, with this structure, the complexity of the system is decreased and the convergence rate of the adaptive algorithms is increased, while preserving most of the stereo effect.

5. MULTI-CHANNEL TIME-DOMAIN ADAPTIVE FILTERS

In this section, we propose a general framework for multi-channel time-domain adaptive filtering. As we will see, the choice of the algorithm is more crucial than the mono-channel case because of the high correlation among all the channels. Standard adaptive algorithms are NLMS, APA, or RLS. For MCAEC, fast versions of the two latter have to be used because of the large number of parameters that has to be adjusted. Straightforward generalizations to the multi-channel case of these algorithms can easily be derived. As discussed in Section 2, the cross-correlation among the channels is a very important issue for multi-channel acoustic echo cancellation. We therefore focus on improved multi-channel versions that take this interchannel correlation into account. By properly factorizing the covariance matrix appearing in (6.4) we can obtain useful insights that lead to sophisticated algorithms that better exploit the cross-correlation among the channels.

5.1 THE CLASSICAL AND FACTORIZED MULTI-CHANNEL RLS

From the normal equation (6.4), we easily derived the classical update equations for the multi-channel RLS:

$$e(n) = y(n) - \hat{\mathbf{h}}^T(n-1)\mathbf{x}(n), \tag{6.11}$$

$$\hat{\mathbf{h}}(n) = \hat{\mathbf{h}}(n-1) + \mathbf{R}^{-1}(n)\mathbf{x}(n)e(n). \tag{6.12}$$

Another way to write the multi-channel RLS is to first factorize the covariance matrix inverse $\mathbf{R}^{-1}(n)$.

Consider the following variables:

$$\begin{aligned}\mathbf{z}_i(n) &= \sum_{j=1}^{P} \mathbf{C}_{i,j}\mathbf{x}_j(n) \\ &= \mathbf{x}_i(n) + \sum_{j=1, j\neq i}^{P} \mathbf{C}_{i,j}\mathbf{x}_j(n) \\ &= \mathbf{x}_i(n) - \hat{\mathbf{x}}_i(n), \ i = 1, ..., P, \qquad (6.13)\end{aligned}$$

with $\mathbf{C}_{i,i} = \mathbf{I}_{L\times L}$ and $\hat{\mathbf{x}}_i(n) = -\sum_{j=1, j\neq i}^{P} \mathbf{C}_{i,j}\mathbf{x}_j(n)$. Matrices $\mathbf{C}_{i,j}$ are the cross-interpolators obtained by minimizing

$$J_{z_i}(n) = \sum_{l=1}^{n} \lambda^{n-l}\mathbf{z}_i^T(l)\mathbf{z}_i(l), \ i = 1, ..., P, \qquad (6.14)$$

and $\mathbf{z}_i(n)$ are the cross-interpolation error vectors.

A general factorization of $\mathbf{R}^{-1}(n)$ can be stated as follows:

Lemma 1:

$$\mathbf{R}^{-1}(n) = \begin{bmatrix} \mathbf{R}_1^{-1}(n) & \mathbf{0}_{L\times L} & \cdots & \mathbf{0}_{L\times L} \\ \mathbf{0}_{L\times L} & \mathbf{R}_2^{-1}(n) & \cdots & \mathbf{0}_{L\times L} \\ \vdots & \vdots & \ddots & \vdots \\ \mathbf{0}_{L\times L} & \mathbf{0}_{L\times L} & \cdots & \mathbf{R}_P^{-1}(n) \end{bmatrix} \times \begin{bmatrix} \mathbf{I}_{L\times L} & \mathbf{C}_{1,2}(n) & \cdots & \mathbf{C}_{1,P}(n) \\ \mathbf{C}_{2,1}(n) & \mathbf{I}_{L\times L} & \cdots & \mathbf{C}_{2,P}(n) \\ \vdots & \vdots & \ddots & \vdots \\ \mathbf{C}_{P,1}(n) & \mathbf{C}_{P,2}(n) & \cdots & \mathbf{I}_{L\times L} \end{bmatrix}, \qquad (6.15)$$

where

$$\mathbf{R}_i(n) = \sum_{j=1}^{P} \mathbf{C}_{i,j}(n)\mathbf{R}_{j,i}(n), \ i = 1, 2, ..., P. \qquad (6.16)$$

Proof: The proof is rather straightforward by multiplying both sides of (6.15) by $\mathbf{R}(n)$ and showing that the result of the right-hand side is equal to the identity matrix with the help of (6.14).

Example: $P = 2$: In this case, we have:

$$\begin{aligned}\mathbf{z}_1(n) &= \mathbf{x}_1(n) + \mathbf{C}_{1,2}\mathbf{x}_2(n), \qquad (6.17)\\ \mathbf{z}_2(n) &= \mathbf{x}_2(n) + \mathbf{C}_{2,1}\mathbf{x}_1(n), \qquad (6.18)\end{aligned}$$

where

$$\begin{aligned}\mathbf{C}_{1,2}(n) &= -\mathbf{R}_{1,2}(n)\mathbf{R}_{2,2}^{-1}(n), \qquad (6.19)\\ \mathbf{C}_{2,1}(n) &= -\mathbf{R}_{2,1}(n)\mathbf{R}_{1,1}^{-1}(n), \qquad (6.20)\end{aligned}$$

are the cross-interpolators obtained by minimizing $\sum_{l=1}^{n} \lambda^{n-l} \mathbf{z}_1^T(l)\mathbf{z}_1(l)$ and $\sum_{l=1}^{n} \lambda^{n-l} \mathbf{z}_2^T(l)\mathbf{z}_2(l)$. Hence:

$$\mathbf{R}^{-1}(n) = \begin{bmatrix} \mathbf{R}_1^{-1}(n) & \mathbf{0}_{L\times L} \\ \mathbf{0}_{L\times L} & \mathbf{R}_2^{-1}(n) \end{bmatrix} \times \begin{bmatrix} \mathbf{I}_{L\times L} & -\mathbf{R}_{1,2}(n)\mathbf{R}_{2,2}^{-1}(n) \\ -\mathbf{R}_{2,1}(n)\mathbf{R}_{1,1}^{-1}(n) & \mathbf{I}_{L\times L} \end{bmatrix}, \quad (6.21)$$

where

$$\mathbf{R}_1(n) = \mathbf{R}_{1,1}(n) - \mathbf{R}_{1,2}(n)\mathbf{R}_{2,2}^{-1}(n)\mathbf{R}_{2,1}(n), \quad (6.22)$$

$$\mathbf{R}_2(n) = \mathbf{R}_{2,2}(n) - \mathbf{R}_{2,1}(n)\mathbf{R}_{1,1}^{-1}(n)\mathbf{R}_{1,2}(n), \quad (6.23)$$

are the cross-interpolation error energy matrices or the Schur complements of $\mathbf{R}$ with respect to $\mathbf{R}_{2,2}$ and $\mathbf{R}_{1,1}$.

From the above result (Lemma 1), we deduce the factorized multi-channel RLS:

$$\hat{\mathbf{h}}_i(n) = \hat{\mathbf{h}}_i(n-1) + \mathbf{R}_i^{-1}(n)\mathbf{z}_i(n)e(n),\ i = 1, 2, ..., P. \quad (6.24)$$

5.2 THE MULTI-CHANNEL FAST RLS

Because RLS has so far proven to perform better than other algorithms in the MCAEC application [14] a fast calculation scheme of a multi-channel version is presented in this section. Compared to standard RLS it has a lower complexity, $6P^2L + 2PL$ multiplications [instead of $O(P^2L^2)$]. This algorithm is a numerically stabilized version of the algorithm proposed in [15]. Some extra stability control has to be added so that the algorithm behaves well for a non-stationary speech signal. The following has to be defined:

$$\chi(n) = [x_1(n)\ x_2(n)\ \dots\ x_P(n)]^T,\ (P \times 1), \quad (6.25)$$

$$\underline{\mathbf{x}}(n) = [\chi^T(n)\ \chi^T(n-1)\ \dots\ \chi^T(n-L+1)]^T,\ (PL \times 1), \quad (6.26)$$

$$\underline{\hat{\mathbf{h}}}(n) = [\hat{h}_{1,0}(n)\ \hat{h}_{2,0}(n)\ \dots\ \hat{h}_{P-1,L-1}(n)\ \hat{h}_{P,L-1}(n)]^T,\ (PL \times 1). \quad (6.27)$$

Note that the channels of the filter- and state-vector [$\underline{\mathbf{x}}(n)$] are interleaved in this algorithm. Defined also:

- $\mathbf{A}(n)$, $\mathbf{B}(n)$ = Forward and backward prediction filter matrices, $(PL \times P)$,
- $\mathbf{E}_\mathrm{A}(n)$, $\mathbf{E}_\mathrm{B}(n)$ = Forward and backward prediction error energy matrices, $(P \times P)$,

- $\mathbf{e}_{\mathrm{A}}(n)$, $\mathbf{e}_{\mathrm{B}}(n)$ = Forward and backward prediction error vectors, $(P \times 1)$,
- $\mathbf{k}(n) = \mathbf{R}^{-1}(n-1)\underline{\mathbf{x}}(n)$ = Kalman gain vector, $(PL \times 1)$,
- $\varphi(n)$ = Maximum likelihood related variable, (1×1),
- $\kappa \in [1.5, 2.5]$, Stabilization parameter, (1×1),
- $\lambda \in (0, 1]$, Forgetting factor, (1×1).

The multi-channel fast RLS (FRLS) is then:

$$\begin{aligned}
& \quad Prediction: \\
\mathbf{e}_{\mathrm{A}}(n) &= \chi(n) - \mathbf{A}^T(n-1)\underline{\mathbf{x}}(n-1), \quad (P \times 1), \\
\varphi_1(n) &= \varphi(n-1) + \mathbf{e}_{\mathrm{A}}^T(n)\mathbf{E}_{\mathrm{A}}^{-1}(n-1)\mathbf{e}_{\mathrm{A}}(n), \quad (1 \times 1), \\
\begin{bmatrix} \mathbf{t}(n) \\ \mathbf{m}(n) \end{bmatrix} &= \begin{bmatrix} \mathbf{0}_{P\times 1} \\ \mathbf{k}(n-1) \end{bmatrix} + \begin{bmatrix} \mathbf{I}_{P\times P} \\ -\mathbf{A}(n-1) \end{bmatrix} \mathbf{E}_{\mathrm{A}}^{-1}(n-1)\mathbf{e}_{\mathrm{A}}(n), \\
& \quad ((PL+P) \times P), \\
\mathbf{E}_{\mathrm{A}}(n) &= \lambda[\mathbf{E}_{\mathrm{A}}(n-1) + \mathbf{e}_{\mathrm{A}}(n)\mathbf{e}_{\mathrm{A}}^T(n)/\varphi(n-1)], \quad (P \times P), \\
\mathbf{A}(n) &= \mathbf{A}(n-1) + \mathbf{k}(n-1)\mathbf{e}_{\mathrm{A}}^T(n)/\varphi(n-1), \quad (PL \times P), \\
\mathbf{e}_{\mathrm{B}_1}(n) &= \mathbf{E}_{\mathrm{B}}(n-1)\mathbf{m}(n), \quad (P \times 1), \\
\mathbf{e}_{\mathrm{B}_2}(n) &= \chi(n-L) - \mathbf{B}^T(n-1)\underline{\mathbf{x}}(n), \quad (P \times 1), \\
\mathbf{e}_{\mathrm{B}}(n) &= \kappa\mathbf{e}_{\mathrm{B}_2}(n) + (1-\kappa)\mathbf{e}_{\mathrm{B}_1}(n), \quad (P \times 1), \\
\mathbf{k}(n) &= \mathbf{t}(n) + \mathbf{B}(n-1)\mathbf{m}(n), \quad (PL \times 1), \\
\varphi(n) &= \varphi_1(n) - \mathbf{e}_{\mathrm{B}_2}^T(n)\mathbf{m}(n), \quad (1 \times 1), \\
\mathbf{E}_{\mathrm{B}}(n) &= \lambda[\mathbf{E}_{\mathrm{B}}(n-1) + \mathbf{e}_{\mathrm{B}_2}(n)\mathbf{e}_{\mathrm{B}_2}^T(n)/\varphi(n)], \quad (P \times P), \\
\mathbf{B}(n) &= \mathbf{B}(n-1) + \mathbf{k}(n)\mathbf{e}_{\mathrm{B}}^T(n)/\varphi(n), \quad (PL \times P). \\
& \quad Filtering: \\
e(n) &= y(n) - \hat{\underline{\mathbf{h}}}^T(n-1)\underline{\mathbf{x}}(n), \quad (1 \times 1), \\
\hat{\underline{\mathbf{h}}}(n) &= \hat{\underline{\mathbf{h}}}(n-1) + \mathbf{k}(n)e(n)/\varphi(n), \quad (PL \times 1).
\end{aligned}$$

5.3 THE MULTI-CHANNEL LMS ALGORITHM

5.3.1 Classical Derivation. The mean-square error criterion is defined as

$$J_2 = E\left\{\left[y(n) - \mathbf{x}^T(n)\hat{\mathbf{h}}\right]^2\right\}, \tag{6.28}$$

where $E\{\cdot\}$ is the statistical expectation operator. Let $\mathbf{f}(\hat{\mathbf{h}})$ denote the value of the gradient vector with respect to $\hat{\mathbf{h}}$. According to the steepest descent method, the updated value of $\hat{\mathbf{h}}$ at time n is computed by using the simple recursive relation [16]:

$$\hat{\mathbf{h}}(n) = \hat{\mathbf{h}}(n-1) + \frac{\mu}{2}\left\{-\mathbf{f}\left[\hat{\mathbf{h}}(n-1)\right]\right\}, \qquad (6.29)$$

where μ is positive step-size constant. Differentiating (6.28) with respect to the filter, we get the following value for the gradient vector:

$$\begin{aligned} \mathbf{f}(\hat{\mathbf{h}}) &= \left[\ \mathbf{f}_1^T(\hat{\mathbf{h}})\ \ \mathbf{f}_2^T(\hat{\mathbf{h}})\ \ \cdots\ \ \mathbf{f}_P^T(\hat{\mathbf{h}})\ \right]^T \\ &= \partial J_2/\partial\hat{\mathbf{h}} = -2\mathbf{r} + 2\mathbf{R}\hat{\mathbf{h}}, \end{aligned} \qquad (6.30)$$

with $\mathbf{r} = E\{y(n)\mathbf{x}(n)\}$ and $\mathbf{R} = E\{\mathbf{x}(n)\mathbf{x}^T(n)\}$. By taking $\mathbf{f}(\hat{\mathbf{h}}) = \mathbf{0}_{LP\times 1}$, we obtain the Wiener-Hopf equation

$$\mathbf{R}\hat{\mathbf{h}} = \mathbf{r} \qquad (6.31)$$

which is similar to the normal equation (6.4) that was derived from a weighted least squares criterion (6.2). Note that we use the same notations for the similar variables that are derived either from the Wiener-Hopf equation or the normal equation.

The steepest-descent algorithm is now:

$$\hat{\mathbf{h}}(n) = \hat{\mathbf{h}}(n-1) + \mu E\{\mathbf{x}(n)e(n)\}, \qquad (6.32)$$

and the classical stochastic approximation (consisting of approximating the gradient with its instantaneous value) [16] provides the multi-channel LMS algorithm,

$$\hat{\mathbf{h}}(n) = \hat{\mathbf{h}}(n-1) + \mu\mathbf{x}(n)e(n), \qquad (6.33)$$

of which the classical mean weight convergence condition under appropriate independence assumption is:

$$0 < \mu < \frac{2}{L\sum_{i=1}^{P}\sigma_{x_i}^2}, \qquad (6.34)$$

where the $\sigma_{x_i}^2$ ($i = 1, 2, ..., P$) are the powers of the input signals. When this condition is satisfied, the weight vector converges in the mean to the optimal Wiener-Hopf solution.

However, the gradient vector corresponding to the filter i is:

$$\mathbf{f}_i(\hat{\mathbf{h}}) = -2\left(\mathbf{r}_i - \sum_{j=1}^{P}\mathbf{R}_{i,j}\hat{\mathbf{h}}_j\right),\ \ i = 1, 2, ..., P, \qquad (6.35)$$

which clearly shows some dependency of $\mathbf{f}_i$ on the full vector $\hat{\mathbf{h}}$. In other words, the filters $\hat{\mathbf{h}}_j$ with $j \neq i$ influence, in a bad direction, the gradient vector $\mathbf{f}_i$ when seeking the minimum, because the algorithm does not take the cross-correlation among all the inputs into account.

5.3.2 Improved Version. We have seen that during the convergence of the multi-channel LMS algorithm, each adaptive filter depends of the others. This dependency must been taken into account. By using this information and Lemma 1, we now differentiate criterion (6.28) with respect to the tap-weight in a different way. The new gradient is obtained by writing that $\hat{\mathbf{h}}_i$ depends of the full vector $\hat{\mathbf{h}}$. We get:

$$\begin{aligned} \mathbf{f}_i(\hat{\mathbf{h}}_i) &= \frac{\partial J_2}{\partial \hat{\mathbf{h}}_i}(\hat{\mathbf{h}}) \\ &= -2E\left\{\mathbf{z}_i(n)\left[y(n) - \mathbf{x}^T(n)\hat{\mathbf{h}}\right]\right\},\ i = 1, 2, ..., P, \end{aligned} \quad (6.36)$$

with

$$\begin{aligned} \mathbf{z}_i(n) &= \sum_{j=1}^{P} [\partial \hat{\mathbf{h}}_j / \partial \hat{\mathbf{h}}_i]^T \mathbf{x}_j(n) \\ &= \sum_{j=1}^{P} \mathbf{C}_{i,j}\mathbf{x}_j(n),\ i = 1, 2, ..., P. \end{aligned} \quad (6.37)$$

We have some interesting orthogonality and decorrelation properties.
Lemma 2:

$$E\{\mathbf{x}_i^T(n)\mathbf{z}_j(n)\} = 0, \quad (6.38)$$

$$E\{\mathbf{z}_i(n)\mathbf{x}_j^T(n)\} = \mathbf{0}_{L\times L},\ \forall i, j = 1, 2, ..., P,\ i \neq j. \quad (6.39)$$

Proof: The proof is straightforward from Lemma 1 (using mathematical expectation instead of weighted least squares).

We can verify by using Lemma 2 that each gradient vector $\mathbf{f}_i$ ($i = 1, 2, ..., P$) now depends only of the corresponding filter $\hat{\mathbf{h}}_i$. In other words, we make the convergence of each $\hat{\mathbf{h}}_i$ independent of the others, which is not the case in the classical gradient algorithm.

Based on the above gradient vector, the improved steepest-descent algorithm is easily obtained, out of which a stochastic approximation leads to the improved multi-channel LMS algorithm:

$$\hat{\mathbf{h}}(n) = \hat{\mathbf{h}}(n-1) + \mu\mathbf{z}(n)e(n) \quad (6.40)$$

with

$$\mathbf{z}(n) = \left[\ \mathbf{z}_1^T(n)\ \ \mathbf{z}_2^T(n)\ \ \cdots\ \ \mathbf{z}_P^T(n)\ \right]^T$$

and

$$0 < \mu < \frac{2}{L\sum_{i=1}^{P}\sigma_{z_i}^2} \quad (6.41)$$

to guaranty the convergence of the algorithm.

5.4 THE MULTI-CHANNEL APA

The affine projection algorithm (APA) [17] has become popular because of its lower complexity compared to RLS while it converges almost as fast. Therefore it is interesting to derive and study the multi-channel version of this algorithm. Like the multi-channel LMS, two versions are derived.

5.4.1 The Straightforward Multi-Channel APA. A simple trick for obtaining the single-channel APA is to search for an algorithm of the stochastic gradient type cancelling N *a posteriori* errors [18]. This requirement results in an underdetermined set of linear equations of which the mininum-norm solution is chosen. In the following, this technique is extended in order to fit our problem [19].

By definition, the set of N *a priori* errors and N *a posteriori* errors are:

$$\mathbf{e}(n) = \mathbf{y}(n) - \mathbf{X}^T(n)\hat{\mathbf{h}}(n-1), \tag{6.42}$$

$$\mathbf{e}_{\mathrm{a}}(n) = \mathbf{y}(n) - \mathbf{X}^T(n)\hat{\mathbf{h}}(n), \tag{6.43}$$

where

$$\mathbf{X}(n) = \begin{bmatrix} \mathbf{X}_1^T(n) & \mathbf{X}_2^T(n) & \cdots & \mathbf{X}_P^T(n) \end{bmatrix}^T$$

is a matrix of size $PL \times N$; the $L \times N$ matrix

$$\mathbf{X}_i(n) = \begin{bmatrix} \mathbf{x}_i(n) & \mathbf{x}_i(n-1) & \cdots & \mathbf{x}_i(n-N+1) \end{bmatrix}$$

is made from the N last input vectors $\mathbf{x}_i(n)$; finally, $\mathbf{y}(n)$ and $\mathbf{e}(n)$ are respectively vectors of the N last samples of the reference signal $y(n)$ and error signal $e(n)$.

Using (6.42) and (6.43) plus the requirement that $\mathbf{e}_{\mathrm{a}}(n) = \mathbf{0}_{N\times 1}$, we obtain:

$$\mathbf{X}^T(n)\Delta\hat{\mathbf{h}}(n) = \mathbf{e}(n), \tag{6.44}$$

where $\Delta\hat{\mathbf{h}}(n) = \hat{\mathbf{h}}(n) - \hat{\mathbf{h}}(n-1)$.

Equation (6.44) (N equations in PL unknowns, $N \leq PL$) is an underdetermined set of linear equations. Hence, it has an infinite number of solutions, out of which the minimum-norm solution is chosen, so that the adaptive filter has smooth variations. This results in [19, 20]:

$$\hat{\mathbf{h}}(n) = \hat{\mathbf{h}}(n-1) + \mathbf{X}(n)\left[\mathbf{X}^T(n)\mathbf{X}(n)\right]^{-1}\mathbf{e}(n). \tag{6.45}$$

However, in this straightforward APA, the normalization matrix $\mathbf{X}^T(n)\mathbf{X}(n) = \sum_{i=1}^{P}\mathbf{X}_i^T(n)\mathbf{X}_i(n)$ does not involve the cross-correlation elements of the P input signals [namely $\mathbf{X}_i^T(n)\mathbf{X}_j(n)$, $i, j = 1, 2, ..., P$, $i \neq j$] and this algorithm may converge slowly.

5.4.2 The Improved Two-Channel APA. A simple way to improve the previous adaptive algorithm is to use the othogonality and decorrelation properties, which will be shown later to appear in this context. Let us derive the improved algorithm by requiring a condition similar to the one used in the improved multi-channel LMS. Just use the constraint that $\Delta\hat{\mathbf{h}}_i$ be orthogonal to $\mathbf{X}_j,\ j \neq i$. As a result, we take into account separately the contributions of each input signal. These constraints read:

$$\mathbf{X}_2^T(n)\Delta\hat{\mathbf{h}}_1(n) = \mathbf{0}_{N\times 1}, \tag{6.46}$$

$$\mathbf{X}_1^T(n)\Delta\hat{\mathbf{h}}_2(n) = \mathbf{0}_{N\times 1}, \tag{6.47}$$

and the new set of linear equations characterizing the improved two-channel APA is:

$$\begin{bmatrix} \mathbf{X}_1^T(n) & \mathbf{X}_2^T(n) \\ \mathbf{X}_2^T(n) & \mathbf{0}_{N\times L} \\ \mathbf{0}_{N\times L} & \mathbf{X}_1^T(n) \end{bmatrix} \begin{bmatrix} \Delta\hat{\mathbf{h}}_1(n) \\ \Delta\hat{\mathbf{h}}_2(n) \end{bmatrix} = \begin{bmatrix} \mathbf{e}(n) \\ \mathbf{0}_{N\times 1} \\ \mathbf{0}_{N\times 1} \end{bmatrix}. \tag{6.48}$$

The improved two-channel APA algorithm is given by the minimum-norm solution of (6.48) which is found as [19],

$$\Delta\hat{\mathbf{h}}_1(n) = \mathbf{Z}_1(n)\left[\mathbf{Z}_1^T(n)\mathbf{Z}_1(n) + \mathbf{Z}_2^T(n)\mathbf{Z}_2(n)\right]^{-1}\mathbf{e}(n), \tag{6.49}$$

$$\Delta\hat{\mathbf{h}}_2(n) = \mathbf{Z}_2(n)\left[\mathbf{Z}_1^T(n)\mathbf{Z}_1(n) + \mathbf{Z}_2^T(n)\mathbf{Z}_2(n)\right]^{-1}\mathbf{e}(n), \tag{6.50}$$

where $\mathbf{Z}_i(n)$ is the projection of $\mathbf{X}_i(n)$ onto a subspace orthogonal to $\mathbf{X}_j(n),\ i \neq j$, i.e.,

$$\mathbf{Z}_i(n) = \left\{\mathbf{I}_{L\times L} - \mathbf{X}_j(n)\left[\mathbf{X}_j^T(n)\mathbf{X}_j(n)\right]^{-1}\mathbf{X}_j(n)\right\}\mathbf{X}_i(n), \quad i, j = 1, 2,\ i \neq j. \tag{6.51}$$

This results in the following orthogonality conditions,

$$\mathbf{X}_i^T(n)\mathbf{Z}_j(n) = \mathbf{0}_{N\times N},\ i \neq j \tag{6.52}$$

which are similar to what appears in the improved multi-channel LMS (Lemma 2).

5.4.3 The Improved Multi-Channel APA. The algorithm explained for two channels is easily generalized to an arbitrary number of channels P. Define the following matrix of size $L \times (P-1)N$:

$$\underline{\mathbf{X}}_i(n) = \begin{bmatrix} \mathbf{X}_1(n) & \cdots & \mathbf{X}_{i-1}(n) & \mathbf{X}_{i+1}(n) & \cdots & \mathbf{X}_P(n) \end{bmatrix}, \quad i = 1, 2, ..., P.$$

The P orthogonality constraints are:

$$\underline{\mathbf{X}}_i^T(n)\Delta\hat{\mathbf{h}}_i(n) = \mathbf{0}_{(P-1)N\times 1},\ i = 1, 2, ..., P, \tag{6.53}$$

and by using the same steps as for $P = 2$, a solution similar to (6.49), (6.50) is obtained [19]:

$$\Delta\hat{\mathbf{h}}_i(n) = \mathbf{Z}_i(n)\left[\sum_{j=1}^{P}\mathbf{Z}_j^T(n)\mathbf{Z}_j(n)\right]^{-1}\mathbf{e}(n),\ i = 1, 2, ..., P, \tag{6.54}$$

where $\mathbf{Z}_i(n)$ is the projection of $\mathbf{X}_i(n)$ onto a subspace orthogonal to $\underline{\mathbf{X}}_i(n)$, i.e.,

$$\begin{aligned}\mathbf{Z}_i(n) &= \left\{\mathbf{I}_{L\times L} - \underline{\mathbf{X}}_i(n)\left[\underline{\mathbf{X}}_i^T(n)\underline{\mathbf{X}}_i(n)\right]^{-1}\underline{\mathbf{X}}_i(n)\right\}\mathbf{X}_i(n), \\ & i = 1, 2, ..., P.\end{aligned} \tag{6.55}$$

Note that this equation holds only under the condition $L \geq (P-1)N$, so that the matrix that appears in (6.55) be invertible.

We can easily see that:

$$\underline{\mathbf{X}}_i^T(n)\mathbf{Z}_i(n) = \mathbf{0}_{(P-1)N\times N},\ i = 1, 2, ..., P. \tag{6.56}$$

6. DISCUSSION

In this chapter we have focused on presenting the main theoretical results that describe and explain the important difference between monophonic from multi-channel echo cancellation and presented a general framework for multi-channel time-domain adaptive filtering. The multi-channel echo canceler has to track echo path changes in both transmission and receiving rooms unless the transmitted multi-channel signals are modified so that the normal equation has a unique solution. The sensitivity can only be reduced by decreasing the interchannel correlation, modifications of the adaptive algorithm itself will not decrease sensitivity. Obviously, there is a limit to the decrease of this interchannel correlation; a strong decorrelation will destroy the stereo effect. Because of the high correlation among the input signals (even after being processed), adaptive algorithms that take this correlation into account should be used, in order to have good performances in terms of convergence rate and tracking of the MCAEC.

Some open research topics within MCAEC are: improving methods for decorrelating the channels. This will always be important since less complex algorithms can be used when the interchannel correlation is "low." What is more important for good performance of an algorithm; ability to spatially or temporally decorrelate the multi-channel signals? To answer this question we need a better characterization of the spatial and temporal correlation for the general far-end case. Factorization of the covariance matrix (Lemma 1) is a good and an important step to this direction but more has to be done. Providing such a theory is challenging and it would be valuable for the design of completely new algorithms.

References

[1] J. Benesty, D. R. Morgan, J. L. Hall, and M. M. Sondhi, "Synthesized stereo combined with acoustic echo cancellation for desktop conferencing," *Bell Labs Tech. J.*, vol. 3, pp. 148-158, July-Sep. 1998.

[2] M. M. Sondhi, D. R. Morgan, and J. L. Hall, "Stereophonic acoustic echo cancellation—An overview of the fundamental problem," *IEEE Signal Processing Lett.*, vol. 2, pp. 148-151, Aug. 1995.

[3] J. Benesty, D. R. Morgan, and M. M. Sondhi, "A better understanding and an improved solution to the specific problems of stereophonic acoustic echo cancellation," *IEEE Trans. Speech Audio Processing*, vol. 6, pp. 156-165, Mar. 1998.

[4] T. Gänsler and P. Eneroth, "Influence of audio coding on stereophonic acoustic echo cancellation," in *Proc. IEEE ICASSP*, 1998, pp. 3649-3652.

[5] A. Gilloire and V. Turbin, "Using auditory properties to improve the behavior of stereophonic acoustic echo cancellers," in *Proc. IEEE ICASSP*, 1998, pp. 3681-3684.

[6] S. Shimauchi, Y. Haneda, S. Makino, and Y Kaneda, "New configuration for a stereo echo canceller with nonlinear pre-processing," in *Proc. IEEE ICASSP*, 1998, pp. 3685-3688.

[7] M. Ali, "Stereophonic echo cancellation system using time-varying all-pass filtering for signal decorrelation," in *Proc. IEEE ICASSP*, 1998, pp. 3689-3692.

[8] Y. Joncour and A. Sugiyama, "A stereo echo canceller with pre-processing for correct echo path identification," in *Proc. IEEE ICASSP*, 1998, pp. 3677-3680.

[9] J. Benesty, P. Duhamel, and Y. Grenier, "Multi-channel adaptive filtering applied to multichannel acoustic echo cancellation," in *Proc. EUSIPCO*, 1996.

[10] B. C. J. Moore, *An Introduction to the Psychology of Hearing*. London: Academic Press, 1989, Ch. 3.

[11] J. Benesty, D. R. Morgan, J. L. Hall, and M. M. Sondhi, "Stereophonic acoustic echo cancellation using nonlinear transformations and comb filtering," *Proc. IEEE ICASSP*, 1998, pp. 3673-3676.

[12] J. Benesty, D. R. Morgan, and M. M. Sondhi, "A hybrid mono/stereo acoustic echo canceler," *IEEE Trans. Speech Audio Processing*, vol. 6, pp. 468-475, Sep. 1998.

[13] F. L. Wightman and D. J. Kistler, "The dominant role of low-frequency interaural time differences in sound localization," *J. Acoust. Soc. Am.*, vol. 91, pp. 1648-1661, Mar. 1992.

[14] J. Benesty, F. Amand, A. Gilloire, and Y. Grenier, "Adaptive filtering algorithms for stereophonic acoustic echo cancellation," in *Proc. IEEE ICASSP*, 1995, pp. 3099-3102.

[15] M. G. Bellanger, *Adaptive Digital Filters and Signal Analysis*. Marcel Dekker, 1987.

[16] S. Haykin, *Adaptive Filter Theory*. New Jersey: Prentice-Hall, Inc, 1996.

[17] K. Ozeki and T. Umeda, "An adaptive filtering algorithm using an orthogonal projection to an affine subspace and its properties," *Elec. Comm. Japan*, vol. J67-A, pp. 126-132, Feb. 1984.

[18] M. Montazeri and P. Duhamel, "A set of algorithms linking NLMS and block RLS algorithms," *IEEE Trans. Signal Processing*, vol. 43, pp. 444-453, Feb. 1995.

[19] J. Benesty, P. Duhamel, and Y. Grenier, "A multi-channel affine projection algorithm with applications to multi-channel acoustic echo cancellation," *IEEE Signal Processing Lett.*, vol. 3, pp. 35-37, Feb. 1996.

[20] S. Shimauchi and S. Makino, "Stereo projection echo canceller with true echo path estimation," in *Proc. IEEE ICASSP*, 1995, pp. 3059-3062.

Chapter 7

MULTI-CHANNEL FREQUENCY-DOMAIN ADAPTIVE FILTERING

Jacob Benesty
Bell Laboratories, Lucent Technologies
jbenesty@bell-labs.com

Dennis R. Morgan
Bell Laboratories, Lucent Technologies
drrm@bell-labs.com

Abstract We derive a new frequency-domain adaptive algorithm by using a frequency-domain recursive least squares criterion, minimizing an error signal in the frequency-domain. A similar criterion was proposed by Mansour and Gray by using mathematical expectations. Here, however, we propose to go one step further and derive an exact adaptive algorithm from the so-called *normal* equation. It is shown that the obtained algorithm is complex to implement, and to reduce the complexity we need to remove a constraint resulting in the unconstrained frequency-domain LMS (UFLMS) algorithm. We also give the optimal adaptation step size for the UFLMS. Most importantly, we generalize all this to the multi-channel case, thereby exploiting the cross-power spectra among all the channels which is very important (for a fast convergence rate) in multi-channel acoustic echo cancellation (AEC), where the input signals are highly correlated.

Keywords: Frequency-Domain, Adaptive Filtering, Echo Cancellation, Multi-Channel, UFLMS

1. INTRODUCTION

Since its first introduction by Dentino *et al.* [1], adaptive filtering in the frequency-domain has progressed very fast, and different sophisticated algorithms have since been proposed. Ferrara [2] was the first to elaborate an efficient frequency-domain adaptive filter algorithm (FLMS) that converges to the optimal (Wiener) solution. Mansour and Gray [3] derived an even more efficient algorithm, the *unconstrained* FLMS (UFLMS), using only three FFT operations per block instead of five for the FLMS, with comparable performances [4]. However, a major handicap with these structures is the delay. Indeed, this delay is equal to the length of the adaptive filter L, which is considerable for some applications like acoustic echo cancellation (AEC) where the number of taps can be easily a thousand or more. A new structure, using the classical overlap save (OLS) method, was proposed in [5, 6] and generalized in [7] where the block processing N was made independent of the filter length L; N can be chosen as small as desired, with a delay equal to N. Although from a complexity point of view, the optimal choice is $N = L$, using smaller block sizes ($N < L$) in order to reduce the delay is still more efficient than the time-domain algorithms. A more general scheme based on weighted overlap and add (WOLA) methods, the *generalized multidelay filter* (GMDFα), was proposed in [8, 9], where α is the overlap factor. The settings $\alpha > 1$ appear to be very useful in the context of adaptive filtering, since the filter coefficients can be adapted more frequently (every N/α samples instead of every N samples in the standard OLS scheme). So this structure introduces one more degree of freedom, but the complexity is increased by roughly a factor α. Taking the block size as large as the delay permits will increase the convergence rate of the algorithm, while taking the overlap factor greater than 1 will increase the tracking abilities of the algorithm.

We first derive a new frequency-domain adaptive algorithm by using a frequency-domain recursive least squares criterion. A similar criterion was proposed in [3] using mathematical expectations instead. Here, however, we propose to go one step further and derive an exact adaptive algorithm from the normal equation. We will see that the obtained algorithm is complex to implement, and to reduce the complexity we will need to remove a constraint and that will give us exactly the UFLMS [3]. We also give the optimal adaptation step for the UFLMS. Most importantly, we generalize all this to the multi-channel case. We will see that the obtained algorithm exploits the cross-power spectra among all the channels, which is very important (for a fast convergence rate) in multi-channel AEC where the input signals are highly correlated [10]. To simplify the presentation, we will derive all the algorithms only for $L = N, \alpha = 1$, and with the OLS method. Generalization to any other case is straightforward.

2. MONO-CHANNEL FREQUENCY-DOMAIN ADAPTIVE FILTERING REVISITED

In the time-domain, the general procedure to derive an adaptive algorithm is to first define an error signal, then to build a cost function based on the error signal, and finally to minimize the cost function with respect to the adaptive filter coefficients [11]. In the context of system identification, the error signal at time n between the system and model filter outputs is given by

$$e(n) = y(n) - \hat{y}(n), \tag{7.1}$$

where

$$\hat{y}(n) = \hat{\mathbf{h}}^T \mathbf{x}(n) \tag{7.2}$$

is an estimate of the output signal $y(n)$,

$$\hat{\mathbf{h}} = \begin{bmatrix} \hat{h}_0 & \hat{h}_1 & \cdots & \hat{h}_{L-1} \end{bmatrix}^T$$

is the model filter, and

$$\mathbf{x}(n) = \begin{bmatrix} x(n) & x(n-1) & \cdots & x(n-L+1) \end{bmatrix}^T$$

is a vector containing the last L samples of the input signal x. Superscript ^{T} denotes transpose of a vector or a matrix. The recursive least squares (RLS) adaptive algorithm is obtained exactly from the normal equation which is derived by minimizing the following time-domain criterion [11]:

$$J_t(n) = (1 - \lambda_t) \sum_{p=0}^{n} \lambda_t^{n-p} e^2(p), \tag{7.3}$$

where λ_t $(0 < \lambda_t < 1)$ is an exponential forgetting factor. In the rest of this section, we will follow the same approach.

We now define the block error signal (of length $N = L$) as:

$$\mathbf{e}(m) = \mathbf{y}(m) - \hat{\mathbf{y}}(m), \tag{7.4}$$

where m is the block time index, and

$$\begin{aligned}
\mathbf{e}(m) &= \begin{bmatrix} e(mL) & \cdots & e(mL+L-1) \end{bmatrix}^T, \\
\mathbf{y}(m) &= \begin{bmatrix} y(mL) & \cdots & y(mL+L-1) \end{bmatrix}^T, \\
\hat{\mathbf{y}}(m) &= \begin{bmatrix} \mathbf{x}(mL) & \cdots & \mathbf{x}(mL+L-1) \end{bmatrix}^T \hat{\mathbf{h}} \\
&= \mathbf{X}^T(m)\hat{\mathbf{h}}.
\end{aligned}$$

It can easily be checked that $\mathbf{X}$ is a Toeplitz matrix of size $(L \times L)$.

It is well known that a Toeplitz matrix $\mathbf{X}$ can be transformed, by doubling its size, to a circulant matrix

$$\mathbf{C} = \begin{bmatrix} \mathbf{X}' & \mathbf{X} \\ \mathbf{X} & \mathbf{X}' \end{bmatrix}$$

where $\mathbf{X}'$ is also a Toeplitz matrix. (The matrix $\mathbf{X}'$ is expressible in terms of the elements of $\mathbf{X}$, except for an arbitrary diagonal.) Using circulant matrices, the block error signal can be re-written equivalently:

$$\begin{bmatrix} \mathbf{0}_{L\times 1} \\ \mathbf{e}(m) \end{bmatrix} = \begin{bmatrix} \mathbf{0}_{L\times 1} \\ \mathbf{y}(m) \end{bmatrix} - \mathbf{W}\hat{\mathbf{y}}'(m), \tag{7.5}$$

where

$$\mathbf{W} = \begin{bmatrix} \mathbf{0}_{L\times L} & \mathbf{0}_{L\times L} \\ \mathbf{0}_{L\times L} & \mathbf{I}_{L\times L} \end{bmatrix}$$

and

$$\hat{\mathbf{y}}'(m) = \mathbf{C}(m) \begin{bmatrix} \hat{\mathbf{h}} \\ \mathbf{0}_{L\times 1} \end{bmatrix}. \tag{7.6}$$

It is also well known that a circulant matrix is easily decomposed as follows: $\mathbf{C} = \mathbf{F}^{-1}\mathbf{D}\mathbf{F}$, where $\mathbf{F}$ is the Fourier matrix [of size $(2L \times 2L)$] and $\mathbf{D}$ is a diagonal matrix whose elements are the discrete Fourier transform of the first column of $\mathbf{C}$. If we multiply (7.5) by $\mathbf{F}$, we get the error signal in the frequency domain:

$$\begin{aligned} \underline{\mathbf{e}}(m) &= \underline{\mathbf{y}}(m) - \mathbf{G}\underline{\hat{\mathbf{y}}}'(m) \\ &= \underline{\mathbf{y}}(m) - \mathbf{G}\mathbf{D}(m)\underline{\hat{\mathbf{h}}}, \end{aligned} \tag{7.7}$$

where

$$\begin{aligned} \underline{\mathbf{e}}(m) &= \mathbf{F}\begin{bmatrix} \mathbf{0}_{L\times 1} \\ \mathbf{e}(m) \end{bmatrix}, \\ \underline{\mathbf{y}}(m) &= \mathbf{F}\begin{bmatrix} \mathbf{0}_{L\times 1} \\ \mathbf{y}(m) \end{bmatrix}, \\ \mathbf{G} &= \mathbf{F}\mathbf{W}\mathbf{F}^{-1}, \\ \underline{\hat{\mathbf{y}}}'(m) &= \mathbf{F}\hat{\mathbf{y}}'(m), \\ \underline{\hat{\mathbf{h}}} &= \mathbf{F}\begin{bmatrix} \hat{\mathbf{h}} \\ \mathbf{0}_{L\times 1} \end{bmatrix}. \end{aligned}$$

Having derived a frequency-domain error signal, we now define a frequency-domain criterion which is similar to (7.3):

$$J_{\mathrm{f}}(m) = (1 - \lambda_{\mathrm{f}}) \sum_{p=0}^{m} \lambda_{\mathrm{f}}^{m-p} \underline{\mathbf{e}}^H(p)\underline{\mathbf{e}}(p), \tag{7.8}$$

where ^{H} denotes conjugate transpose. Let ∇ be the gradient operator (with respect to $\hat{\underline{\mathbf{h}}}$). Applying the operator ∇ to the cost function J_f, we obtain (noting that $\mathbf{G}^H\mathbf{G} = \mathbf{G}^2 = \mathbf{G}$) the complex gradient vector:

$$\begin{aligned}\nabla J_\mathrm{f}(m) &= \frac{\partial J_\mathrm{f}(m)}{\partial \hat{\underline{\mathbf{h}}}(m)} \\ &= -(1-\lambda_\mathrm{f})\sum_{p=0}^{m}\lambda_\mathrm{f}^{m-p}\mathbf{D}(p)\mathbf{G}^*\underline{\mathbf{y}}^*(p) \\ &+ (1-\lambda_\mathrm{f})\left[\sum_{p=0}^{m}\lambda_\mathrm{f}^{m-p}\mathbf{D}(p)\mathbf{G}^*\mathbf{D}^*(p)\right]\hat{\underline{\mathbf{h}}}^*(m),\end{aligned} \tag{7.9}$$

where * denotes complex conjugate. By setting the gradient of the cost function equal to zero, conjugating, and noting that $\mathbf{G}\underline{\mathbf{y}}(p) = \underline{\mathbf{y}}(p)$, we obtain the so-called *normal* equation:

$$\mathbf{S}(m)\hat{\underline{\mathbf{h}}}(m) = \mathbf{s}(m), \tag{7.10}$$

where

$$\begin{aligned}\mathbf{S}(m) &= (1-\lambda_\mathrm{f})\sum_{p=0}^{m}\lambda_\mathrm{f}^{m-p}\mathbf{D}^*(p)\mathbf{G}\mathbf{D}(p) \\ &= \lambda_\mathrm{f}\mathbf{S}(m-1) + (1-\lambda_\mathrm{f})\mathbf{D}^*(m)\mathbf{G}\mathbf{D}(m)\end{aligned} \tag{7.11}$$

and

$$\begin{aligned}\mathbf{s}(m) &= (1-\lambda_\mathrm{f})\sum_{p=0}^{m}\lambda_\mathrm{f}^{m-p}\mathbf{D}^*(p)\underline{\mathbf{y}}(p) \\ &= \lambda_\mathrm{f}\mathbf{s}(m-1) + (1-\lambda_\mathrm{f})\mathbf{D}^*(m)\underline{\mathbf{y}}(m).\end{aligned} \tag{7.12}$$

It can be shown that if the covariance matrix of the input signal is of rank L then the matrix $\mathbf{S}(m)$ is nonsingular [3, 4]. In this case, the normal equation has a unique solution which is the optimal Wiener solution.

Enforcing the normal equation at block time indices m and $m-1$, and using (7.11) and (7.12), we easily derive an exact adaptive algorithm:

$$\underline{\mathbf{e}}(m) = \underline{\mathbf{y}}(m) - \mathbf{G}\mathbf{D}(m)\hat{\underline{\mathbf{h}}}(m-1) \tag{7.13}$$

$$\hat{\underline{\mathbf{h}}}(m) = \hat{\underline{\mathbf{h}}}(m-1) + (1-\lambda_\mathrm{f})\mathbf{S}^{-1}(m)\mathbf{D}^*(m)\underline{\mathbf{e}}(m). \tag{7.14}$$

This will be called the *constrained* algorithm. Note that our definition is different from the original frequency-domain adaptive algorithm proposed by Ferrara [2]. (The constraint here is on the update of the matrix $\mathbf{S}$ while in Ferrara's algorithm the constraint is on the update of the coefficients of the filter.) It can be shown that the convergence of the proposed algorithm for stationary signals does not depend on the statistics of the input signal, which is of course a very nice feature. Frequency-domain adaptive algorithms were first introduced to reduce the arithmetic complexity of the LMS algorithm [2]. Unfortunately, the matrix **S** is not diagonal, so the proposed algorithm has a high complexity and may not be very useful in practice. We will argue here that 2**G** can be well approximated by the identity matrix, and we then obtain the following unconstrained algorithm:

$$\mathbf{S}_\mathrm{u}(m) = \lambda_\mathrm{f}\mathbf{S}_\mathrm{u}(m-1) + (1-\lambda_\mathrm{f})\mathbf{D}^*(m)\mathbf{D}(m) \tag{7.15}$$

$$\hat{\underline{\mathbf{h}}}(m) = \hat{\underline{\mathbf{h}}}(m-1) + \mu_\mathrm{u}\mathbf{S}_\mathrm{u}^{-1}(m)\mathbf{D}^*(m)\underline{\mathbf{e}}(m) \tag{7.16}$$

where $\mathbf{S}_\mathrm{u}$ is now a diagonal matrix and $\mu_\mathrm{u} = 2(1-\lambda_\mathrm{f})$ is a positive number. This algorithm is exactly the *unconstrained* frequency-domain adaptive filter proposed by Mansour and Gray [3] and since $\mathbf{S}_\mathrm{u}$ is diagonal, this algorithm becomes very attractive from a complexity point of view. Now the questions are the following. Is this approximation justified? And what will be the optimum value of μ_u?

Let's examine the structure of the matrix **G**. We have: $\mathbf{G}^* = \mathbf{F}^{-1}\mathbf{W}\mathbf{F}$; since **W** is a diagonal matrix, $\mathbf{G}^*$ is a circulant matrix. Therefore, inverse transforming the diagonal of **W** gives the first column of $\mathbf{G}^*$,

$$\begin{aligned} \mathbf{g}^* &= \begin{bmatrix} g_0^* & g_1^* & \cdots & g_{2L-1}^* \end{bmatrix}^T \\ &= \mathbf{F}^{-1}[0 \ldots 0\ 1 \ldots 1]^T. \end{aligned}$$

The elements of vector **g** can be written explicitly as:

$$\begin{aligned} g_k &= \frac{1}{2L}\sum_{l=L}^{2L-1} \exp(-i2\pi kl/2L) \\ &= \frac{(-1)^k}{2L}\sum_{l=0}^{L-1} \exp[-i\pi kl/L], \end{aligned} \tag{7.17}$$

where $i^2 = -1$. Since g_k is the sum of a geometric progression, we have:

$$\begin{aligned} g_k &= \begin{cases} 0.5 & k=0 \\ \dfrac{(-1)^k}{2L}\dfrac{1-\exp(-i\pi k)}{1-\exp(-i\pi k/L)} & k \neq 0 \end{cases} \\ &= \begin{cases} 0.5 & k=0 \\ 0 & k \text{ even} \\ \dfrac{-1}{2L}[1 - i\cot(\dfrac{\pi k}{2L})] & k \text{ odd}, \end{cases} \end{aligned} \tag{7.18}$$

where $L-1$ elements of vector $\mathbf{g}$ are equal to zero. Moreover, since $\mathbf{G}^H\mathbf{G} = \mathbf{G}$, then $\mathbf{g}^H\mathbf{g} = g_0 = 0.5$ and we have

$$\mathbf{g}^H\mathbf{g} - g_0^2 = \sum_{l=1}^{2L-1} |g_l|^2 = 2\sum_{l=1}^{L-1} |g_l|^2 = \frac{1}{4}. \tag{7.19}$$

We can see from (7.19) that the first element of vector $\mathbf{g}$, i.e. g_0, is dominant, in a mean-square sense, and from (7.18) that the L first elements of $\mathbf{g}$ decrease rapidly to zero as k increases. Because of the conjugate symmetry, some of the last elements of $\mathbf{g}$ are not negligible, but this is of little concern since $\mathbf{G}$ is circulant with $\mathbf{g}$ as its first column and its other columns have those non-negligible elements shifted in such a way that they are concentrated around the main diagonal. To summarize, we can say that only the very first (few) off-diagonals of $\mathbf{G}$ will be non-negligible while the others can be completely neglected. Thus, approximating $\mathbf{G}$ by a diagonal matrix, i.e. $\mathbf{G} \approx g_0\mathbf{I} = \mathbf{I}/2$, is reasonable, and in this case we will have $\mu_{\mathrm{u}} \approx (1 - \lambda_{\mathrm{f}})/g_0 = 2(1 - \lambda_{\mathrm{f}})$ for an optimal convergence rate. Note that this is in agreement with previous derivations [2] that give an optimal step size of $1 - \lambda_{\mathrm{f}}$ divided by a power normalizing factor, which for our assumed unit-power signal with L-sample zero padding is equal to $1/2$.

3. GENERALIZATION TO THE MULTI-CHANNEL CASE

The generalization to the multi-channel case is rather straightforward. Therefore, in this section we only highlight some important steps and state the algorithms. For convenience, we will use the same notation as previously employed. Let J be the number of channels. Our definition of multi-channel is that we have a system with J input signals x_j, $j = 1, 2, \ldots, J$ and one output signal y. Now the block error signal is defined as:

$$\mathbf{e}(m) = \mathbf{y}(m) - \sum_{j=1}^{J} \mathbf{X}_j^T(m)\hat{\mathbf{h}}_j, \tag{7.20}$$

where $\mathbf{e}$ and $\mathbf{y}$ are vectors of e_j and y_j respectively, all matrices $\mathbf{X}_j$ are Toeplitz of size $(L \times L)$, and $\hat{\mathbf{h}}_j$ is the estimated impulse response of the jth channel. In the frequency-domain, we have [c.f. (7.7)]:

$$\begin{aligned} \underline{\mathbf{e}}(m) &= \underline{\mathbf{y}}(m) - \mathbf{G}\sum_{j=1}^{J} \mathbf{D}_j(m)\underline{\hat{\mathbf{h}}}_j \\ &= \underline{\mathbf{y}}(m) - \mathbf{G}\mathbf{D}(m)\underline{\hat{\mathbf{h}}}, \end{aligned} \tag{7.21}$$

where $\mathbf{D} = \begin{bmatrix} \mathbf{D}_1 & \mathbf{D}_2 & \cdots & \mathbf{D}_J \end{bmatrix}$ is a $(2L \times 2LJ)$ matrix containing all the J diagonal matrices $\mathbf{D}_j$ and $\hat{\underline{\mathbf{h}}} = \begin{bmatrix} \hat{\underline{\mathbf{h}}}_1^T & \hat{\underline{\mathbf{h}}}_2^T & \cdots & \hat{\underline{\mathbf{h}}}_J^T \end{bmatrix}^T$ is the $(2LJ \times 1)$ vector of concatenated, transformed, zero-padded estimated impulse responses. Minimizing the criterion defined in (7.8), we obtain the normal equation for the multi-channel case:

$$\mathbf{S}(m)\hat{\underline{\mathbf{h}}}(m) = \mathbf{s}(m), \tag{7.22}$$

where

$$\begin{aligned} \mathbf{S}(m) &= (1-\lambda_\mathrm{f})\sum_{p=0}^{m}\lambda_\mathrm{f}^{m-p}\mathbf{D}^H(p)\mathbf{G}\mathbf{D}(p) \\ &= \lambda_\mathrm{f}\mathbf{S}(m-1) + (1-\lambda_\mathrm{f})\mathbf{D}^H(m)\mathbf{G}\mathbf{D}(m) \end{aligned} \tag{7.23}$$

is a $(2LJ \times 2LJ)$ matrix and

$$\begin{aligned} \mathbf{s}(m) &= (1-\lambda_\mathrm{f})\sum_{p=0}^{m}\lambda_\mathrm{f}^{m-p}\mathbf{D}^H(p)\underline{\mathbf{y}}(p) \\ &= \lambda_\mathrm{f}\mathbf{s}(m-1) + (1-\lambda_\mathrm{f})\mathbf{D}^H(m)\underline{\mathbf{y}}(m) \end{aligned} \tag{7.24}$$

is a $(2LJ \times 1)$ vector. Using the same approach and definitions as in Section 2, we get the multi-channel constrained frequency-domain adaptive algorithm:

$$\underline{\mathbf{e}}(m) = \underline{\mathbf{y}}(m) - \mathbf{G}\mathbf{D}(m)\hat{\underline{\mathbf{h}}}(m-1) \tag{7.25}$$

$$\hat{\underline{\mathbf{h}}}(m) = \hat{\underline{\mathbf{h}}}(m-1) + (1-\lambda_\mathrm{f})\mathbf{S}^{-1}(m)\mathbf{D}^H(m)\underline{\mathbf{e}}(m) \tag{7.26}$$

and the multi-channel unconstrained frequency-domain adaptive algorithm:

$$\mathbf{S}_\mathrm{u}(m) = \lambda_\mathrm{f}\mathbf{S}_\mathrm{u}(m-1) + (1-\lambda_\mathrm{f})\mathbf{D}^H(m)\mathbf{D}(m) \tag{7.27}$$

$$\hat{\underline{\mathbf{h}}}(m) = \hat{\underline{\mathbf{h}}}(m-1) + \mu_\mathrm{u}\mathbf{S}_\mathrm{u}^{-1}(m)\mathbf{D}^H(m)\underline{\mathbf{e}}(m). \tag{7.28}$$

Now $\mathbf{S}_\mathrm{u}$ is not a diagonal matrix but a block matrix containing J^2 diagonal matrices that are estimates of the power spectra and cross-power spectra of all the input signals.

Particular case: The two-channel unconstrained frequency-domain adaptive algorithm

We easily deduce the algorithm from (7.27) and (7.28):

$$\underline{\mathbf{e}}(m) = \underline{\mathbf{y}}(m) - \mathbf{G}[\mathbf{D}_1(m)\hat{\underline{\mathbf{h}}}_1(m-1) + \mathbf{D}_2(m)\hat{\underline{\mathbf{h}}}_2(m-1)] \tag{7.29}$$

$$\begin{aligned} \hat{\underline{\mathbf{h}}}_1(m) &= \hat{\underline{\mathbf{h}}}_1(m-1) \\ &+ \mu_\mathrm{u}\mathbf{S}_1^{-1}(m)[\mathbf{D}_1^*(m) - \mathbf{S}_{1,2}(m)\mathbf{S}_{2,2}^{-1}(m)\mathbf{D}_2^*(m)]\underline{\mathbf{e}}(m) \end{aligned} \tag{7.30}$$

$$\begin{aligned} \hat{\underline{\mathbf{h}}}_2(m) &= \hat{\underline{\mathbf{h}}}_2(m-1) \\ &+ \mu_\mathrm{u}\mathbf{S}_2^{-1}(m)[\mathbf{D}_2^*(m) - \mathbf{S}_{2,1}(m)\mathbf{S}_{1,1}^{-1}(m)\mathbf{D}_1^*(m)]\underline{\mathbf{e}}(m) \end{aligned} \tag{7.31}$$

where $\mathbf{S}_{j,l}$ are the (diagonal) sub-matrices of $\mathbf{S}_{\mathrm{u}}$,

$$\mathbf{S}_j(m) = \mathbf{S}_{j,j}(m)[\mathbf{I}_{2L\times 2L} - \mathbf{U}^H(m)\mathbf{U}(m)], \ j = 1, 2, \tag{7.32}$$

and

$$\mathbf{U}(m) = [\mathbf{S}_{1,1}(m)\mathbf{S}_{2,2}(m)]^{-1/2}\mathbf{S}_{1,2}(m) \tag{7.33}$$

is the coherence matrix. We can verify that this algorithm is exactly the same as in [17, 18], exploiting the coherence between the two channels in order to improve the convergence rate of the adaptive filter.

For simplicity, we have derived the multi-channel case assuming $N = L$ and no overlap ($\alpha = 1$). It is easy to generalize this for $\alpha > 1$ by simply computing the FFT's using overlapped data, which we do in the next section for the application example ($\alpha = 4$). Furthermore, it is straightforward, although tedious, to generalize to the case of $N < L$ [8].

4. APPLICATION TO ACOUSTIC ECHO CANCELLATION AND SIMULATIONS

Multi-channel acoustic echo cancellation (AEC) is typically intended for use in high quality teleconferencing systems and in multiparticipant desktop conferencing, implementing sound transmission through at least two channels [12]. Multi-channel AEC can be viewed as a straightforward generalization of the single-channel acoustic echo cancellation principle [13]. Figure 7.1 shows this technique, in the two-channel (stereo) case, for one microphone in the *receiving* room (which is represented by the two echo paths h_1 and h_2 between the two loudspeakers and the microphone). The two reference signals x_1 and x_2 from the transmission room are obtained from two microphones in the case of teleconferencing. These signals are derived by filtering from a common source, and this gives rise to a non-uniqueness problem that does not arise for the single-channel AEC [13, 14, 15]. As a result, the usual adaptive algorithms converge to solutions that depend on the impulse responses in the *transmission* room. This means that for good echo cancellation one must track not only the changes in the receiving room, but also the changes in the transmission room (for example, when one person stops talking and another person starts). The same problem occurs for multi-channel desktop conferencing, where multi-channel sound is synthesized from the single-microphone signals of all the participants [12].

In [14, 16], a simple but efficient solution was proposed that overcomes the above problem by adding a small nonlinearity into each channel. The distortion due to the nonlinearity (NL), shown in Fig. 7.1, is hardly perceptible for speech (and does not affect the stereo effect), yet it reduces interchannel coherence, thereby allowing reduction of misalignment to a low level. However, this solution is fruitful only when combined with the multi-channel FRLS algorithm. This requirement implies a high level of computational complexity, so a

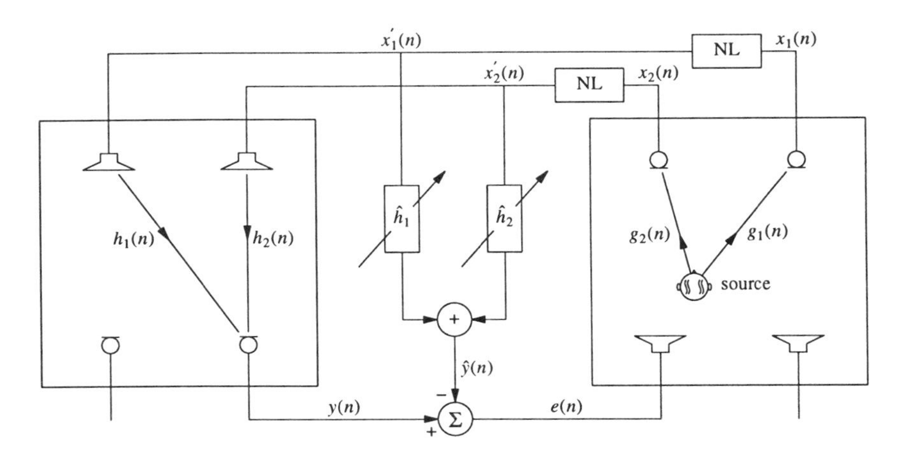

Figure 7.1 Schematic diagram of stereophonic acoustic echo cancellation with nonlinear transformations of the two input signals.

real-time implementation of this algorithm is difficult. That is why an efficient multi-channel frequency-domain adaptive filter can be a very good alternative.

We now show, by way of simulation using two channels, that the proposed unconstrained frequency-domain adaptive filter is a good alternative to some classical algorithms, i.e., the two-channel NLMS and the two-channel FRLS. The signal source s in the transmission room is a 10 s speech signal. The two microphone signals x_1 and x_2 were obtained by convolving s with two impulse responses g_1, g_2 of length 4096, as measured in an actual room. The microphone output signal y in the receiving room is obtained by summing the two convolutions $(h_1 * x_1)$ and $(h_2 * x_2)$, where h_1 and h_2 were also measured in an actual room as 4096-point responses. A white noise signal with 45 dB SNR is added to y. The sampling rate is 16 kHz. The length of the two adaptive filters is taken as $L = 1024$. In all of our simulations, we added a half-wave rectifier nonlinearity (with gain of 0.5) [14] to the signals x_1 and x_2. For the proposed algorithm, we used the following parameters: $N = 1024$, $\alpha = 4$ (that implies an overall delay equal to 1024 samples, i.e., 64 ms). With these values of N and α, the proposed algorithm is 7 times less complex than the two-channel NLMS and 40 times less complex than the two-channel FRLS.

Figures 7.2, 7.3, and 7.4 show the convergence of the mean square error (MSE) (a) and misalignment (b) respectively for the two-channel NLMS, the two-channel FRLS, and the proposed algorithm. For the purpose of smoothing the curves, error and misalignment samples are averaged over 128 points. For the FRLS algorithm, we have chosen a value $\lambda_{\text{RLS}} = 1 - 1/(20L)$ where $L = 1024$ in this case; accordingly for the block frequency-domain algorithm, we have chosen $\lambda_{\text{f}} = \lambda_{\text{RLS}}^L = 0.95$ in order to have the same effective window

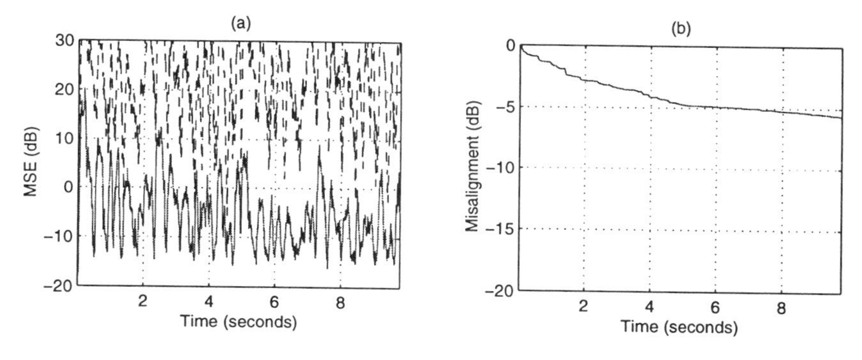

Figure 7.2 Performance of the two-channel NLMS, $L = 1024$. (a) MSE (–) as compared to original echo level (– –). (b) misalignment.

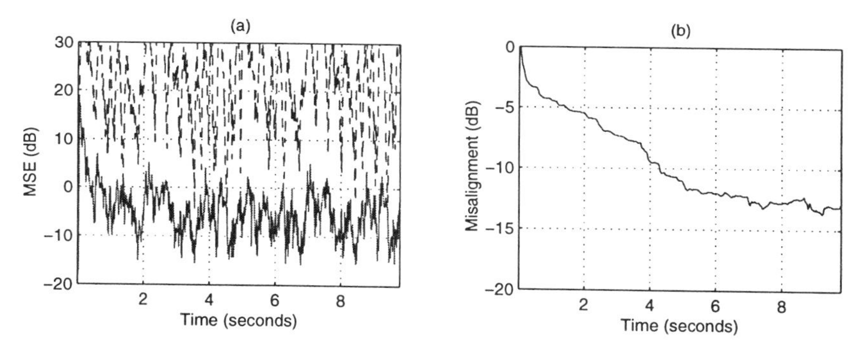

Figure 7.3 Performance of the two-channel FRLS with $\lambda_{\mathrm{RLS}} = 1 - 1/(20L)$, $L = 1024$. (a) MSE (–) as compared to original echo level (– –). (b) misalignment.

length. We can see that the proposed algorithm outperforms the other two as far as the misalignment is concerned. Also, the steady-state attenuation of the MSE for the frequency-domain algorithm appears to be as good as that for the FRLS (and somewhat better than NLMS) as confirmed by informal listening tests. However, the initial convergence rate of the proposed algorithm is somewhat slower than that of the FRLS. This can be improved by using a smaller exponential forgetting factor as shown in Fig. 7.5 with $\lambda_{\mathrm{f}} = 0.9$, but the misalignment is then increased somewhat. The optimal tradeoff between convergence rate and misalignment is very subjective and application dependent; we do not attempt to perform this tradeoff here.

5. CONCLUSIONS

We have derived a class of multi-channel frequency-domain adaptive algorithms from a frequency-domain recursive least squares criterion. The con-

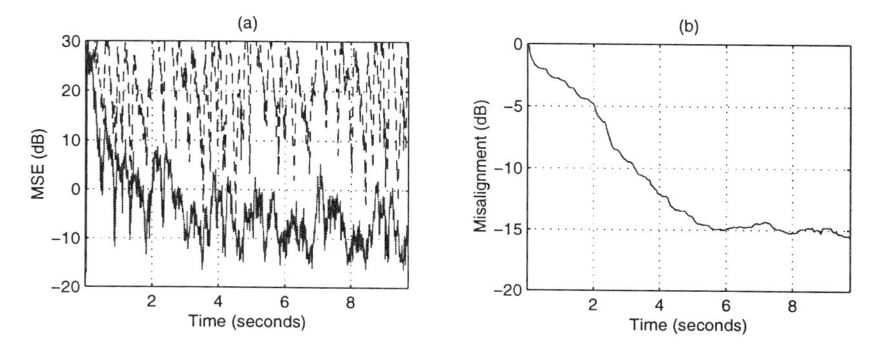

Figure 7.4 Performance of the proposed algorithm (unconstrained version) with $\lambda_f = 0.95$, $L = 1024$. (a) MSE (–) as compared to original echo level (– –). (b) misalignment.

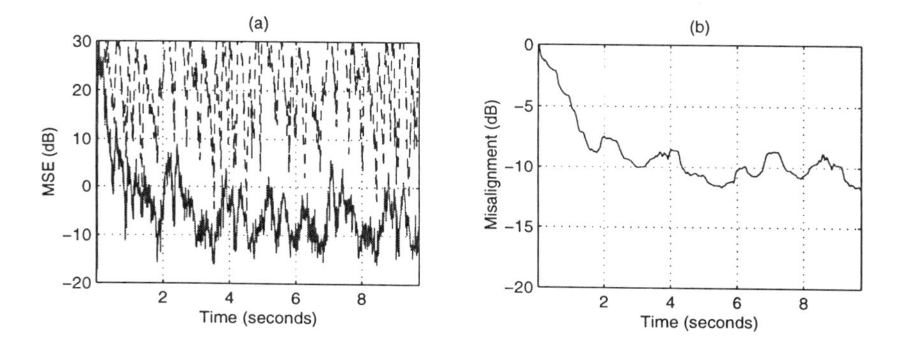

Figure 7.5 Same as in Fig. 7.4 with $\lambda_f = 0.9$.

strained algorithm was deduced directly and exactly from the normal equation and in this sense it is optimal, while an unconstrained version is developed as a good approximation. Most importantly, both algorithms exploit the cross-power spectra (or equivalently the cross-correlations in the time-domain) among all the channels and this feature is fundamental for the algorithms to converge rapidly to the Wiener solution, especially for applications like multi-channel AEC where the channels are highly correlated.

References

[1] M. Dentino, J. McCool, and B. Widrow, "Adaptive filtering in the frequency domain," *Proc. IEEE*, vol. 66, pp. 1658-1659, Dec. 1978.

[2] E. R. Ferrara, Jr., "Fast implementation of LMS adaptive filter," *IEEE Trans. Acoust., Speech, Signal Processing*, vol. ASSP-28, pp. 474-475, Aug. 1980.

[3] D. Mansour and A. H. Gray, JR., "Unconstrained frequency-domain adaptive filter," *IEEE Trans. Acoust., Speech, Signal Processing*, vol. ASSP-30, pp. 726-734, Oct. 1982.

[4] J. C. Lee and C. K. Un, "Performance analysis of frequency-domain block LMS adaptive digital filters," *IEEE Trans. Circuits Syst.*, vol. CAS-36, pp. 173-189, Feb. 1989.

[5] J.-S. Soo and K. K. Pang, "Multidelay block frequency domain adaptive filter," *IEEE Trans. Acoust., Speech, Signal Processing*, vol. ASSP-38, pp. 373-376, Feb. 1990.

[6] J. Benesty and P. Duhamel, "Fast constant modulus adaptive algorithm," *IEE Proc.-F*, vol. 138, pp. 379-387, Aug. 1991.

[7] J. Benesty and P. Duhamel, "A fast exact least mean square adaptive algorithm," *IEEE Trans. Signal Processing*, vol. 40, pp. 2904-2920, Dec. 1992.

[8] E. Moulines, O. Ait Amrane, and Y. Grenier, "The generalized multidelay adaptive filter: structure and convergence analysis," *IEEE Trans. Signal Processing*, vol. 43, pp. 14-28, Jan. 1995.

[9] J. Prado and E. Moulines, "Frequency-domain adaptive filtering with applications to acoustic echo cancellation," *Ann. Télécommun.*, vol. 49, pp. 414-428, 1994.

[10] J. Benesty, F. Amand, A. Gilloire, and Y. Grenier, "Adaptive filtering algorithms for stereophonic acoustic echo cancellation," in *Proc. IEEE ICASSP*, 1995, pp. 3099-3102.

[11] M. G. Bellanger, *Adaptive Digital Filters and Signal Analysis*. Marcel Dekker, 1987.

[12] J. Benesty, D. R. Morgan, J. L. Hall, and M. M. Sondhi, "Synthesized stereo combined with acoustic echo cancellation for desktop conferencing," *Bell Labs Tech. J.*, vol. 3, pp. 148-158, Jul.-Sep. 1998.

[13] M. M. Sondhi, D. R. Morgan, and J. L. Hall, "Stereophonic acoustic echo cancellation—An overview of the fundamental problem," *IEEE Signal Processing Lett.*, vol. 2, pp. 148-151, Aug. 1995.

[14] J. Benesty, D. R. Morgan, and M. M. Sondhi, "A better understanding and an improved solution to the specific problems of stereophonic acoustic echo cancellation," *IEEE Trans. Speech Audio Processing*, vol. 6, pp. 156-165, Mar. 1998.

[15] A. Gilloire, "Current issues in stereophonic and multi-channel acoustic echo cancellation," in *Proc. IWAENC*, 1997, pp. K5-K8.

[16] J. Benesty, D. R. Morgan, and M. M. Sondhi, "A hybrid mono/stereo acoustic echo canceler," *IEEE Trans. Speech Audio Processing*, vol. 6, pp. 468-475, Sep. 1998.

[17] J. Benesty, A. Gilloire, and Y. Grenier, "A frequency-domain stereophonic acoustic echo canceler exploiting the coherence between the channels and using nonlinear transformations," in *Proc. IWAENC*, Sep. 1999, pp. 28-31.

[18] J. Benesty, A. Gilloire, and Y. Grenier, "A frequency domain stereophonic acoustic echo canceler exploiting the coherence between the channels," *J. Acoust. Soc. Am.*, vol. 106, pp. L30-L35, Sep. 1999.

Chapter 8

A REAL-TIME STEREOPHONIC ACOUSTIC SUBBAND ECHO CANCELER

Peter Eneroth
Department of Applied Electronics
Lund University
peter.eneroth@tde.lth.se

Steven L. Gay
Bell Laboratories, Lucent Technologies
slg@bell-labs.com

Tomas Gänsler
Bell Laboratories, Lucent Technologies
gaensler@bell-labs.com

Jacob Benesty
Bell Laboratories, Lucent Technologies
jbenesty@bell-labs.com

Abstract Teleconferencing systems employ acoustic echo cancelers to reduce echoes that results from the coupling between loudspeaker and microphone. To enhance the sound realism, two-channel audio is advantageous. However, stereophonic acoustic echo cancellation (SAEC) is more difficult to solve as is shown in Chapter 6 of this book. In this chapter, a wideband stereophonic acoustic echo canceler, based on a subband structure of the two-channel fast recursive least squares algorithm, is presented. The structure has been used in a real-time implementation, with which experiments have been performed. In this chapter, all fundamental blocks of the echo canceler are described, and simulation results using the implementation, on real life recordings, are studied. The results clearly verify that the theoretic fundamental problem of SAEC also applies to real-life situations.

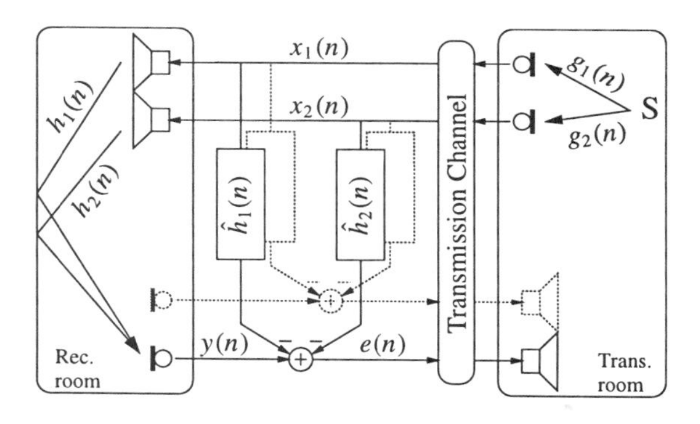

Figure 8.1 A stereophonic echo canceler.

Keywords: Stereophonic Acoustic Echo Canceler, Subband, Two-Channel Fast RLS, NLMS, Real-Time Implementation

1. INTRODUCTION

In conferencing systems, such as teleconferencing and desktop conferencing, acoustic echo cancelers (AECs) are needed to reduce the echo that results from the acoustic coupling between the loudspeaker and the microphone. The AEC identifies the echo path and simultaneously reduces the echo by means of adaptive filtering. If the conferencing system has dual audio channels in each direction, classical monophonic AECs will not provide sufficient echo suppression, and more sophisticated stereophonic acoustic echo cancelers (SAECs) are needed. In this chapter, a real-time implementation of a SAEC is described and studied.

In stereophonic conferencing systems, spatial audio information is transmitted. Not only will the listener experience a more realistic sound image, but she will also be able to aurally localize the speaker at the other end. Studies have shown that this improves perception, especially when speech from several speakers overlap [1]. However, there are now four acoustic echo paths to identify, two to each microphone, as illustrated in Fig. 8.1. This will not only cause increased calculation complexity, but also a new fundamental problem of non-uniqueness due to the correlation between the two channels, as shown in Chapter 6.

Several different channel decorrelation methods have been suggested, [2, 3, 4, 5, 6, 7], which reduce the effects of this new fundamental problem. However, even when these are used, the normal equation to be solved is ill conditioned. Moreover, the standard normalized least mean square (NLMS) adaptive algorithm is known to converge slowly in these situations. Therefore more sophisticated algorithms such as the affine projection algorithm (APA) or the recursive least squares (RLS), that are less affected by a high condition number, are preferred in SAEC's. The combination of four adaptive filters per SAEC and sophisticated adaptive algorithms results in high calculation complexity, and a well known way to reduce calculation complexity is to perform the adaptive filtering in a subband structure. The complexity reduction is possible since the subband signals are downsampled.

In the following, we will propose a structure for a high-performance SAEC, that has been verified in a real-time implementation. We will also show and discuss results from real-life recordings using this structure.

2. ACOUSTIC ECHO CANCELER COMPONENTS

As can be seen in Fig. 8.1, there are four different echo paths between the two loudspeakers and the two microphones in the receiving room. Therefore four adaptive filters are needed in order to estimate these impulse responses, also depicted in Fig. 8.1. Due to the fundamental problem in stereophonic acoustic echo cancellation (Chapter 6) these four adaptive filters may also need to have a faster convergence rate than what is provided by the normalized least mean square algorithm. Algorithms such as recursive least squares should therefore be used and implemented in a subband scheme. Another component needed in a SAEC is a residual echo suppressor, which suppress remaining echoes after the adaptive filter.

All the above components, depicted in Fig. 8.2, are described in this section.

2.1 ADAPTIVE ALGORITHM

The most central part of an echo canceler is of course the adaptive filter, which identifies the room impulse response. The performance and the calculation complexity of the echo canceler are highly dependent upon the choice of adaptive algorithm.

In traditional mono echo cancelers, the NLMS [8] algorithm is commonly used. It has several positive characteristics as stable robust adaptation and straightforward easy implementation. In SAECs, the convergence rate of the NLMS algorithm is low due to the correlation between the channels (Chapter 6).

The RLS algorithm has a significantly faster convergence rate than the NLMS algorithm, especially in the SAEC situation with highly correlated channels. But the calculation complexity, even in its fast version, fast RLS (FRLS), is several times higher than for the NLMS. A general analysis of the RLS algorithm

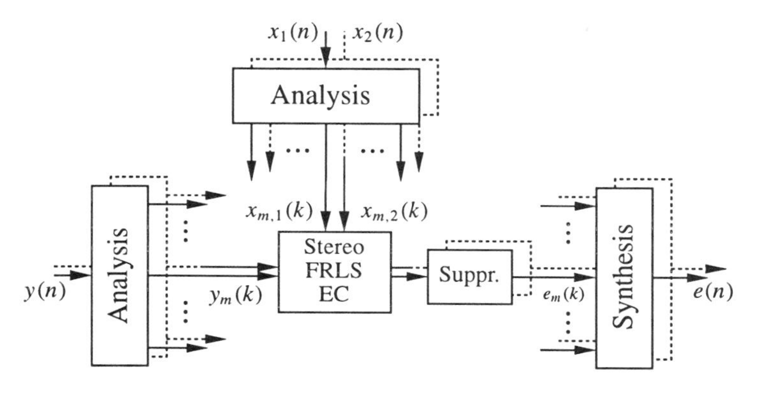

Figure 8.2 Subband stereophonic acoustic echo canceler. The transmission room and receiving room signals, $x_{(\cdot)}(n)$ and $y(n)$ respectively, is decomposed by four analysis filterbanks. The subband signals are then processed by the adaptive filters and residual echo suppressors, and finally the return signals are formed by two synthesis filterbanks.

can be found in [8] and the stabilized two-channel FRLS is described in Chapter 6 and in [9, 10].

In a subband echo canceler, several independent adaptive filters will be used in parallel, one for each subband. The main reason for applying the adaptive filters in a subband scheme, as we will see in Section 2.4, is to reduce the computational complexity of the SAEC. On the other hand it also gives us the freedom to use different adaptive filter parameters, such as adaptive filter lengths, in different frequency regions, as well as the freedom to use different types of adaptive algorithms in different subbands. In Section 3, examples with two-channel FRLS in the lower frequency regions and NLMS in the higher frequency regions are given.

Non-stationary input signals such as speech, may cause the FRLS algorithm to become unstable. However, the stability can increased by monitoring the state of the algorithm, and restarting it if the algorithm is determined to be unstable [9]. Alternatively, the FRLS algorithms can be restarted, one subband at the time, on a regularly interval [11]. During the time between restart and until the algorithm has reconverged, echo cancellation may be poor. A two-path structure, presented next, can improve the performance in these situations.

2.1.1 Two-path Adaptive Filtering. In situations with large disturbances, for example double-talk, or if the adaptive filter becomes unstable, the filter may diverge from a good estimate. Under these conditions, it would be better to use an earlier filter estimate until the adaptive filter has reconverged. This is the purpose of the two-path adaptive filter structure [12], described in Chapter 5.

It should be noted that the two-path structure is adapted in the subbands, and that the decisions made in one subband is independent of the states in all other subbands.

2.1.2 Channel Decorrelation. As presented in Chapter 6, the performance of the adaptive filters in a stereophonic echo canceler will suffer when the two channels are highly correlated. Therefore some form of channel decorrelation is needed in implementations of SAECs.

To reduce the perceived distortion it would be preferable if the decorrelating signal is similar to the original signal. But the core problem in the SAEC is that the two channels are linearly related, i.e. adding a signal that is linearly related to the original signal will not reduce the correlation between the two channels. In [2], it is suggested that a non-linearly processed source signal should be added to the source signal itself. It was found that the half-wave rectifier performed well in addition to having a simple and low-complexity structure. The half-wave rectifier is discussed in Chapter 6.

Audio coders in the transmission path between the transmission and the receiving rooms will decorrelate the channels. In [3], the influence of a perceptual audio coder, the MPEG layer III coder [14], has on a SAEC is analyzed. It is shown that the coder can decorrelate the signal because non-perceivable quantization noise is added to the source signal. How efficient decorrelator the coder is depends on what features that are used for compression. For example, advanced stereo coders usually operate in a joint stereo mode, where two correlated channels are coded jointly. This can actually increase the correlation between the channels, and should not be used if the coder also has to serve as decorrelator.

2.2 FILTERBANK DESIGN

The main reason for using a subband scheme is the reduction of calculation complexity, but other positive effects include increased stability of the adaptive filter, due to that fewer taps are adapted in each subband, and a structure that allow for efficient implementations on parallel systems. The two biggest disadvantages are the transmission path delay that is introduced, and possible aliasing due to downsampling. In [15] it has been shown that if critically downsampling is used, i.e. if we have the same downsampling factor r as number of subbands M, aliasing will significantly decrease the performance of the adaptive filters. Therefore non-critical downsampling, i.e. $r < M$, in conjunction with filters that have good stopband attenuation is commonly used. General discussions of filterbanks can be found in numerous books and articles [16, 17, 18, 19], but since the emphasis has been on critical downsampled filterbanks, an efficient structure for non-critical downsampling is given in this section. Methods how to design the prototype filters are also discussed.

As is shown in Fig. 8.2, four analysis filterbanks are used to decompose the transmission room and receiving room signals, and each signal is decomposed into M subband signals. After the adaptive filters, two synthesis filterbanks reconstruct the two echo canceled fullband signals.

2.2.1 Analysis Filterbank. The subband signal for subband m, $x_m(k)$, is calculated by bandpass filtering and downsampling the fullband input signal $x(n)$,

$$x_m(k) = \sum_{l=0}^{K-1} f_m(l)x(rk-l), \tag{8.1}$$

where the subband filter for subband m of length K is denoted $f_m(n)$. The subband filters are modulated versions of a low-pass prototype filter $f(n)$. If M denotes the number of subbands in the filterbank, the individual subband filters can be expressed with the following modulation,

$$\mathbf{f}_m^T = \mathbf{w}_m^T \mathbf{F}, \tag{8.2}$$

where

$$\mathbf{f}_m = \begin{bmatrix} f_m(0) & f_m(1) & \dots & f_m(K-1) \end{bmatrix}^T, \tag{8.3}$$

$$\mathbf{w}_m = \begin{bmatrix} 1 & e^{-j\frac{2\pi m}{M}(1)} & \dots & e^{-j\frac{2\pi m}{M}(M-1)} \end{bmatrix}^T, \tag{8.4}$$

$$\mathbf{F} = \begin{bmatrix} \text{diag}(\tilde{\mathbf{f}}_0) & \text{diag}(\tilde{\mathbf{f}}_M) & \dots & \text{diag}(\tilde{\mathbf{f}}_{K-M}) \end{bmatrix}. \tag{8.5}$$

The prototype filter matrix $\mathbf{F}$ is of size $M \times K$, and $\text{diag}(\tilde{\mathbf{f}}_i)$ denotes a diagonal matrix with elements from the prototype filter as,

$$\text{diag}(\tilde{\mathbf{f}}_i) = \begin{bmatrix} f(i) & & \mathbf{0} \\ & \ddots & \\ \mathbf{0} & & f(i+M-1) \end{bmatrix}. \tag{8.6}$$

Due to the downsampling, r new input samples are needed for each new subband sample, and we denote r new samples a *frame*. Now the output sample in subband m, frame k can be expressed as $x_m(k) = \mathbf{w}_m^T \mathbf{F}\mathbf{x}(k)$, where the fullband input vector $\mathbf{x}(k)$, consists of r new and $K-r$ old samples,

$$\mathbf{x}(k) = \begin{bmatrix} x(rk) & x(rk-1) & \dots & x(rk-K+1) \end{bmatrix}^T. \tag{8.7}$$

By exchanging the modulation vector $\mathbf{w}_m$ with the modulation matrix,

$$\mathbf{W} = \begin{bmatrix} \mathbf{w}_0 & \mathbf{w}_1 & \dots & \mathbf{w}_{M-1} \end{bmatrix}^T, \tag{8.8}$$

a vector containing all subband samples for frame k can be calculated in one step as,

$$\begin{bmatrix} x_0(k) & x_1(k) & \dots & x_{M-1}(k) \end{bmatrix}^T = \mathbf{WFx}(k). \tag{8.9}$$

The computational efficient polyphase filterbank structure can be seen in (8.9). Since $\mathbf{F}$ is a real sparse matrix, $\mathbf{Fx}(k)$ can be calculated with K real multiplications. Secondly, the modulation matrix, $\mathbf{W}$, is also known as the discrete Fourier transform (DFT) matrix. Therefore, the calculation complexity of (8.9) can be reduced by using efficient fast Fourier transforms,

$$\begin{bmatrix} x_0(k) & x_1(k) & \dots & x_{M-1}(k) \end{bmatrix}^T = \text{FFT}\left\{\mathbf{Fx}(k)\right\}. \tag{8.10}$$

Let us return to (8.8) once again. Due to the symmetry in the modulation vector (8.4), the following relation between rows can be seen in (8.8),

$$\mathbf{w}_i = \mathbf{w}_{M-i}^* \qquad 1 \le i \le \frac{M}{2} - 1 \tag{8.11}$$

where * denotes the conjugate operator. Consequently, $x_{M-i}(k) = x_i^*(k)$ as long as the input signal $x(n)$ and the elements of the prototype filter matrix $\mathbf{F}$ are real valued. Therefore, it is only necessary to calculate the first $M/2+1$ subbands, and also only necessary to apply the adaptive filters, the echo canceler, in the $M/2+1$ lowest subbands. It should be noted that complex valued adaptive filters are needed.

2.2.2 Synthesis Filterbank. The synthesis filterbank reconstructs the fullband signal, $e(n)$, by a summation of the up-sampled and filtered subband signals $e_m(k)$,

$$e(n) = \sum_{m=0}^{M-1} \sum_{l=0}^{K-1} g_m(l)\tilde{e}_m(n-l), \tag{8.12}$$

$$\tilde{e}_m(n) = \begin{cases} e_m(\frac{n}{r}) & \frac{n}{r} \in \mathcal{Z}, \\ 0 & \frac{n}{r} \notin \mathcal{Z}. \end{cases} \tag{8.13}$$

In contrast to the analysis filterbank, where the bandpass filter is used to suppress aliasing, the filter, g_m, is needed to suppress imaging resulting from the up-sampling of the subband residual echo signal. In order to find a structure where the number of arithmetic operations can be reduced, the filtering operation in (8.12) is described by a state vector, depicted in Fig. 8.3. As can be seen in the figure, the state-vector is updated by shifting the vector one position to the right, and by adding the modulated prototype filter vector $\mathbf{g}_m$ multiplied with the up-sampled residual echo signal for subband m. Let us now define the state

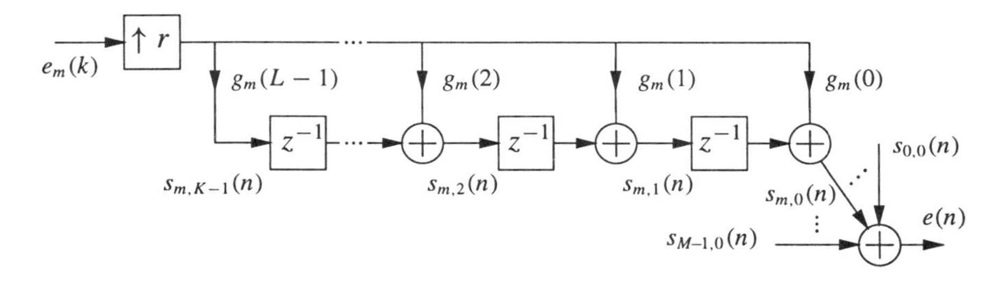

Figure 8.3 State representation of the synthesis filterbank.

vector as one upper and one lower vector,

$$\mathbf{s}_m(k) = \begin{bmatrix} \mathbf{s}_m^u(k) \\ \mathbf{s}_m^l(k) \end{bmatrix}, \tag{8.14}$$

$$\mathbf{s}_m^u(k) = \begin{bmatrix} s_{m,0}(kr) & s_{m,1}(kr) & \ldots & s_{m,r-1}(kr) \end{bmatrix}^T, \tag{8.15}$$

$$\mathbf{s}_m^l(k) = \begin{bmatrix} s_{m,r}(kr) & s_{m,r+1}(kr) & \ldots & s_{m,K-1}(kr) \end{bmatrix}^T. \tag{8.16}$$

Due to the interpolation, every subband sample $e_m(rk)$ is followed by $r - 1$ zeros. For these zeros, the state vectors are updated with pure shifts. The output signal, $e(n)$, can therefore be calculated for r samples, i.e. one frame, at the time,

$$\begin{bmatrix} e(kr-r+1) & e(kr-r+2) & \ldots & e(kr) \end{bmatrix}^T = \sum_{m=0}^{M-1} \mathbf{s}_m^u(k). \tag{8.17}$$

Similar to the analysis filters, the synthesis filters $g_m(n)$ are modulated versions of the low-pass prototype filter $g(n)$. Using the modulation vector in (8.4), $\mathbf{g}_m = \mathbf{G}\mathbf{w}_m$, where $\mathbf{G}$ is a sparse prototype matrix of size $K \times M$,

$$\mathbf{G} = \begin{bmatrix} \mathrm{diag}(\tilde{\mathbf{g}}_0) & \mathrm{diag}(\tilde{\mathbf{g}}_M) & \ldots & \mathrm{diag}(\tilde{\mathbf{g}}_{K-M}) \end{bmatrix}^T. \tag{8.18}$$

As mentioned above, the state vectors can be updated on a frame basis. The state vectors are shifted r positions and the input subband sample, multiplied with $\mathbf{g}_m$, is added,

$$\mathbf{s}_m(k) = \begin{bmatrix} \mathbf{s}_m^l(k-1) \\ \mathbf{0}_{r \times 1} \end{bmatrix} + \mathbf{G}\mathbf{w}_m e_m(k). \tag{8.19}$$

When reconstructing the output signal $e(n)$, it is enough to know the sum of the state vectors, as is shown in (8.17). Therefore, the state vectors do not need

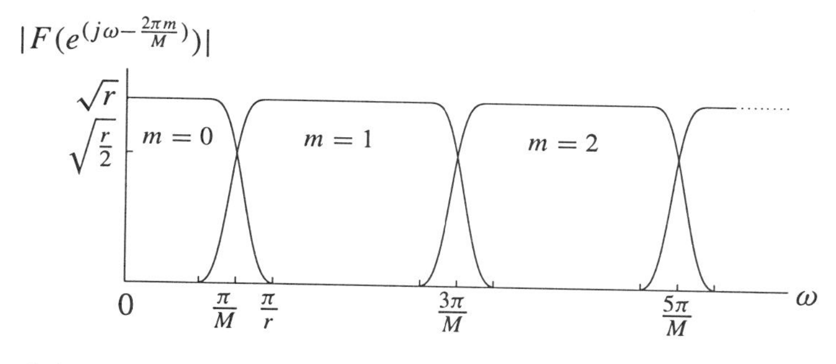

Figure 8.4 An example of a filterbank designed by solving (8.28). The filters prevent aliasing by sufficient stopband attenuation, see filter $m = 0$ for $f > \frac{1}{2r}$.

to be updated individually, instead it is sufficient to update the sum of the state vectors, as

$$\mathbf{s}(k) = \begin{bmatrix} \mathbf{s}^l(k-1) \\ \mathbf{0}_{r\times 1} \end{bmatrix} + \mathbf{G}\mathbf{W}\mathbf{e}^{\text{sub}}(k), \tag{8.20}$$

where $\mathbf{W}$ is defined in (8.8) and $\mathbf{e}^{\text{sub}}(k) = \begin{bmatrix} e_0(k) & e_1(k) & \dots & e_{M-1}(k) \end{bmatrix}^T$. Finally, (8.20) can be calculated with a computationally efficient FFT algorithm,

$$\mathbf{s}(k) = \begin{bmatrix} \mathbf{s}^l(k-1) \\ \mathbf{0}_{r\times 1} \end{bmatrix} + \mathbf{G} \cdot \text{FFT}\left\{\mathbf{e}^{\text{sub}}(k)\right\}, \tag{8.21}$$

and the synthesis filterbank output for frame k is then the upper r elements of the state vector, in reverse order,

$$\begin{bmatrix} e(kr-r+1) & e(kr-r+2) & \dots & e(kr) \end{bmatrix}^T = \mathbf{s}^u(k). \tag{8.22}$$

2.2.3 Prototype Filter Requirements and Design. Since aliasing in the filterbank will drastically decrease the performance of the SAEC, aliasing suppression is an important property in the design of the prototype filter. This in combination with non-critical down-sampling, allow us to neglect aliasing cancellation in the filterbank. The only requirement, neglecting small amounts of aliasing errors, is to ensure that the filterbank, when seen as an input output device, is a pure delay,

$$\frac{1}{r}\sum_{m=0}^{M-1} F(e^{(j\omega-\frac{2\pi m}{M})})G(e^{(j\omega-\frac{2\pi m}{M})}) = e^{-j\omega K}, \tag{8.23}$$

where K is the prototype filter length, $F(z) = \sum_{k=0}^{K-1} f(k)z^{-k}$ the analysis filter, and $G(z) = \sum_{k=0}^{K-1} g(k)z^{-k}$ the synthesis filter. In order to guarantee linear phase, we let $G(z) = z^{-K}F(z^{-1})$. If we define the non-causal filter $R(z) = F(z)F(z^{-1})$, which differ from the total filterbank, $F(z)G(z)$, only by a delay, an equivalent requirement to (8.23) would be,

$$\frac{1}{r}\sum_{m=0}^{M-1} R(e^{j(\omega-\frac{2\pi m}{M})}) = 1. \tag{8.24}$$

Now, we can formulate the filter design problem as a minimization problem. First we need to make sure that the passband amplification is sufficient flat, by minimizing

$$\Phi_1 = \int_0^{\frac{2\pi}{M}-\frac{\pi}{r}} (R(e^{j\omega}) - r)^2 d\omega. \tag{8.25}$$

In the region where two adjacent bands overlap, see Fig. 8.4, the sum of the two bands should also be flat, by minimizing

$$\Phi_2 = \int_{\frac{2\pi}{M}-\frac{\pi}{r}}^{\frac{\pi}{r}} (R(e^{j\omega}) + R(e^{j(\omega-\frac{2\pi}{M})}) - r)^2 d\omega. \tag{8.26}$$

Finally we need to minimize aliasing by enforcing good stopband attenuation,

$$\Phi_3 = \int_{\frac{\pi}{r}}^{\pi} (R(e^{j\omega}))^2 d\omega. \tag{8.27}$$

The minimization problem can now be expressed as

$$\min_{R(z)} \quad \alpha_1\Phi_1 + \alpha_2\Phi_2 + \alpha_3\Phi_3, \tag{8.28}$$

under the constraints,

$$r(0) \geq r(n) \qquad n \neq 0, \tag{8.29}$$

$$r(n) = r^*(-n) \qquad \forall n, \tag{8.30}$$

where $\alpha_i \geq 0$ are trade-off parameters and $R(z) = \sum_{k=-K+1}^{K-1} r(k)z^{-k}$. Quadratic programming [20] has been found to be a fast and stable method to solve this minimization problem. And in [21], a method how to calculate $F(z)$ out of $R(z)$ is presented. An alternative method to design filters that satisfies (8.23) can be found in [22]. This method has more stringent requirements, forcing Φ_1 and Φ_2 to be equal to zero. Because of this, longer filters are needed in order to achieve as good stopband attenuation as with filters designed according to (8.28).

2.2.4 Noncausal Subband Impulse Responses. When a signal is filtered with a bandpass filter, it will always become spread in the time domain. This is also the case for the subbands signals in a subband echo canceler. Due to that the signals that are used for estimation of the impulse response are spread in time, the impulse response to be estimated will also be spread in time. The perhaps surprisingly, but well-known fact, is that these impulse responses are be better modeled if a couple of non-causal taps are included in the model. The topic has not been well described in the literature, but it is discussed in [23].

In the practical situation it usually enough to know that approximately 5 to 10 non-causal taps in each subband should be included. This can easily be performed by the delaying the loud-speaker, $x(n)$, or microphone signal, $y(n)$, left of the echo canceler in Fig. 8.1, with the corresponding number of taps. In most applications there is a natural transmission delay between the loud-speaker and microphone signal, usually denoted flat delay. This delay reduces the need to artificially delaying the microphone signal, and may in some applications be a sufficient delay.

2.3 RESIDUAL ECHO SUPPRESSION

The residual echo suppressor contains two different types of suppressors and a comfort noise fill device, depicted in Fig. 8.5. Each subband has its own residual echo suppressor that operate independently of the other subbands.

The first is a short-time transmission room energy based suppressor. It multiplies the residual echo signal with an attenuation factor, $\lambda_{A,m}(k)$, that is roughly inversely proportional to the square root of the short-time energy of the transmission room signal, $\sigma^2_{x,m}(k)$. The attenuation factor has a lower and an upper bound, T_{lo} and T_{hi}, and can specifically be written as

$$\lambda_{A,m}(k) = \begin{cases} 1 & \sigma_{x,m}(k) < T_{\text{lo}}, \\ T_{\text{lo}}/\sigma_{x,m}(k) & T_{\text{lo}} < \sigma_{x,m}(k) < T_{\text{hi}}, \\ T_{\text{lo}}/T_{\text{hi}} & T_{\text{hi}} < \sigma_{x,m}(k). \end{cases} \tag{8.31}$$

The second suppressor, an echo path gain based suppressor [24], can be viewed as a mild form of center clipping in that if the residual echo is very strong, it is left unaffected, but when it is below a threshold, it is attenuated by an amount roughly proportional to the residual echo signal. The attenuation factor, $\lambda_{B,m}(k)$, is a function of $\sigma^2_{x,m}(k)$ and the short-time energy of the signal after the first suppressor, $\sigma^2_{e_B,m}(k)$. It is upper bounded by one, and can be expressed as,

$$\lambda_{B,m}(k) = \min\left\{ \varpi_m \frac{\sigma^2_{e_B,m}(k)}{\sigma^2_{x,m}(k)}, 1 \right\}, \tag{8.32}$$

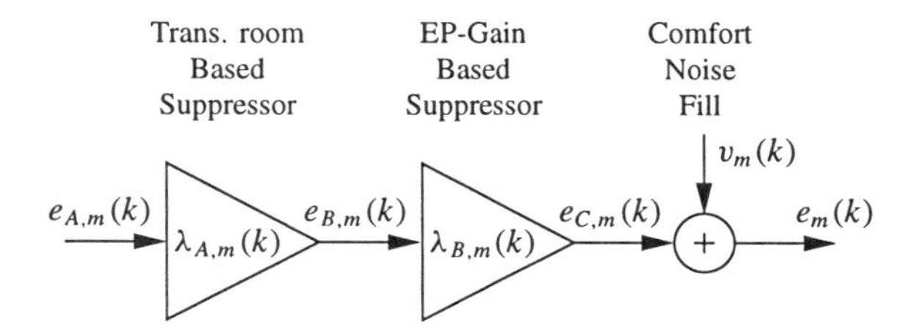

Figure 8.5 Suppression and comfort noise fill. (Elaboration of the block denoted "Suppr." in Fig. 8.2).

where the constant ϖ_m determines the amount of suppression. Finally, comfort noise is added to the residual echo signal.

2.4 COMPUTATIONAL COMPLEXITY

In this section we calculate the number of real valued multiplications and additions the adaptive filter and the filterbanks need per fullband sample period. The Fourier transform in the filterbank and most parts of the adaptive filter (FRLS) are performed with complex arithmetic. In this analysis multiplication between two complex numbers are counted as four real multiplications and two real additions.

The number of multiplications needed for the real valued two channel FRLS is $32L$, and the number of additions is $32L$, where L is the length of the adaptive filter. This includes calculation of the residual signals for both channels. The subband signals are complex valued, and the complex version of the two channel FRLS needs $128L_{\text{sub}}$ real multiplications and $128L_{\text{sub}}$ real additions, where L_{sub} denotes the length of the adaptive filter in the subbands. Echo cancellation in the subband structure described in previous sections, needs $M/2+1$ adaptive filters but due to downsampling they are updated $1/r$ times the fullband rate. The length of the adaptive filters are $L_{\text{sub}} = L/r + C_{\text{nc}}$, where C_{nc} compensates for the non-causal taps, Section 2.2 and [23]. The total number of real multiplications per fullband sample is then $\frac{M/2+1}{r}(128L_{\text{sub}})$ and the number of additions $\frac{M/2+1}{r}(128L_{\text{sub}})$.

The analysis, (8.10), and synthesis, (8.21), filterbanks include two parts, polyphase filtering and fast Fourier transformation. The polyphase filtering uses K multiplications and K additions per filterbank, where K is the length of the prototype filter. Since the input signal of the Fourier transform in the analysis filterbank is real valued, and output signal from the Fourier transform in the synthesis filterbank also is real valued, only two Fourier transforms are needed for the four analysis filterbanks and one Fourier transform for the two synthesis filterbanks. $2M-4$ extra additions per Fourier transform are needed

Table 8.1 Calculation complexity comparison given as number of real valued mult/add per full-band sample period. Corresponding fullband impulse response length $L = 3168$, downsampling rate $r = \frac{3}{4}M$, number of non-causal taps per subband $C_{nc} = 5$ [23].

Number of subbands, M	1	16	32	64	128
Prototype filter length, K	–	207	415	831	1663
Filterbank add.	–	182	194	207	219
Filterbank mult.	–	110	117	125	132
NLMS add./mult.	25344	6456	3105	1562	823
FRLS add./mult.	101376	25824	12421	6248	3293

to separate the two real- valued transforms [25]. If the Fourier transforms are implemented with a Radix 2 structure, they each need $2M \log_2 M - 7M + 12$ real multiplications and $3M \log_2 M - 3M + 4$ additions [25]. Also the filterbanks need to be updated $1/r$ times the fullband rate. The total number of multiplications of the four analysis filterbanks is $\frac{1}{r}(4K + 4M \log_2 M - 14M + 24)$ and the total number of additions is $\frac{1}{r}(4K + 6M \log_2 M - 2M)$. In the synthesis filterbank, K extra additions are needed to update the state vector in each filterbank, (8.21). The total number of multiplications for the two synthesis filterbanks is $\frac{1}{r}(2K + 2M \log_2 M - 7M + 12)$ and the number of additions $\frac{1}{r}(4K + 3M \log_2 M - M)$. In Table 8.1, complexity examples for different number of subbands are given.

2.5 IMPLEMENTATION ASPECTS

In the previous section we have seen that, by using a subband scheme, it is possible to reduce the calculation complexity of the adaptive filters. This is of course a major advantage when an algorithm is to be implemented in a real-time environment. But a subband scheme offers other advantages to an implementation as well. For example, if the system is to be implemented on a hardware with parallel processing units, the subband structure makes it easy to fully utilize all units. Typically the adaptive filters represent the majority of the calculation complexity, and a straight-forward way to utilize several processors would be to divide the adaptive filters, each filter corresponding to a specific subband, over several processors. This will require data to be transferred between the different units, but the amount of data is limited, since the adaptive filters operate independently of each other. A disadvantage is that the adaptive filters may not finish in time to deliver the data to the synthesis filterbank, introducing more delay to signal path. This can though be avoided by using the two-path structure described in Section 2.1. Using this structure, the adaptive filters could even be stopped for short times, in a worst case scenario.

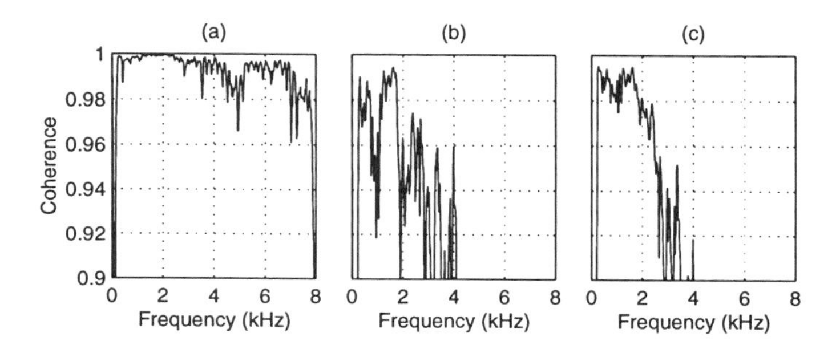

Figure 8.6 Magnitude coherence between the right and left channel in the transmission room. (a) Unprocessed transmission room signals. (b) Signals pre-processed with the half-wave rectifier, $\alpha = 0.5$. (c) Signal coded/decoded with an MPEG layer III audio coder.

With a subband structure, the total processing power needed can in a simple manor be adjusted, to the amount of available processing power. Not only the filter lengths can be adjusted, but also the number of subbands and type of adaptive filter in each subband can be altered. This could be especially advantageous in desktop conferencing, where the echo canceler is operating on the local workstation.

3. SIMULATIONS

In this section, simulations will highlight the difficulties of stereophonic acoustic echo cancellation, and also indicate the performance that can be expected from a subband SAEC based on the two-channel FRLS algorithm. The data was recorded in a typical office size room, with very low background noise level (HuMaNet room B [26]). Increased background noise level in the transmission room would reduce the correlation between the channels, and that way improve the ability for the SAEC to converge to the correct solution. First a recording representing the transmission room was performed. In this recording, the excitation signal was a high-quality speech signal recorded in an anechoic chamber. In order to create signals with spatial changes in the transmission room, two loudspeakers were used, representing two speakers at different positions. Also, signals representing the receiving room were recorded, using the transmission room signal above as excitation signal. The signals were recorded at 16 kHz sample rate and the average SNR (echo-to-noise ratio) was approximately 40 dB. In the simulations in this section, the subband echo canceler had $M = 64$ subbands, the downsampling factor was $r = 48$ and adaptive filter length in each subband was $L_{\text{sub}} = 66$, corresponding to 198 ms. The residual echo suppressor, described in Section 2.3, was deactivated.

As shown in Chapter 6, the echo canceler problem has an infinite number of solutions when the two input signals, x_1 and x_2 in Fig. 8.1, are linearly related. The magnitude coherence function,

$$\gamma(f) = \frac{|S_{x_1x_2}(f)|}{\sqrt{S_{x_1x_1}(f)S_{x_2x_2}(f)}}, \tag{8.33}$$

is a measure of how correlated the two signals are [2], where $\gamma(f) = 1$ shows that the two signals are completely linearly related to each other. That is, the SAEC will have difficulties to converge to the correct solution in those frequency regions where the magnitude coherence function is close to one. In Fig. 8.6, the magnitude coherence function is shown for speech signals recorded as described above. The power spectral estimates, $S_{x(\cdot)x(\cdot)}$, were calculated using the Welch method [27] with a Hanning window of 8196 samples on the signals recorded at 16 kHz. Figure 8.6(a) shows the coherence between the left and right channels for an unprocessed transmission room speech recording. The correlation between the channels can be reduced by pre-processing the signals. In Fig. 8.6(b) the coherence for signals pre-processed with half-wave rectifiers, Section 2.1 [2], with $\alpha = 0.5$ is shown. The signals can also be decorrelated by the use of a coder and a decoder [3]. In Fig. 8.6(c) the coherence function is shown for a signal that has been coded/decoded by an MPEG Layer III coder [14, 28], with joint stereo coding disabled and the bitrate 32 kbit/s per channel.

One way to show the effectiveness of the decorrelation is to study how the performance of the echo canceler decreases after a position change of the transmission room speaker. As performance index the mean square error (MSE) energy of the residual echo signal is used. The MSE is given by,

$$\text{MSE} = \frac{P_{e-w}}{P_{y-w}}, \tag{8.34}$$

$$P_{e-w} = \text{LPF}\{e(n) - w(n)\}^2, \tag{8.35}$$

where w denotes the receiving room background noise signal and LPF denotes a lowpass filter; in this case it has a single pole in 0.999. P_{y-w} is analogously calculated. In all the following examples, the background noise signal w is unknown, and cannot be subtracted according to (8.35). This will somewhat increase the MSE. In the simulation, the speaker moves from a position close to the left microphone, to a position closer to the right microphone after 5.1 seconds. The transmission room signal (left channel) is shown in Fig. 8.7(a). In Fig. 8.7(b) the MSE resulting from the use of an unprocessed signal is plotted with a solid line. Especially notice the severe increase of MSE after the transmission room speaker position change at 5.1 seconds. In the same figure, but plotted with a dotted line, is the mean square error for the same excitation signal

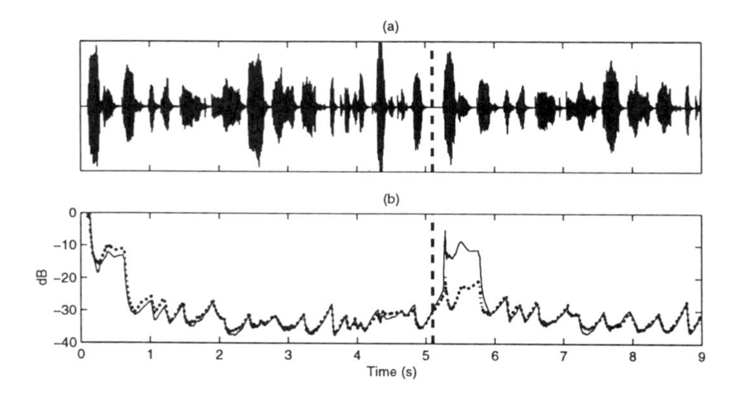

Figure 8.7 Mean square error convergence of the SAEC. (a) Excitation signal. (b) Mean square error for unprocessed signal (solid line) and signal processed with the half-wave rectifier (dotted line).

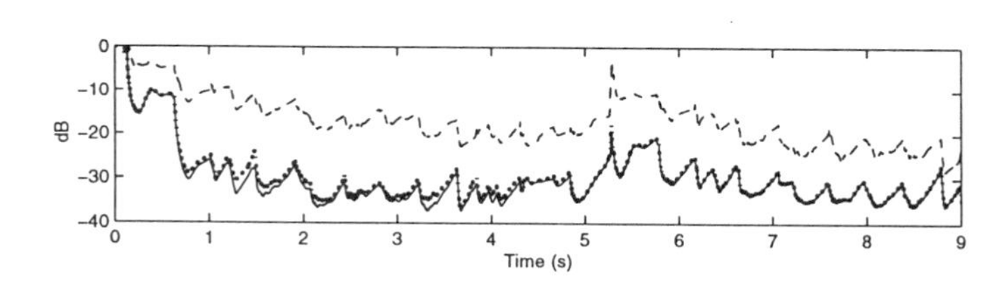

Figure 8.8 Mean square error convergence of the SAEC. Comparison between two-channel FRLS (solid line), NLMS (dashed line), and a SAEC with FRLS in the lower subbands and NLMS in the higher subbands (dotted line).

but processed with the half-wave rectifier, $\alpha = 0.5$. Under these conditions, the SAEC converges towards the true solution, and is therefore independent of echo path changes in the transmission room. Because of this, the MSE is less affected by the the transmission room speaker position change at 5.1 seconds. Simulations have shown similar behavior for the coded/decoded signal used in Fig. 8.6(c) as for signals processed with the half-wave rectifier, $\alpha = 0.5$. This shows that the coherence function is an effective measure of how ill-conditioned the correlation matrix, $\mathbf{R}_{xx}(n)$, is and how this affects the performance of the SAEC.

Other simulations have shown that when a background noise source, in our case fan noise from a personal computer, was emitted in the transmission room, the correlation between the channels was reduced. Convergence of the adaptive filters were improved, especially in the higher frequency regions. Channel decorrelation is though still considered needed in normal office environments, especially in the lower frequency regions.

In the previous examples, the two-channel FRLS algorithm is used in all subbands. In order to reduce the calculation complexity, it is possible to use an NLMS algorithm in the upper subbands without significantly reducing the performance of the echo canceler. In Fig. 8.8, the dotted line is generated with a SAEC with two-channel FRLS in the lower half of the subbands ($\approx$ 0–4 kHz), and NLMS in the upper half ($\approx$ 4–8 kHz).

References

[1] J. Benesty, D. R. Morgan, J. Hall, and M. M. Sondhi, "Synthesized stereo combined with acoustic echo cancellation for desktop conferencing," *Bell Labs Tech. J.*, vol. 3, pp. 148-158, July-Sep. 1998.

[2] J. Benesty, D. R. Morgan, and M. M. Sondhi, "A better understanding and an improved solution to the specific problems of stereophonic acoustic echo cancellation," *IEEE Trans. Speech Audio Processing*, vol. 6, pp. 156-165, Mar. 1998.

[3] T. Gänsler and P. Eneroth, "Influence of audio coding on stereophonic acoustic echo cancellation," in *Proc. IEEE ICASSP*, 1998, pp. 3649-3652.

[4] A. Gilloire and V. Turbin, "Using auditory properties to improve the behavior of stereophonic acoustic echo cancellers," in *Proc. IEEE ICASSP*, 1998, pp. 3681-3684.

[5] S. Shimauchi, Y. Haneda, S. Makino, and Y. Kaneda, "New configuration for a stereo echo canceller with nonlinear pre-processing," in *Proc. IEEE ICASSP*, 1998, pp. 3685-3688.

[6] M. Ali, "Stereophonic echo cancellation system using time-varying all-pass filtering for signal decorrelation," in *Proc. IEEE ICASSP*, 1998, pp. 3689-3692.

[7] Y. Joncour and A. Sugiyama, "A stereo echo canceller with pre-processing for correct echo path identification," in *Proc. IEEE ICASSP*, 1998, pp. 3677-3680.

[8] S. Haykin, *Adaptive Filter Theory*. Chapter 9, Prentice Hall International, 3rd edition, 1996.

[9] J. Benesty, F. Amand, A. Gilloire, and Y. Grenier, "Adaptive filtering algorithms for stereophonic acoustic echo cancellation," in *Proc. IEEE ICASSP*, 1995, pp. 3099-3102.

[10] M. G. Bellanger, *Adaptive Digital Filters and Signal Analysis*. Marcel Dekker, 1987.

[11] B. Hätty, "Block recursive least squares adaptive filters using multirate systems for cancellation of acoustical echoes," in *IEEE Conf. Record of the ASSP Workshop on Application of Digital Signal Processing to Audio and Acoustics*, 1989.

[12] K. Ochiai, T. Araseki, and T. Ogihara, "Echo cancellation with two path models," *IEEE Trans. on Commun.*, vol. COM-25, pp. 589-595, June 1977.

[13] M. M. Sondhi, D. R. Morgan, and J. L. Hall, "Stereophonic acoustic echo cancellation—an overview of the fundamental problem," *IEEE Signal Processing Lett.*, vol. 2, pp. 148-151, Aug. 1995.

[14] B. G. Haskell, A. Puri, and A. N. Netravali, *Digital Video: An Introduction to* MPEG-2. Chapter 4, pp. 55-79, Digital Multimedia Standards Series. Chapman & Hall, 1st edition, 1997.

[15] A. Gilloire and M. Vetterli, "Adaptive filtering in subbands with critical sampling: Analysis, experiments, and application to acoustic echo cancellation," *IEEE Trans. on Signal Processing*, vol. 40, pp. 1862-1875, Aug. 1992.

[16] M. Vetterli and J. Kovačević, *Wavelets and Subband Coding*. Prentice Hall PTR, 1995.

[17] G. Strang and T. Nguyen, *Wavelet and Filter Banks.* Wellesley-Cambridge Press, 1996.

[18] P. P. Vaidyanathan, *Multirate Systems and Filter Banks.* Prentice Hall PTR, 1993.

[19] N. J. Fliege, *Multirate Digital Signal Processing.* John Wiley & Sons, 1994.

[20] Coleman and Y. Li, "A reflective newton method for minimizing a quadratic function subject to bounds on some of the variables," *SIAM Journal on Optimization*, vol. 6, pp. 1040-1058, 1996.

[21] R. Boite and H. Leich, "A new procedure for the design of high order minimum phase FIR digital or CCD filters," *Signal Processing*, pp. 101-108, 1981.

[22] G. Wackersreuther, "On the design of filters for ideal QMF and polyphase filter banks," *AEÜ*, vol. 39, pp. 123-130, 1985.

[23] W. Kellermann, *Zur Nachbildung physikalischer Systeme durch parallelisierte digitale Erzatzsysteme im Hinblick auf die Kompensation akustischer Echos.* Ph.D. thesis, Darmstadt, 1989.

[24] E. J. Diethorn, "An algorithm for subband echo suppression in speech communications," Private Communication, 1998.

[25] H. Sorensen, D. Jones, M. Heideman, and S. Burrus, "Real-values fast fourier transform algorithms," *IEEE Transactions on Acoustics, Speech, and Signal Processing*, vol. ASSP-35, pp. 849-863, June 1987.

[26] D. A. Berkley and J. L. Fanagan, "HuMaNet: an experimental human-machine communications network based on ISDN wideband audio," *AT&T Tech. J.*, vol. 69, pp. 87-99, Sep.-Oct. 1990.

[27] P. D. Welch, "The use of fast Fourier transform for the estimation of power spectra: a method based on time averaging over short, modified periodograms," *IEEE Trans. on Audio and Electroacoustics*, vol. 15, pp. 70-73, June 1967.

[28] Fraunhofer IIS, "MPEG-1 LAYER III shareware audio coder," 1995, Am Weichselgarten 3 D-91058 Erlangen Germany, Encoder and decoder code: http://www.iis.fhg.de/amm/techinf/layer3/index.html, Public domain decoder source code (ANSI c): ftp://ftp.fhg.de/pub/iis/layer3/public_c/.

III

NOISE REDUCTION TECHNIQUES WITH A SINGLE MICROPHONE

Chapter 9

SUBBAND NOISE REDUCTION METHODS FOR SPEECH ENHANCEMENT

Eric J. Diethorn
Microelectronics and Communications Technologies, Lucent Technologies
eric.diethorn@lucent.com

Abstract Digital noise reduction processing is used in many telecommunications applications to enhance the quality of speech. This investigation focuses on the class of single-channel noise reduction methods employing the technique of short-time spectral modification, a class that includes the popular method of spectral subtraction. The simplicity and relative effectiveness of these subband noise reduction methods has resulted in explosive growth in their use for a variety of speech communications applications. The most commonly used forms of the short-time spectral modification method are discussed, including the Wiener filter, magnitude subtraction, power subtraction, and generalized parametric subtraction. Because of its importance to the subjective performance of any noise reduction method, the subject of real-time signal- and noise-level estimation is also reviewed. A low-complexity noise reduction algorithm is also presented and its implementation is discussed.

Keywords: Noise Reduction, Wiener Filtering, Spectral Subtraction, Short-Time Fourier Analysis, Subband Filter Banks, Implementation

1. INTRODUCTION

Noise enters speech communications systems in many ways. In traditional wire-line telephone calls, one or both parties may be speaking within an environment having high levels of background noise. Calls made from public telephone booths located near roadways, transportation stations, and shopping areas serve as examples. Similarly, cellular, or wireless, telephones permit users to place calls from virtually any location, and it is common for such communications to be degraded by noise of varied origin. In room teleconferencing applications, in which the acoustical characteristics of the environment are generally assumed to be controlled (quiet), it is not uncommon for heating, ventilation and air-conditioning systems to contribute substantial levels of noise. Noise originates not only from acoustical sources, however. Circuit noise, generated electrically within the telephone network, is still prevalent throughout the global telecommunications system.

When present at small or moderate additive levels, noise degrades the subjective quality of speech communications. Listening tests broadly show that people grow less tolerant of, and less attentive to, listening material as the signal-to-noise (SNR) ratio of the material decreases. This phenomenon is known as listener fatigue. When the SNR of speech material is very low, say less than 10 dB, the intelligibility of speech is affected.

Even traditionally low levels of noise can present a problem, especially when multiple speech channels are combined as in conferencing or bridging. In multiparty, or multipoint, teleconferencing, the background noise present at the microphone(s) of each point of the conference combines additively at the network bridge with the noise processes from all other points. The loudspeaker at each location of the conference therefore reproduces the combined sum of the noise processes from all other locations. This problem becomes serious as the number of conferencing points increases. Consider a three-point conference in which the room noise at all locations is stationary and independent with power P. Each loudspeaker receives noise from the other two locations, resulting in a total received noise power of $2P$, or 3 dB greater than that of a two-point conference. With N points, each side receives a total noise power that is $10\log(N-1)$ dB greater than P. For example, in a conference with 10 participating locations, the received noise power at each point is about 10 dB greater than that of the two-party case. Because a 10 dB increase in sound power level roughly translates to a doubling of perceived loudness, the noise level perceived by each participant is twice as loud as that of the two-party case. The benefits of noise reduction processing for cases such as this are clearly evident.

A variety of approaches have been proposed to reduce noise for purposes of speech enhancement. Included are: classic (static) Wiener filtering [6]; dynamic comb filtering (see citations in [14]), in which a linear filter is adapted to pass only the harmonic components of voiced speech as derived from the pitch period; dynamic, linear all-pole and pole-zero modeling of speech (see citations in [14]), in which the coefficients of the (noise-free) model are estimated from the noisy speech; short-time spectral modification techniques, in which the magnitude of the short-time Fourier transform is attenuated at frequencies where speech is absent [1]–[5], [11]–[20], [23]–[25]; and hidden Markov modeling [21, 22], a technique also employing time-varying models for speech but where the evolution of model coefficients is governed by transition probabilities associated with model states.

Predominately, speech noise reduction systems are used to improve the subjective quality of speech, lessening the degree to which listener fatigue limits the perceived quality of speech communications. Though some work [14, 24], has shown that intelligibility improvement is possible through digital noise-reduction processing, these results apply to carefully designed listening tests.

All the above noise reduction methods share the property that they operate on a single channel of noisy speech. They are blind techniques, in the sense that only the noise-corrupt speech is known to the algorithm. Thus, in order to enhance the speech-signal-to-noise ratio, the algorithms must form bootstrap estimates of the signal and noise. When multiple channels containing either the same noisy speech source or noise source alone are available, a wide range of spatial acoustic processing technologies applies. Included are adaptive beamformers and adaptive noise cancelers. Adaptive noise cancellation methods are coherent noise reduction processors, exploiting the phase coherency among multiple time series channels to cancel the noise, whereas the noise reduction methods listed above are incoherent processors.

This work focuses solely on the class of single-channel noise reduction methods employing short-time spectral modification techniques. The simplicity and relative effectiveness of these methods has resulted in explosive growth in their use for a variety of speech communications applications. Today, noise reduction processors appear in a variety of commercial products, including cellular telephone handsets; cellular hands-free, in-the-car telephone adjuncts; room teleconferencing systems; in-network speech processors, such as bridges and echo cancelers; in-home telephone appliances, including speakerphones and cordless phones; and hearing aid and protection devices. Frequently, noise reduction processors are commonly used in conjunction with other audio and speech enhancement devices. In room teleconferencing systems, for example, noise reduction is often combined with acoustic echo cancellation and microphone-array processing (beamforming). In summary, the diversity and complexity of modern communications systems present ample opportunity to apply methods of digital noise reduction processing.

The organization of this chapter is as follows. We begin in Section 2 with a brief review of Wiener filtering because of its fundamental relation to all past and modern noise reduction methods that employ spectral modification. Section 3 reviews the technique of short-time Fourier analysis and discusses its use in noise reduction. The short-time Wiener filter is discussed, and a variety of commonly used variations on the Wiener filter are reviewed. Techniques for estimating the signal and noise envelopes are reviewed in Section 4. Last, Section 5 presents a low-complexity implementation of a noise reduction processor for speech enhancement.

2. WIENER FILTERING

Consider the problem of recovering a signal $s(n)$ that is corrupted by additive noise. Let

$$y(n) = s(n) + v(n) \tag{9.1}$$

represent the noisy signal and let the power spectrum of the noise source $v(n)$ be known or, at least, accurately estimated. When $s(n)$ and $v(n)$ are uncorrelated stationary random processes, the power spectrum of the noise-corrupt signal, $P_y(\omega)$, is simply the sum of the power spectrums of the signal and noise:

$$P_y(\omega) = P_s(\omega) + P_v(\omega). \tag{9.2}$$

Under these circumstances the power spectrum of the signal is easily recovered by exploiting (9.2) to subtract the power spectrum of the noise from that of the noisy observation, that is,

$$P_s(\omega) = P_y(\omega) - P_v(\omega). \tag{9.3}$$

Though trivial in concept, this fundamental spectral power subtraction relation forms the basis for the noise reductions methods discussed throughout this chapter.

Of course, (9.3) only provides recovery of the power spectrum of the random process to which sample function $s(n)$ is associated. To estimate $s(n)$ we rely upon classical linear estimation theory. The estimate $\hat{s}(n)$ of $s(n)$ that minimizes the mean-squared error $\left\|s(n) - \hat{s}(n)\right\|^2$ is given by [6]

$$\hat{S}_W(\omega) = H_W(\omega)Y(\omega) \tag{9.4}$$

where $\hat{S}_W(\omega)$ is the Fourier transform corresponding to the optimum $\hat{s}(n)$, $Y(\omega)$ is the Fourier transform of $y(n)$, and

$$H_W(\omega) = \frac{P_s(\omega)}{P_s(\omega) + P_v(\omega)} \tag{9.5}$$

is the (noncausal) Wiener filter frequency response function derived by Norbert Wiener many years ago. Thus, the least-mean-squares estimate of the signal is acquired simply by applying a frequency dependent gain function to the spectrum of the noisy signal. Note that by using (9.2) and (9.3) in (9.5) and expanding (9.4) we also have

$$\hat{S}_W(\omega) = \left[\frac{P_y(\omega) - P_v(\omega)}{P_y(\omega)}\right] Y(\omega). \tag{9.6}$$

Equation (9.6) illustrates a form of the Wiener recovery method that utilizes a spectral subtraction operation.

To apply Wiener's theory we must make several approximations, which in practice do not limit the usefulness of Wiener's result. When $P_y(\omega)$ and $P_v(\omega)$ are not known they can be estimated from the observed signals. Using the principle of ensemble averaging for stationary signals, the power spectrums of the signal and noise are given by the expected value of the squared-modulous of their respective Fourier transforms. With $E\{\cdot\}$ denoting expectation, we have $P_s(\omega) = E\left\{|S(\omega)|^2\right\}$, $P_v(\omega) = E\left\{|V(\omega)|^2\right\}$ and, consequently, $P_y(\omega) = E\left\{|Y(\omega)|^2\right\}$. Substituting expected values into (9.2) and (9.3), (9.5) gives

$$H_{\mathrm{W}}(\omega) = \frac{E\left\{|Y(\omega)|^2\right\} - E\left\{|V(\omega)|^2\right\}}{E\left\{|Y(\omega)|^2\right\}}. \tag{9.7}$$

When the ensemble averages themselves are unknown we may go one step further. The Fourier transforms of the observed signals can be used as sample estimates of the ensemble averages, leading to

$$H_{\mathrm{W}}(\omega) \cong \frac{|Y(\omega)|^2 - |V(\omega)|^2}{|Y(\omega)|^2} \tag{9.8}$$

as an estimate of the Wiener filter. This form of the Wiener filter serves as the basis for the large majority of spectral-based nose reduction techniques in use today.

3. SPEECH ENHANCEMENT BY SHORT-TIME SPECTRAL MODIFICATION

Wiener filter theory applies to stationary signals and their power spectrums. Speech is, of course, not stationary; its spectral content evolves with time. Further, although in many applications noise source $v(n)$ is accurately modeled as a stationary process, its power spectrum is not known exactly and must be estimated. Under these circumstances Wiener's theory can be applied in a block-processing arrangement using short-time Fourier analysis. In the next several sections we review the short-time Wiener filter method of noise reduction as well as a few of the most commonly used variants.

3.1 SHORT-TIME FOURIER ANALYSIS AND SYNTHESIS

Any time series $x(n)$, stationary or otherwise, can be represented by its short-time Fourier transform (STFT) [5, 8, 9]

$$X(k, m) = \sum_{n=0}^{N-1} h(n)x(m-n)e^{-j2\pi kn/K}, \quad k = 0, \ldots, K-1 \tag{9.9}$$

where m is the time index about which the short-time spectrum is computed, k is the discrete frequency index, $h(n)$ is an analysis window, N dictates the duration over which the transform is computed, and K is the number of frequency bins at which the STFT is computed. For stationary signals the magnitude-squared of the STFT provides a sample estimate of the power spectrum of the underlying random process.

A key motivation for using the STFT in speech enhancement applications is that there exist synthesis formulae by which a time series can be exactly reconstructed from its STFT representation. As a result, if the noise can be eliminated from the STFT of $y(n)$, the signal estimate $\hat{s}(n)$ can be recovered through the appropriate synthesis procedure. A STFT analysis and synthesis structure is a type of subband filter bank, and filter bank theory specifies criteria under which perfect reconstruction is possible [5]. Proper synthesis of the time series from its STFT is key to the performance of any noise reduction method. Improper or ad hoc synthesis results in audible artifacts in the reconstructed speech time series, artifacts which reduce the quality of the very material for which enhancement is desired.

In typical noise reduction applications N in (9.9) falls within the range 32–256 for speech time series sampled at 8 kHz. The number of frequency bins, or subband time series channels, K is often within this same range. Analysis windows $h(n)$ for subband filter banks can be designed for particular characteristics [5]. Alternatively, any of the common data windows (e.g., Hamming) can be used, though such traditional windows place constraints on filter bank structure that must be considered by the designer.

Filter bank theory is not discussed further in this chapter, though the example implementation in Section 5 includes a description of a subband filter bank. The reader is directed to [5] for a complete treatment.

3.2 SHORT-TIME WIENER FILTER

The STFT representation can be used to define a short-time Wiener filter. Replacing the Fourier transforms in (9.8) with their corresponding STFTs yields the short-time Wiener filter

$$H_{\mathrm{W}}(k,m) = \frac{|Y(k,m)|^2 - |V(k,m)|^2}{|Y(k,m)|^2}. \tag{9.10}$$

Replacing the Fourier transforms in (9.4) with their corresponding STFT representations and using (9.10) gives

$$\hat{S}_{\mathrm{W}}(k,m) = \frac{|Y(k,m)|^2 - |V(k,m)|^2}{|Y(k,m)|^2} Y(k,m) \tag{9.11}$$

as the estimate of the STFT of the desired signal. As discussed above, the desired full-band speech time series estimate is recovered from $\hat{S}_{\mathrm{W}}(k,m)$ by using the appropriate synthesis procedure associated with the subband filter bank structure being used.

The short-time Wiener filter method of noise reduction was studied in [12, 13, 14]. Though the Wiener filter was not the first of the spectral modification method to be investigated for noise reduction, its form is basic to nearly all the noise reduction methods investigated over the last forty years.

In any implementation of (9.11), or any other noise reduction method discussed herein, $Y(k, m)$ is computed at every time index m while $V(k, m)$ is computed only for m for which speech is absent. Otherwise, corruption of the noise envelope estimate occurs. When speech is present, $V(k, m)$ must be estimated from past samples. This subject is discussed further in Section 4.

3.3 POWER SUBTRACTION

An alternative estimate of $S(k, m)$ is arrived at by departing from Wiener's theory. $S(\omega)$ can be represented in terms of its magnitude and phase components, namely,

$$S(\omega) = |S(\omega)|e^{j\phi_s(\omega)}. \tag{9.12}$$

Thus, $S(\omega)$ can be estimated if estimates of its magnitude and phase can be found. Consider, first, stationary $s(n)$ and $v(n)$. As above, if the sample functions $|S(\omega)|^2$ and $|V(\omega)|^2$ are used in place of the power spectrums $P_s(\omega)$ and $P_v(\omega)$, (9.3) becomes

$$|S(\omega)|^2 = |Y(\omega)|^2 - |V(\omega)|^2. \tag{9.13}$$

The square root of (9.13) therefore provides an estimate of the signal's magnitude spectrum. Concerning the signal's phase, if the signal-to-noise ratio of the noisy signal is reasonably high, the phase of the noisy signal, $\phi_y(\omega)$, can be used in place of $\phi_s(\omega)$. Using these magnitude and phase estimates (9.12) yields

$$\hat{S}_{\text{PS}}(\omega) = \sqrt{|Y(\omega)|^2 - |V(\omega)|^2}\; e^{j\phi_y(\omega)} \tag{9.14}$$

as the spectrum estimate of the desired signal. Using STFT quantities in place of the power spectrums in (9.14) yields

$$\hat{S}_{\text{PS}}(k, m) = \sqrt{|Y(k, m)|^2 - |V(k, m)|^2}\; e^{j\phi_y(k,m)} \tag{9.15}$$

as the short-time spectrum estimate. The form in (9.14) and (9.15) is referred to as the power subtraction method of noise reduction. Power subtraction was studied in [4, 12, 13, 14, 17].

Note that (9.14) provides a consistent estimate of the magnitude-squared of $S(\omega)$. Assuming the signal and noise are uncorrelated, we have

$$\begin{aligned} E\{|\hat{S}_{\mathrm{PS}}(\omega)|^2\} &= E\{|Y(\omega)|^2\} - E\{|V(\omega)|^2\} \\ &= E\{|S(\omega)|^2\} \\ &= P_s(\omega). \end{aligned} \tag{9.16}$$

Like the Wiener filter, the power subtraction method also can be shown to be optimal from within the estimation theoretic framework, albeit from a different formulation of the problem. Let $S(k, m)$ and $V(k, m)$ be realizations of independent, stationary Gaussian random processes. Also, without loss of generality, assume $S(k, m)$ and $V(k, m)$ are real. If σ_S^2 and σ_V^2 are the variances of the signal and noise STFTs, the probability density function of the observation $Y(k, m)$ given the known signal and noise variances is given by [7, 12]

$$p\left(Y(k, m) \mid \sigma_S^2, \sigma_V^2\right) = \frac{1}{\pi\left(\sigma_S^2 + \sigma_V^2\right)} e^{-\frac{Y^2(k,m)}{\sigma_S^2 + \sigma_V^2}}. \tag{9.17}$$

The maximum likelihood estimate of the signal variance is the estimate $\hat{\sigma}_S^2$ that maximizes the likelihood of the noisy observation occurring. Maximizing (9.17) with respect to the signal variance gives

$$\hat{\sigma}_s^2 = Y^2(k, m) - \sigma_V^2 \tag{9.18}$$

as the best estimate of the signal variance. This estimate suggests the power subtraction estimator (9.15) when variances are replaced by sample approximations, that is, by the corresponding magnitude-squared STFT quantities. Thus, the power subtraction estimator results from the optimum maximum likelihood signal-variance estimator. In comparison, the Wiener estimator results from the optimum minimum mean-squared error estimate of the signal spectrum, or equivalently, the signal time series.

McAulay and Mulpass [13] show that the time domain dual of (9.17), in which $y(n)$ replaces $Y(k, m)$ and the variances are of the signal and noise time series, also yields the power subtraction form when maximized with respect to the unknown signal variance.

3.4 MAGNITUDE SUBTRACTION

Yet another estimate of $S(k, m)$ is suggested by the magnitude-only form of (9.3). Consider

$$|\hat{S}(k, m)| = |Y(k, m)| - |V(k, m)| \tag{9.19}$$

as an estimate of the magnitude $S(k, m)$. Using (9.19) in (9.12), and appending the phase of the noisy signal as done in Section 3.3, gives

$$\hat{S}_{\mathrm{MS}}(k, m) = \big[|Y(k, m)| - |V(k, m)|\big] e^{j\phi_y(k,m)} \tag{9.20}$$

as the short-time spectrum estimate of the signal. The form (9.20) is referred to as the magnitude subtraction method of noise reduction. Like power subtraction, magnitude subtraction is also a consistent estimator of $|S(\omega)|^2$, assuming the signal and noise are stationary and independent.

The magnitude subtraction method of noise reduction was popularized by Boll [11], but was suggested by Weiss et al. [15] and even earlier by Schroeder [1, 2, 4].

3.5 PARAMETRIC WIENER FILTERING

The Wiener, power subtraction and magnitude subtraction schemes are closely related. Several authors have exploited this fact to define a general class of noise reduction methods. To see this, first note that the power subtraction estimate (9.15) can be rewritten as

$$\begin{aligned} \hat{S}_{\mathrm{PS}}(k, m) &= \sqrt{|Y(k, m)|^2 - |V(k, m)|^2}\; e^{j\phi_y(k,m)} \\ &= H_{\mathrm{PS}}(k, m) Y(k, m), \end{aligned} \tag{9.21}$$

where

$$H_{\mathrm{PS}}(k, m) = \left[\frac{|Y(k, m)|^2 - |V(k, m)|^2}{|Y(k, m)|^2}\right]^{1/2}. \tag{9.22}$$

Similarly, the magnitude subtraction estimate (9.20) can be rewritten as

$$\begin{aligned} \hat{S}_{\mathrm{MS}}(k, m) &= \big[|Y(k, m)| - |V(k, m)|\big] e^{j\phi_y(k,m)} \\ &= H_{\mathrm{MS}}(k, m) Y(k, m), \end{aligned} \tag{9.23}$$

where

$$H_{\mathrm{MS}}(k, m) = \frac{|Y(k, m)| - |V(k, m)|}{|Y(k, m)|}. \tag{9.24}$$

Both forms result from the generalized estimate

$$\hat{S}_{\mathrm{G}}(k, m) = H_{\mathrm{G}}(k, m) Y(k, m), \tag{9.25}$$

where

$$H_{\mathrm{G}}(k, m) = \left[1 - \left(\frac{|V(k, m)|}{|Y(k, m)|}\right)^{\gamma}\right]^{\beta}. \tag{9.26}$$

This has been referred to as parametric Wiener filtering [14] or parametric spectral subtraction [17]. Power subtraction results from using $(\gamma, \beta) = (2, 1/2)$, magnitude subtraction from $(\gamma, \beta) = (1, 1)$, and the Wiener estimate from $(\gamma, \beta) = (2, 1)$.

For general (γ, β) the parametric form (9.25) has not been demonstrated to satisfy optimality criteria, though this fact does not in any way diminish the usefulness of the generalized form for choosing noise reduction gain formulae.

Figure 9.1 shows a plot of the Wiener (9.10), power subtraction (9.22) and magnitude subtraction (9.24) noise reduction gain functions as a function of *a priori* SNR. The *a priori* SNR is the ratio σ_s^2/σ_v^2. Also, for purposes of evaluating each gain function, the magnitude-squared sample spectrums have been replaced by their respective variances. The curves show the attenuation of each gain function as input SNR decreases. Note that all three gain functions provide an attenuation of no greater than 6 dB at an input SNR of 0 dB. That is, in any subband k in which the signal and noise powers are equal, the contribution of that subband to the reconstructed speech time series is no less than half its input amplitude. A fourth gain curve appearing in the figure, namely, that of a voice-activity-detection-based method, will be discussed in Section 5.

Gain curves such as Fig. 9.1 provide some insight into the nature of a given noise reduction algorithm, but provide little indication of the subjective speech quality of the resulting noise-reduced signal. More important to the subjective quality of a noise reduction algorithm is the manner in which the speech and noise envelopes are estimated; this subject is discussed in Section 4.

3.6 REVIEW AND DISCUSSION

3.6.1 Schroeder's Noise Reduction Device. The first use of spectral gain modification methods in speech noise reduction is described in a little-known U.S. patent issued in 1965 to M. R. Schroeder [1]–[4], who at the time was working for AT&T Bell Laboratories. A block diagram of Schroeder's noise reduction system is shown in Fig. 9.2. This diagram is modified from its original form for this discussion, and incorporates elements presented by Schroeder in a related 1968 patent [2] and also by Schroeder's colleagues in a subsequent published work [4].

Schroeder's system was a purely analog implementation of spectral magnitude subtraction. As shown in the figure, a bank of bandpass filters separates the noisy signal into K different frequency bands. The bandwidth of each filter is about 300 Hz. Ten individual filters therefore cover the 300 to 3300 Hz range necessary for telephony grade speech applications. The noise-reduction processing performed in each band is identical. First, the output of each filter bank is rectified and averaged using a low-pass filter to produce a short-time estimate of the noisy speech envelope for the band. The lowpass filter has a

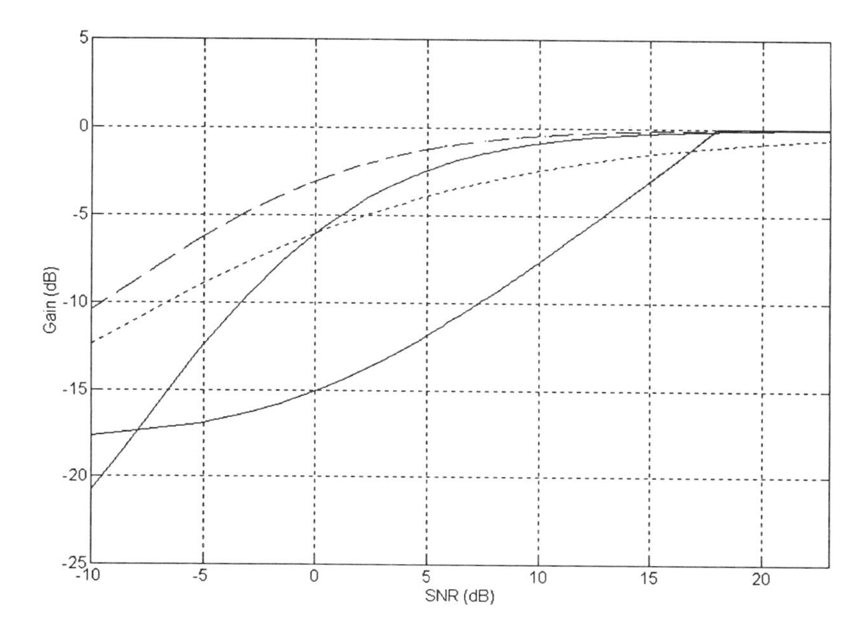

Figure 9.1 Gain functions for the Wiener (upper-most solid), spectral power subtraction (dash), spectral magnitude subtraction (large dash), and *a posteriori* SNR voice activity detection (lower-most solid) methods of noise reduction as a function of *a priori* input signal-to-noise ratio.

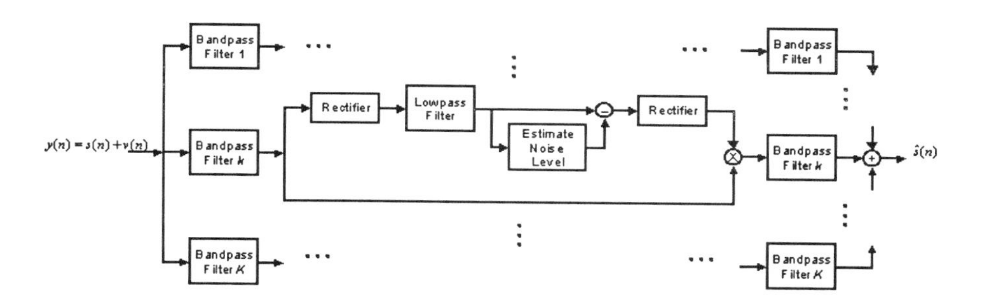

Figure 9.2 Schroeder's noise reduction system. After M. R. Schroeder [1, 2].

cutoff of between 0 and 10 Hz. The noisy speech envelope is then subtracted from an estimate of the noise-only envelope. To estimate the noise, the noise level estimator uses a series of resistors, capacitors and diodes to produce a running estimate of the minima of the noisy speech envelope. The decay time of this noise estimator is instantaneous while the rise time is very large, on the order of seconds. Between speech utterances the noisy speech envelope contains only noise, and the noise level estimator quickly decays to meet the level of the noise. During utterances the noise level estimate changes very little. Thus, the output of the subtraction block is an estimate of the noise-free signal envelope for the band, or $|\hat{S}(k, m)|$ in the current notation. A second rectification is performed on the output to accommodate negative results from the difference node (negative estimates are simply set to zero). Finally, the noise-free signal envelope is used as a multiplier with the unmodified output of the bandpass filter for the band, and the result is summed with the results from all bands to form the reconstructed full-band time series $\hat{s}(n)$.

It is interesting to note that Schroeder's implementation was a purely analog one, employing bandpass filters and rectification and averaging circuitry. Other aspects of the design preceded presentations by later authors. By rectifying the output of the signal envelope estimator, Schroeder's system could correct for negative estimates.

3.6.2 Literature Review. Following Schroeder's work, it was not until the mid-1970's that interest in noise reduction systems grew, presumably because of the availability of digital computers and analog processors that could be controlled by digital decision logic. With the unification of digital multirate processing theory in the late 1970's and 1980's came the realization that Wiener-filter processing, among other operations, could be accomplished efficiently using digital subband architectures [5, 8, 9].

Digital noise reduction processing for speech enhancement was popularized by the technique of spectral subtraction. This renewed interest appears to have been sparked in a 1974 paper by Weiss, Aschkenasy, and Parsons [15]. Their paper describes a "spectrum shaping" method that used amplitude clipping, or gating, in filter banks to remove low-level excitation, presumably noise. A few years later, Boll [11], in an often-sited reference, was apparently the first to reintroduce the spectral subtraction method that Schroeder had identified nearly 20 years earlier. Boll was perhaps the first to cast the magnitude subtraction method in the framework of digital short-time Fourier analysis, which had earlier been under development by, among others, Allen [9] and Protnoff [8]. Shortly after, McAulay and Malpass [13] presented one of the first treatments of the spectral subtraction method from within a framework of optimal estimation, which included Wiener filter theory. They described a class of spectral subtraction estimators, including power subtraction and magnitude subtraction, from within an estimation theoretic framework [7]. Coincident with [13], Lim and Oppenheim [14] presented one of the first comprehensive treatments of methods of speech enhancement and noise reduction. The spectral subtraction methods are discussed, also within a framework of optimal estimation, and are compared to other methods of speech enhancement.

Lim and Oppenheim recognize Weiss, Aschkenasy, and Parsons as originators of the spectral subtraction technique, as Schroeder's work was likely unknown to them at the time. Also in 1980, Sondhi, Schmidt, and Rabiner [4] published results from a series of implementation studies that grew from Schroeder's work in the 1960's. This work was the first published reference of Schroeder's work, other than the patents [1, 2], but likely was not widely know itself because it was published in the Bell System Technical Journal.

Yet another spectral noise reduction method has been proposed by Ephraim and Malah [12]. They derive a related spectral noise reduction method based on optimum short-time Fourier amplitude estimation. Differing from a variance estimator, that is, power subtraction, the amplitude estimator is optimum in the sense of providing the best minimum mean-squared error estimate of the spectral amplitude. The Fourier amplitude estimator converges to the Wiener estimator at high input signal-to-noise ratios.

Recently, noise reduction processing incorporating psychoacoustic perceptual models has been proposed. Tsoukalas and Mourjopoulos [24] present a spectral gain modification technique that uses perceptual models to suppress only those components of the noise that are above audibility thresholds. These thresholds are dynamic, changing with the changing spectral character of the speech itself. A reported 40% gain in intelligibility can be achieved when precise information about the noise power level is known.

3.6.3 Musical Noise. Much of the work in noise reduction in the last 20 years has been directed toward implementation issues associated with spectral subtraction methods, particularly in understanding and eliminating a host of processing artifacts commonly referred to as musical noise. Musical noise is a processing artifact that has plagued all spectral modification methods. This artifact is perceived by many as the sound made by an ensemble of low-amplitude tonal components, the frequencies of which are changing rapidly over time. The amplitude of these components is usually small, on the order of the noise power itself. First, because the STFTs in (9.26) are computed over short intervals of time, and because the noise envelope estimate is made only during periods of silence while the noisy signal envelope estimate is always made, the difference in (9.26) can actually be negative. In such cases the common approach is set the difference to zero or to invert the sign of the difference and use the result. Such harsh action creates sudden discontinuities in the trajectories of spectral amplitudes, and this induces the artifact. Second, artifacts are induced by improper synthesis of the full-band time series. Ad hoc "FFT processing" results in filter banks that do not possess the quality of perfect reconstruction and, moreover, cause aliasing of the subband time series in both frequency and time.

The works of a variety of authors have shown that artifacts such as musical noise can be nearly eliminated by taking proper action in chiefly three aspects of the processing. First, careful design of the filter bank is required. Second, the proper use of time-averaging techniques, in conjunction with appropriate decision criteria, is necessary to produce stable estimates and obviate the need for rectification following spectral subtraction. Third, the complementary technique of augmenting the traditional gain function with a soft-decision, voice-activity-detection (VAD) statistic has proven extremely successful. This later technique supplants the traditional gain-based noise reduction form in (9.25) with

$$\hat{S}_G(k, m) = H_G(k, m) P\big(H_1 \mid Y(k, m)\big) Y(k, m), \tag{9.27}$$

where H_1 denotes the hypothesis that the signal is present in the observation and $P\big(H_1 \mid Y(k, m)\big)$ is the probability that the signal is present conditioned on the observation. $P\big(H_1 \mid Y(k, m)\big)$ acts as a gain function itself. At very low input SNRs, $P\big(H_1 \mid Y(k, m)\big)$ further suppresses fluctuations in the traditional gain function, fluctuations caused by the statistical volatility of the signal and noise envelope estimates.

Boll [11] was one of the first to augment a traditional gain function (magnitude subtraction) with a VAD. Boll's method integrates the estimated *a priori* SNR[1] across all frequency bins and uses the resulting scaler in a binary VAD which, if satisfied, applies a fixed additional attenuation to the right-hand side of (9.25). McAulay and Malpass [13] derived a "soft-decision" VAD from within the detection theoretic framework [7]. Unlike a binary VAD, a soft VAD varies continuously within $(0, 1)$ as a function of $Y(k, m)$. Later, Ephraim and Malah [12] incorporated a measure of the "signal presence uncertainty" into their Fourier amplitude estimator. Clearly, $H_G(\cdot)$ and $P(\cdot)$ in (9.27) can be combined into a single gain function; this idea is pursued in the implementation presented in Section 5.

3.6.4 A Word About Phase Augmentation. It is often noted that the practice of phase augmentation – that is, using the phase of the noisy signal as an estimate of the signal's phase – is acceptable because the ear is relatively insensitive to phase corruption. More is known, however. Within the context of noise reduction, Vary [20] has shown that as long as the SNR is at least 6 dB in any subband k for which the gain function is near unity, the resulting distortion is generally imperceptible. In other words, because the noisy phase contributes to $\hat{S}(k, m)$ only at those k for which the input SNR is positive, the phase of $\hat{S}(k, m)$ itself is essentially noise-free. Additionally, Ephraim and Malah [12] show that the noisy phase is a good choice because it has the property of not corrupting the envelope of the optimum short-time Fourier amplitude estimator that forms the basis of their method.

4. AVERAGING TECHNIQUES FOR ENVELOPE ESTIMATION

Proper estimation of both the noisy signal envelope and noise envelope is paramount to the performance of any noise reduction technique. Improper estimation of either envelope will result in an unacceptably high level of audible processing artifacts, such as musical noise. Up to this point, instantaneous envelope quantities have been used in the gain formulae of the noise reduction methods discussed. To combat musical noise, however, averaged, or smoothed, envelopes are used. In the next few sections a variety of commonly used time averaging techniques are reviewed and discussed in the context of noise reduction.

4.1 MOVING AVERAGE

One of the simplest ways to improve the stability of the noise estimate is to replace it with an arithmetic average computed over its recent past. For each time index m, $|V(k,m)|$ is replace by a smoothed version $\overline{V}(k,m)$ given by

$$\overline{V}(k,m) = \frac{1}{M}\sum_{l=0}^{M-1} |V(k,m-l)|, \tag{9.28}$$

where M is the number of samples used in the average. The over-bar notation in (9.28) shall be used throughout to denote any averaged magnitude quantity, regardless of the averaging technique actually used. In the computation of (9.28) the reader should note that $V(k,m)$ is simply $Y(k,m)$ under the assumption that speech is absent.

Because of statistical variation in the noisy signal, it is also fortuitous to use a smoothed version, $\overline{Y}(k,m)$, in place of $|Y(k,m)|$ in the gain formulae. Though the amount of smoothing necessary depends upon several aspects of the implementation, such as the subband filter bank structure, $\overline{Y}(k,m)$ is generally smoothed much less than $\overline{V}(k,m)$. This is because variations in $\overline{Y}(k,m)$ and $\overline{V}(k,m)$ induce different artifacts in the signal estimate $\hat{S}_G(k,m)$. Referring to (9.26), positive fluctuations in $|V(k,m)|$ can cause the difference in (9.26) to be negative, requiring rectification. Consequently, it is beneficial to average the noise magnitude over long intervals (assuming stationary noise). Similarly, positive fluctuations in $|Y(k,m)|$ resulting from the statistical variability of the noise component reduce the effectiveness of noise reduction because $H_G(k,m)$ is larger for larger $|Y(k,m)|$. Thus, some smoothing of the noisy speech envelope is also beneficial. Excessive smoothing of the noisy speech envelope, however, degrades the speech quality of the signal estimate because $S(k,m)$ is not stationary. Excessive smoothing of $|Y(k,m)|$, and therefore $H_G(k,m)$, disperses $H_G(k,m)$ to the point that it is no longer well matched to the speech component of the noisy observation $Y(k,m)$.

Early on, Boll [11] described the use of arithmetic averaging to reduce the presence of artifacts. For the STFT filter bank implementation used, Boll applied the same sized average to both the noise and noisy speech envelopes (about 38 ms). McAulay and Malpass [13] also discuss arithmetic averaging.

4.2 SINGLE-POLE RECURSION

The arithmetic average requires an M-length history of the data. Further, each sample in the average receives the same weight, although it is trivial to include a tapered weighting window in (9.28) if desired. An alternative to arithmetic averaging is recursive averaging. Using a single-pole recursive average the noise envelope estimate becomes

$$\overline{V}(k, m) = \alpha \overline{V}(k, m-1) + (1-\alpha)|V(k, m)|, \tag{9.29}$$

where α, $0 < \alpha < 1$, is the coefficient of smoothing. Equation (9.29) defines a first-order lowpass filter and so the variance of $\overline{V}(k, m)$ is less than the variance of $|V(k, m)|$ itself.

Recursive averaging has been by far the most popular method of averaging used in the spectral noise reduction methods. This is due to its simplicity and efficiency, requiring only a single memory location for state variable storage. Also, because the impulse response corresponding to (29) decays as α^n, $n \geq 0$, the recursive average weights the recent past more heavily than the distant past. This characteristic has been found beneficial to noise reduction processing. Indeed, the first investigations into the use of averaging techniques employed recursive averaging. Schroeder's method (Fig. 9.2) incorporates an analog version of (9.29) in the signal path common to both the noise and noisy signal envelope estimators. Sondhi et al. [4] experimented with variations on the recursive average for both the power subtraction and magnitude subtraction methods. For computing $\overline{V}(k, m)$, cutoff frequencies of between 10 and 30 Hz, or greater, were found to be effective [4]. For estimating $\overline{Y}(k, m)$, cutoff frequencies of between 1 and 10 Hz were sufficient. Other proponents of the recursive average include McAulay and Malpass [13], Ephraim and Malah [12], and Cappé [16].

4.3 TWO-SIDED SINGLE-POLE RECURSION

An alternative to the classic single-pole recursive filter involves choosing α in (9.29) based upon the magnitude of $|V(k, m)|$ relative to $\overline{V}(k, m-1)$. Consider the so-called two-sided single-pole recursion in which α in (9.29) is given by

$$\alpha = \begin{cases} \alpha_{\mathrm{a}}, & \text{if } |V(k, m)| \geq \overline{V}(k, m-1) \\ \alpha_{\mathrm{d}}, & \text{if } |V(k, m)| < \overline{V}(k, m-1) \end{cases}, \tag{9.30}$$

where α_a is the "attack" coefficient and α_d is the "decay" coefficient. The two-sided recursive average employs two different filter response times, depending on whether the input is increasing or decreasing in magnitude relative to the current average. This property can be advantageous. Consider, first, the computation of $\overline{V}(k, m)$. Although it is desirable to update $V(k, m)$ only when speech is absent, it is not always possible to determine when speech is present and when it is not. If speech or other transient phenomena are present in $Y(k, m)$ and (9.29) is updated, $\overline{V}(k, m)$ will become corrupt. This problem can be reduced by choosing $\alpha_a > \alpha_d$. In this case increases in $|V(k, m)|$ change $\overline{V}(k, m)$ much less than decreases in $|V(k, m)|$, and therefore $\overline{V}(k, m)$ is less perturbed by transient phenomenon that are not components of the stationary noise.

The two-sided single-pole recursion can also be used to update $\overline{Y}(k, m)$. For this purpose it is common to choose $\alpha_a < \alpha_d$, in which case $\overline{Y}(k, m)$ is more responsive to the sudden onset of speech energy than to the end of an utterance when speech energy decays. This characteristic improves the response of $H_G(k, m)$ in (9.26) to the onset of speech.

Etter and Moschytz [17] and Diethorn [19] used the two-sided single-pole recursion in the context of noise reduction; it is also used in the implementation discussed in Section 5. The technique has it origins in speakerphone technology, where it is used for voice activity detection; for example, see [26].

As a variation on (9.29)–(9.30), Etter and Moschytz [17] also proposed using

$$\overline{V}(k, m) = \begin{cases} \alpha_a \overline{V}(k, m-1), & \text{if } a_a \overline{V}(k, m-1) < |V(k, m)| \\ \alpha_d \overline{V}(k, m-1), & \text{if } a_d \overline{V}(k, m-1) > |V(k, m)| \\ |V(k, m)|, & \text{otherwise} \end{cases}, \quad (9.31)$$

where $\alpha_a > 1$ and $0 < \alpha_d < 1$. In subjective listening tests, this so-called two-slope limitation filter reportedly performs better than (9.29)–(9.30) for some material [17].

4.4 NONLINEAR DATA PROCESSING

To improve the stability of the noise estimate further, Sondhi et al. [4] also experimented with a scheme to post-process the noise envelope $\overline{V}(k, m)$ based upon a short-term histogram of its past values. This early rank ordering technique provided a means to prune wild points from the noise envelope estimate. Median filtering and other rank-order statistical filtering can be used to post-process $\overline{V}(k, m)$ and $\overline{Y}(k, m)$ following any of the averaging techniques described above; see [10] for an early reference on such methods. More recently, Plante et al. [25] have described a noise reduction method using reassignment methods to replace envelope estimates that are deemed erroneous. In general,

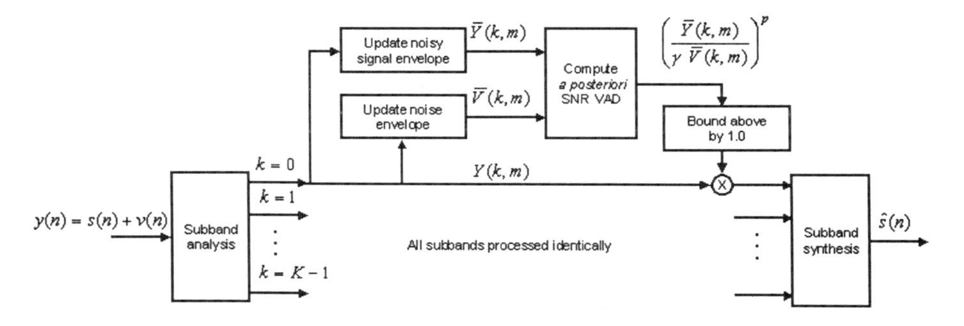

Figure 9.3 Noise reduction system based on *a posteriori* SNR voice activity detection.

nonlinear data processing techniques can provide improved noise reduction performance, although the behavior of such methods is sometimes difficult to analyze analytically.

5. EXAMPLE IMPLEMENTATION

Noise reduction systems need not be complicated to produce acceptable results, as the implementation described in this section demonstrates. The method described is a variation on that first presented in [19].

Figure 9.3 shows a signal-flow diagram of the noise reduction system. The algorithm consists of four key processes: subband analysis, envelope estimation, gain computation, and subband synthesis. Each of these components is described below.

5.1 SUBBAND FILTER BANK ARCHITECTURE

The subband architecture implements a perfect reconstruction filter bank using the uniform discrete Fourier transform (DFT) filter bank method [5]. This filter bank is one of a sub-class of so-called polyphase filter banks, but is somewhat simpler in computational structure.

The subband filter bank implements an overlap-add process. At the start of each processing epoch, a block of L new time-series samples is shifted into an N-sample shift register. Here, $L = 16$ and $N = 64$. The shift-register data are multiplied by a length-N analysis window (the filter bank's prototype FIR filter) and transformed via an N-point DFT. Each frequency bin output from the DFT represents one new complex time-series sample for the subband frequency range corresponding to that bin. The subband sampling rate is equal to the full-band sampling rate divided by L. The bandwidth of each subband is the ratio of the full-band sampling rate to N. Thus, for 8 kHz sampling, the subband sampling rate and bandwidth are, respectively, 500 Hz and 125 Hz (indicative of an oversampled-by-4 filter bank architecture). Following subband analysis, the vector of subband time series is presented to the envelope estimators. Next, the noise reduction gain is computed. To reconstruct the noise-reduced full-band time series, the subband synthesizer first transforms the gain-modified vector of subband time series using an inverse DFT. The synthesis window (same as analysis window) is applied, and the result is overlapped with, and added to, the contents of an N-sample output accumulator. Last, a block of L processed samples is produced at the output of the synthesizer.

The prototype analysis and synthesis window, input-output block size L and transform block size N are chosen to maintain these properties of the filter bank:

- no time-domain aliasing at the subband level,
- no frequency domain aliasing at the subband level, and
- perfect reconstruction (unity transfer function) when analysis is followed directly by synthesis (no intermediate processing).

5.2 A-POSTERIORI-SNR VOICE ACTIVITY DETECTOR

The gain function of the noise reduction algorithm is based on the idea of a composite gain function and soft-acting voice activity detector as discussed in Section 3.6.3.

5.2.1 Envelope Computations. As shown in Fig. 9.3, the time series output from each subband k of the analysis filter bank is used to update estimates of the noisy speech and noise-only envelopes, respectively, $\overline{Y}(k, m)$ and $\overline{V}(k, m)$. These estimates are generated using the two-sided single-pole recursion described in Section 4.3. Specifically,

$$\overline{Y}(k, m) = \beta \overline{Y}(k, m-1) + (1-\beta)|Y(k, m)|, \tag{9.32}$$

and

$$\overline{V}(k, m) = \alpha \overline{V}(k, m-1) + (1-\alpha)|Y(k, m)|, \tag{9.33}$$

where β takes on attack and decay constants of about 1 ms and 10 ms, respectively, and α takes on attack and decay time constants of about about 4 sec. and 1 ms. Note that, in comparison with (9.29), (9.33) uses $Y(k, m)$ in place of $V(k, m)$. This substitution is possible because the long attack time (4 sec.) of the noise envelope estimate is used in place of the logic that would otherwise be needed to discern the speech/no-speech condition. This approach further simplifies the implementation.

5.2.2 Gain Computation. The envelope estimates are used to compute a gain function that incorporates a type of voice activity likelihood function.

This function consists of the *a posteriori* SNR normalized by the threshold of speech activity detection, γ. Specifically, the noise reduction formula is

$$\hat{S}(k, m) = H(k, m)Y(k, m), \tag{9.34}$$

where gain function $H(k, m)$ is given by

$$H(k, m) = \min\left[1, \left(\frac{\overline{Y}(k, m)}{\gamma \overline{V}(k, m)}\right)^p\right]. \tag{9.35}$$

Threshold γ specifies the *a posteriori* SNR level at which the certainty of speech is declared and p, a positive integer, is the gain expansion factor. Typical values for the detection threshold fall in the range $5 \leq \gamma \leq 20$, though the (subjectively) best value depends on the characteristics of the filter bank architecture and the time constants used to compute the envelope estimates, among other things. The expansion factor p controls the rate of decay of the gain function for *a posteriori* SNRs below unity. With $p = 1$, for example, the gain decays linearly with *a posteriori* SNR. Factor p also governs the amount of noise reduction possible by controlling the lower bound of (9.35); larger p results in a smaller lower bound. The $\min(\cdot)$ operator insures the gain reaches a value no greater than unity.

Looking at (9.35), subband time series whose *a posteriori* SNR exceeds the speech detection threshold are passed to the synthesis bank with unity gain. Subband time series whose *a posteriori* SNR is less than the threshold are passed to the synthesis bank with a gain that is proportional to the SNR raised to the power p.

Note in particular that (9.35) does not involve a spectral subtraction operation. This has the benefit of circumventing the problem of a negative argument, as occurs with the parametric form in (9.26). A disadvantage of (9.35) is that the gain function, and therefore noise reduction level, is bounded below by the reciprocal of the detection threshold. That is, as the *a priori* SNR goes to zero we have (for $p = 1$)

$$\begin{aligned} \frac{|Y(k, m)|}{\gamma |V(k, m)|} &= \frac{|S(k, m) + V(k, m)|}{\gamma |V(k, m)|} \\ &\approx \frac{|V(k, m)|}{\gamma |V(k, m)|} \\ &= \frac{1}{\gamma}. \end{aligned} \tag{9.36}$$

For example, with $\gamma = 10$ the system provides no more than 20 dB of noise reduction.

A variation on the above technique incorporates for each subband k both the per-band, or narrowband, normalized *a posteriori* SNR and a k-wise arithmetic average of the *a posteriori* SNRs from neighboring bands. This narrowband-broadband hybrid gain function can provide improved noise reduction performance for wideband speech utterances, such as fricatives. The reader is referred to [19] for more information.

5.3 EXAMPLE

The time series and spectrogram data in Figs. 9.4, 9.5, and 9.6 show the results of processing a noisy speech sample using the subband noise reduction method presented above. For this example $\gamma = 8$ and expansion factor $p = 1$, resulting in a minimum gain in (9.35) of -18 dB. The lower-most solid line in Fig. 9.1 shows the gain function (9.35) in comparison with the Wiener, magnitude subtraction and power subtraction gain functions.

The upper trace in Fig. 9.4 shows a segment of raw (unprocessed) time series for a series of short utterances (digit counting) recorded in an automobile traveling at highway speeds. The speech was recorded from the microphone channel of a wireless-phone handset and later digitized at an 8 kHz sampling rate. The lower trace in Fig. 9.4 shows the corresponding noise-reduced time series produced by the noise reduction algorithm. Figure 9.5 shows spectrograms corresponding to the time series in Fig. 9.4. The spectrograms show, at least visibly, that the noise reduction method introduces no noticeable distortion. Figure 9.6 shows averaged background noise power spectra for the raw and noise-reduced time series. These spectrums were computed for the [10 12] second interval in Fig. 9.4. As can be seen, the noise floor of the processed time series is about 18 dB below that of the raw time series uniformly across the speech band.

6. CONCLUSION

The subject of noise reduction for speech enhancement is a mature one with a 40-year history in the field of telecommunication. The majority of research has focused on the class of noise reduction methods incorporating the technique of short-time spectral modification. These methods are based upon subband filter bank processing architectures, are relatively simple to implement and can provide significant gains to the subjective quality of noisy speech. The earliest of these methods was developed in 1960 by researchers at Bell Laboratories.

Noise reduction processing has its roots in classical Wiener filter theory. Reviewed in this chapter were the most commonly used noise reduction formulations, including the short-time Wiener filter, spectral magnitude subtraction, spectral power subtraction, and the generalized parametric Wiener filter. When implemented digitally, these methods frequently suffer from the presence of

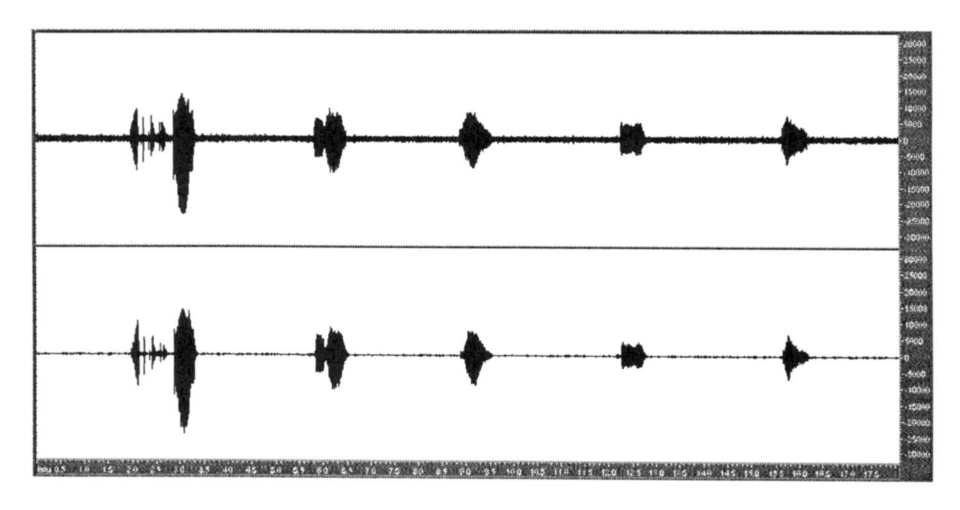

Figure 9.4 Speech time series for the noise reduction example. Original (top) and noise-reduced (bottom) time series. Samples are 18 seconds in duration. Abscissa shows 0.5 second graduation between tic mark labels.

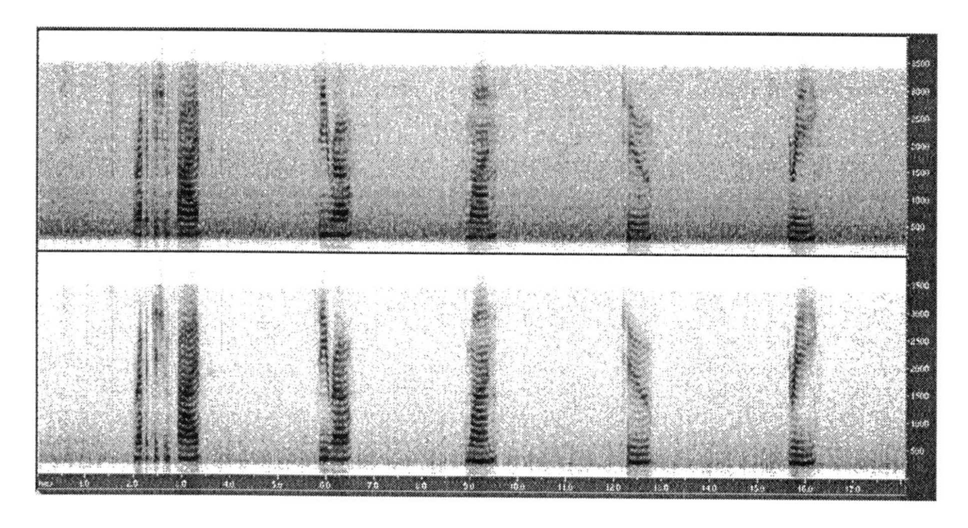

Figure 9.5 Spectrograms corresponding to speech time series in Fig. 9.3. Original (top) and noise-reduced (bottom) samples. Spectrograms computed with a display range of 120 dB. Abscissa shows 1 second graduation between tic mark labels; ordinate, 500 Hz.

processing artifacts, a phenomenon known as musical noise. The origins of musical noise were reviewed, as were approaches to combating the problem. The subject of speech envelope estimation was presented in detail and several averaging techniques for computing envelope estimates were reviewed. A

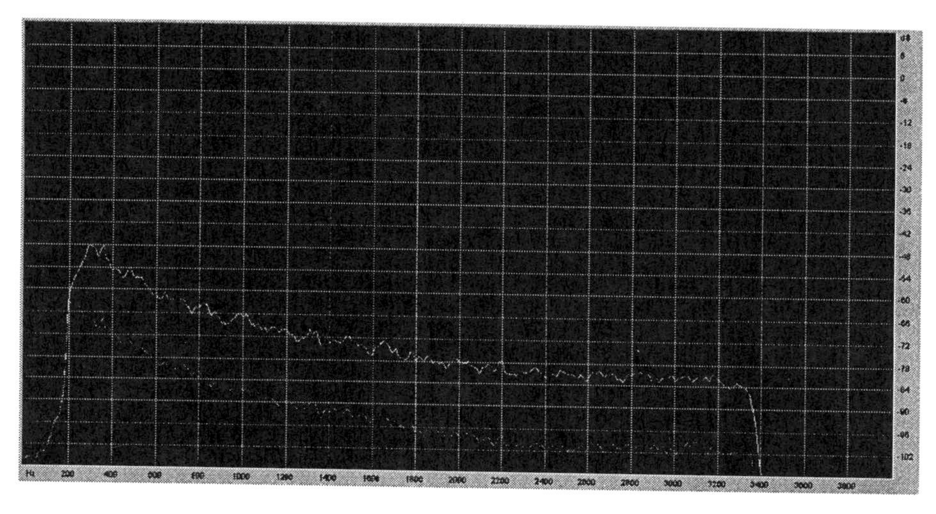

Figure 9.6 Noisy (top) and noise-reduced (bottom) power spectrums corresponding to the time series in Fig. 9.3. Block-averaged power spectrums computed over the interval [10 12] seconds. Abscissa shows 200 Hz graduation between tic mark labels; ordinate, 6 dB.

low-complexity noise reduction algorithm was presented and demonstrated by example.

Notes

1. The estimated instantaneous *a priori* SNR is the ratio $|\hat{S}(k,m)|^2/|V(k,m)|^2$.

References

[1] M. R. Schroeder, U.S. Patent No. 3,180,936, filed Dec. 1, 1960, issued Apr. 27, 1965.

[2] M. R. Schroeder, U.S. Patent No. 3,403,224, filed May 28, 1965, issued Sep. 24, 1968.

[3] M. M. Sondhi and S. Sievers, AT&T Bell Laboratories Internal Report (unpublished), Dec. 1964.

[4] M. M. Sondhi, C. E. Schmidt, and L. R. Rabiner, "Improving the quality of a noisy speech signal," *Bell Syst. Techn. J.*, vol. 60, Oct. 1981.

[5] R. E. Crochiere and L. R. Rabiner, *Multirate Digital Signal Processing*. Englewood Cliffs, NJ: Prentice-Hall, 1983.

[6] N. Wiener, *Extrapolation, Interpolation and Smoothing of Stationary Time Series with Engineering Applications*. New York: Wiley, 1949.

[7] H. L. Van Trees, *Detection, Estimation, and Modulation Theory, Part I*. New York: John Wiley & Sons, 1968.

[8] M. R. Portnoff, "Time-frequency representation of digital signals and systems based on short-time Fourier analysis," *IEEE Trans. Acoust. Speech Signal Process.*, vol. ASSP-28, pp. 55-69, Feb. 1980.

[9] J. B. Allen, "Short-time spectral analysis, synthesis and modification by discrete Fourier transform," *IEEE Trans. Acoust. Speech Signal Process.*, vol. ASSP-25, pp. 235-238, June 1977.

[10] L. R. Rabiner, M. R. Sambur, and C. E. Schmidt, "Applications of a nonlinear smoothing algorithm to speech processing," *IEEE Trans. Acoust. Speech Signal Process.*, vol. ASSP-23, Dec. 1975.

[11] S. F. Boll, "Suppression of acoustic noise in speech using spectral subtraction," *IEEE Trans. Acoust., Speech, Signal Proc.*, vol. ASSP-27, Apr. 1979.

[12] Y. Ephraim and D. Malah, "Speech enhancement using a minimum mean-square error short-time spectral amplitude estimator," *IEEE Trans. Acoust., Speech, Signal Process.*, vol. ASSP-32, Dec. 1984.

[13] R. J. McAulay and M. L. Malpass, "Speech enhancement using a soft-decision noise suppression filter," *IEEE Trans. Acoust., Speech, Signal Process.*, vol. ASSP-28, Apr. 1980.

[14] J. S. Lim and A. V. Oppenheim, "Enhancement and bandwidth compression of noisy speech," *Proc. of the IEEE*, vol. 67, Dec. 1979.

[15] M. R. Weiss, E. Aschkenasy, and T. W. Parsons, "Processing speech signals to attenuate interference," in *Proc. IEEE Symposium on Speech Recognition*, Carnegie-Mellon Univ., Apr. 15-19, 1974, pp. 292-295.

[16] O. Cappé, "Elimination of the musical noise phenomenon with the Ephraim and Malah noise suppressor," *IEEE Trans. Speech Audio Process.*, vol. 2, Apr. 1994.

[17] W. Etter and G. S. Moschytz, "Noise reduction by noise-adaptive spectral magnitude expansion," *J. Audio Eng. Soc.*, vol. 42, May 1994.

[18] B. M. Helf and P. L. Chu, "Reduction of background noise for speech enhancement," U.S. Patent No. 5,550,924, Mar. 13, 1995.

[19] E. J. Diethorn, "A subband noise-reduction method for enhancing speech in telephony & teleconferencing," in *IEEE Workshop on Applications of Signal Processing to Audio and Acoustics*, Mohonk Mountain House, New Paltz, NY, Oct. 19-22, 1997.

[20] P. Vary, "Noise suppression by spectral magnitude estimation – mechanism and theoretical limits," *Signal Processing*, vol. 8, pp. 387-400, July 1985.

[21] Y. Ephraim, D. Malah, and B. H. Juang, "On the application of hidden Markov models for enhancing noisy speech," *IEEE Trans. Acoust., Speech, Signal Process.*, vol. ASSP-37, pp. 1846-1856, Dec. 1989.

[22] H. Sameti, H. Sheikhzadeh, and Li Deng, "HMM-based strategies for enhancement of speech signals embedded in nonstationary noise," *IEEE Trans. Speech Audio Process.*, vol. 6, Sep. 1998.

[23] B. L. Sim, Y. C. Tong, and J. S. Chang, "A parametric formulation of the generalized spectral subtraction method," *IEEE Trans. Speech Audio Process.*, vol. 6, July 1998.

[24] D. E. Tsoukalas and J. N. Mourjopoulos, "Speech enhancement based on audible noise suppression," *IEEE Trans. Speech Audio Process.*, vol. 5, Nov. 1997.

[25] F. Plante, G. Meyer, and W. A. Ainsworth, "Improvement of speech spectrogram accuracy by the method of reassignment," *IEEE Trans. Speech Audio Process.*, vol. 6, May 1998.

[26] R. H. Erving, W. A. Ford, and R. Miller, U.S. Patent No. 5,007,046, filed Dec. 28, 1988, issued Apr. 9, 1991.

IV

MICROPHONE ARRAYS

Chapter 10

SUPERDIRECTIONAL MICROPHONE ARRAYS

Gary W. Elko
Bell Labs, Lucent Technologies
gwe@bell-labs.com

Abstract Noise and reverberation can seriously degrade both the microphone reception and the loudspeaker transmission of speech signals in hands-free telecommunication. Directional loudspeakers and microphone arrays can be effective in combating these problems. This chapter covers the design and implementation of differential arrays that are small compared to the acoustic wavelength. Differential arrays are therefore also superdirectional arrays since their directivity is higher than that of a uniformly summed array with the same geometry. Aside from the small size, another beneficial feature of these differential arrays is that their directivity is independent of frequency. Derivations are included for several optimal differential arrays that may be useful for teleconferencing and speech pickup in noisy and reverberant environments. Novel expressions and design details covering multiple-order hypercardioid and supercardioid-type differential arrays are given. Also, the design of Dolph-Chebyshev equi-sidelobe differential arrays is covered for the general multiple-order case. The results shown here should be useful in designing and selecting directional microphones for a variety of applications.

Keywords: Acoustic Arrays, Beamforming, Directional Microphones, Differential Microphones, Room Acoustics

1. INTRODUCTION

Noise and reverberation can seriously degrade both the microphone reception and the loudspeaker transmission of speech signals in hands-free telecommunication. The use of small directional microphones and loudspeakers can be effective in combating these problems. First-order differential microphones have been in existence now for more than 50 years. Due to their directional and close-talking properties, they have proven essential for the reduction of feedback in public address systems. In telephone applications, such as speakerphone teleconferencing, directional microphones are very useful but at present are seldom utilized. Since small differential arrays can offer significant improvement in typical teleconferencing configurations, it is expected that they will become more prevalent in years to come.

Work on various differential microphone arrays has been ongoing in the Acoustics Research Department at Bell Labs for many years. The intent of the present work is to fill in some of the missing information and to develop some of the necessary analytical expressions in differential microphone array design. Included are several new results of potential importance to a designer of such microphone systems. Various design configurations of multiple-order differential arrays, that are optimal under various criteria, are discussed.

Generally, designs and applications of differential microphones are illustrated. Since transduction and transmission of acoustic waves are generally reciprocal processes the results are also applicable to loudspeakers. However, the loudspeaker implementation is difficult because of the large volume velocities required to approximate ideal differential microphones. The reasons are twofold: first, the source must be small compared to the acoustic wavelength; second, the real-part of the radiation impedance becomes very small for differential operation. Another additional factor that must be carefully accounted for in differential loudspeaker array design is the mutual radiation impedance between array elements.

2. DIFFERENTIAL MICROPHONE ARRAYS

The term *first-order* differential applies to any array whose sensitivity is proportional to the first spatial derivative of the acoustic pressure field. The term n^{th}*-order* differential is used for arrays that have a response proportional to a linear combination of the spatial derivatives up to, and including n. The classification *superdirectional* is applied to an array whose directivity factor (to be defined later) is higher than that of an array of the same geometry with uniform amplitude weighting. The systems that are discussed in this chapter, respond to finite-differences of the acoustic pressure that closely approximate the pressure differentials for general order. Thus, the interelement spacing of the array microphones is much smaller than the acoustic wavelength and the arrays are therefore superdirectional. Typically, these arrays combine the outputs of closely-spaced microphones in an alternating sign fashion. These differential arrays are therefore also commonly referred to as *pattern-differencing* arrays.

Before we discuss the various implementations of nth-order finite-difference systems, we develop expressions for the nth-order spatial acoustic pressure derivative in a direction $\mathbf{r}$ (the bold-type indicates a vector quantity). Since realizable differential arrays approximate the true acoustic pressure differentials, the equations for the general order differentials provide significant insight into the operation of these systems. To begin, we examine the case for a propagating acoustic plane-wave. The acoustic pressure field for a propagating acoustic plane-wave can be written as,

$$p(k, \mathbf{r}, t) = P_o e^{j(\omega t - \mathbf{k}^T \mathbf{r})} = P_o e^{j(\omega t - kr\cos\theta)}, \tag{10.1}$$

where P_o is the plane-wave amplitude, ω is the angular frequency, T is the transpose operator, $\mathbf{k}$ is the acoustic wavevector ($\|\mathbf{k}\| = k = \omega/c = 2\pi/\lambda$ where λ is the acoustic wavelength), c is the speed of sound, and $r = \|\mathbf{r}\|$ where $\mathbf{r}$ is the position vector relative to the selected origin, and θ is the angle between the position vector $\mathbf{r}$ and the wavevector $\mathbf{k}$. Dropping the time dependence and taking the nth-order spatial derivative along the direction of the position vector $\mathbf{r}$ yields,

$$\frac{d^n}{dr^n} p(k, r) = P_o(-jk\cos\theta)^n e^{-jkr\cos\theta}. \tag{10.2}$$

The plane-wave solution is valid for the response to sources that are "far" from the microphone array. By "far" we mean distances that are many times the square of the relevant source dimension divided by the acoustic wavelength. Using (10.2) we can conclude that the nth-order differential has a bidirectional pattern with the shape of $(\cos\theta)^n$. We can also see that the frequency response of a differential microphone is a high-pass system with a slope of $6n$ dB per octave. If we relax the far-field assumption and examine the response of the differential system to a point source located at the coordinate origin, then

$$p(k, r) = P_o \frac{e^{-j(kr\cos\theta)}}{r}. \tag{10.3}$$

The nth-order spatial derivative in the direction r is

$$\frac{d^n}{dr^n} p(k, r, \theta) = P_o \frac{n!}{r^{n+1}} e^{-jkr\cos\theta} (-1)^n \sum_{m=0}^{n} \frac{(jkr\cos\theta)^m}{m!}, \tag{10.4}$$

where r is the distance to the source. The interesting thing to notice in (10.4) is that for $kr\cos\theta$ small, the microphone is independent of frequency. Another fundamental property is that the general n^{th}-order differential response is a weighted sum of bidirectional terms of the form $\cos^m\theta$. We use this property in later sections. First, though, we consider the effects of the finite-difference approximation of the spatial derivative.

After expanding the acoustic pressure field into a Taylor series, we must keep zero and first-order terms to express the first derivative. The resulting equation is nothing other than the finite-difference approximation to the first-order derivative. As long as the spacing is small compared to the acoustic wavelength, the higher-order terms (namely, the higher-order derivatives) become insignificant over a desired frequency range. Likewise, the second and third-order derivatives can be incorporated by retaining the second and third-order terms in the expansion. The resulting approximation can be expressed as the exact spatial derivative multiplied by a bias error term. For the first-order case in a plane-wave acoustic field [(10.2)], the pressure derivative is

$$\frac{dp(k,r,\theta)}{dr} = -jkP_o \cos\theta \, e^{-jkr\cos\theta}. \tag{10.5}$$

The finite-difference approximation for the first-order system is defined as

$$\begin{aligned} \frac{\Delta\, p(k,r,\theta)}{\Delta\, r} &\equiv \frac{p(k, r+d/2, \theta) - p(k, r-d/2, \theta)}{d} \\ &= \frac{-j2P_o \sin(k\,d/2\cos\theta) e^{-jkr\cos\theta}}{d}, \end{aligned} \tag{10.6}$$

where d the distance between the two microphones. If we now define the amplitude bias error ϵ_1 as

$$\epsilon_1 = \frac{\Delta p/\Delta r}{dp/dr}, \tag{10.7}$$

then on-axis ($\theta = 0$),

$$\epsilon_1 = \frac{\sin kd/2}{kd/2} = \frac{\sin \pi d/\lambda}{\pi d/\lambda}. \tag{10.8}$$

Figure 10.1 shows the amplitude bias error ϵ_1 between the true pressure differential and the approximation by the two closely-spaced omnidirectional (zero-order) microphones. The bias error is plotted as a nondimensional function of microphone spacing divided by the acoustic wavelength (d/λ). From Fig. 10.1, it can be seen that for less than 1 dB error, that the element spacing must be less than 1/4 of the acoustic wavelength. Similar equations can be written for higher-order arrays. The bias function for these systems is low-pass in nature if the frequency range is limited to the small kd range.

In general, to realize an array that is sensitive to the nth derivative of the incident acoustic pressure field, we require m p^{th}-order microphones, where, $m + p - 1 = n$. For example, a first-order differential microphone requires two zero-order microphones. The linearized Euler's equation for an ideal (no viscosity) fluid states that

$$-\nabla p = \rho \frac{\partial \mathbf{v}}{\partial t}, \tag{10.9}$$

where ρ is the fluid density and $\mathbf{v}$ is the acoustic particle velocity. The time derivative of the particle velocity is proportional to the pressure-gradient. For

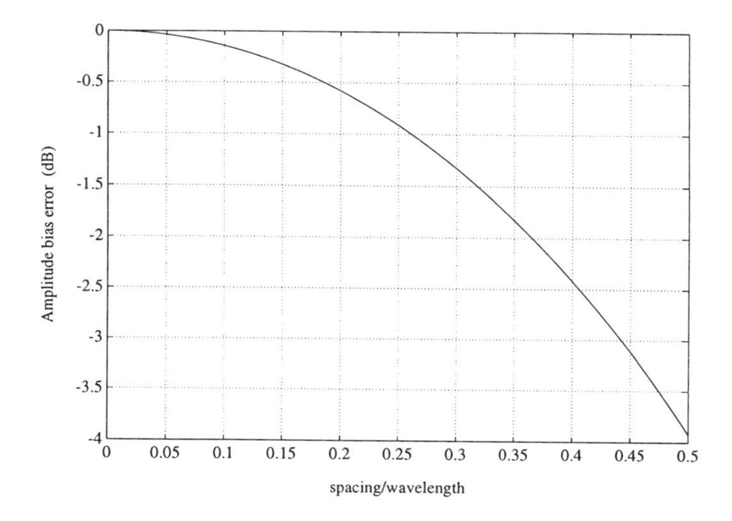

Figure 10.1 Finite-difference amplitude bias error in dB for a plane-wave propagating along the microphone axis, as a function of element spacing divided by the acoustic wavelength.

an axial component of a sinusoidal velocity vector, the output is proportional to the pressure differential along that axis. More typically, a first-order differential dipole microphone is designed as a diaphragm that is open to the sound field on both sides. The motion of the diaphragm is dependent on the net force difference (pressure-difference) across the diaphragm and is therefore an acoustic particle velocity microphone. Thus, by proper design of a microphone, it is easy to construct a first-order acoustic microphone. The design of higher order microphones can be formed by combinations of lower-order microphones where the sum of all of component microphone orders is equal to the desired order differential microphone.

For a plane-wave with amplitude P_o and wavenumber k incident on a two-element array, as shown in Fig. 10.2, the output can be written as

$$E_1(k, \theta) = P_o \left(1 - e^{-jkd \cos \theta}\right), \tag{10.10}$$

where d is the interelement spacing and the subscript indicates a first-order differential array. Note again that the explicit time dependence factor is neglected for the sake of compactness. If it is now assumed that the spacing is much smaller than the acoustic wavelength, we can write

$$E_1(k, \theta) \approx P_o kd \cos \theta. \tag{10.11}$$

As expected, the first-order array has the factor $\cos \theta$ that resolves the component of the acoustic particle velocity along the microphone axis.

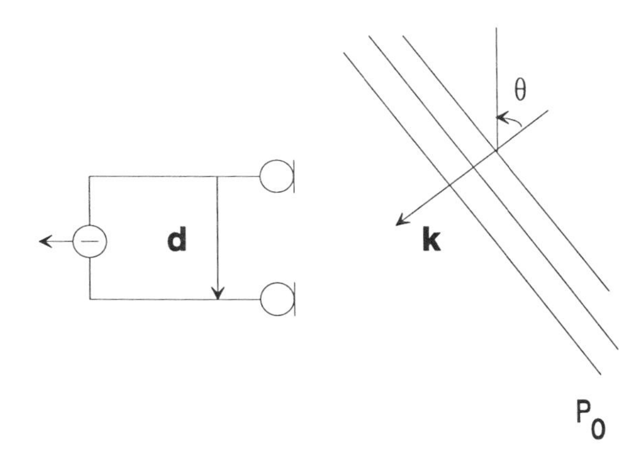

Figure 10.2 Diagram of first-order microphone composed of two zero-order (omnidirectional) microphones.

We now examine the case where a delay is introduced between these two zero-order microphones. For a plane-wave incident on this new array, we can write

$$E_1(\omega, \theta) = P_o \left(1 - e^{-j\omega(\tau + d \cos \theta / c)}\right), \tag{10.12}$$

where τ is equal to the delay applied to the signal from one microphone and we have made the substitution $k = \omega/c$. If we again assume a small spacing ($kd \ll \pi$ and $\omega\tau \ll \pi$),

$$E_1(\omega, \theta) \approx P_o\omega \left(\tau + d/c \cos \theta\right). \tag{10.13}$$

One thing to notice about (10.13), is that the first-order array has a first-order high-pass frequency dependence. The term in the parentheses in (10.13) contains the array directional response. In the design of differential arrays, the array directivity function is the quantity that is of interest. To simplify further analysis for the directivity of the first-order array, let us define a_0, a_1, and α_1, such that

$$\alpha_1 = a_0 = \frac{\tau}{\tau + d/c} \tag{10.14}$$

and

$$1 - \alpha_1 = a_1 = \frac{d/c}{\tau + d/c}. \tag{10.15}$$

Then

$$a_0 + a_1 = 1. \tag{10.16}$$

Thus, the normalized directional response is

$$E_{N_1}(\theta) = a_0 + a_1 \cos\theta = \alpha_1 + (1 - \alpha_1)\cos\theta, \tag{10.17}$$

where the subscript N denotes the normalized response of a first-order system, i.e., $E_{N_1}(0) = 1$. The normalization of the array response has effectively factored out the term that defines the normalized directional response of the microphone array. The most interesting thing to notice in (10.17) is that the first-order differential array directivity function is *independent* of frequency within the region where the assumption of small spacing compared to the acoustic wavelength holds. Note that we have substituted the dependent variable α_1 which is itself a function of the variables d and τ.

The magnitude of (10.17) is the parametric expression for the "limaçon of Pascal" algebraic curve. The two terms in (10.17) can be seen to be the sum of a zero-order microphone (first-term) and a first-order microphone (second term), which is the general form of the first-order array. Early unidirectional microphones were actually constructed by summing the outputs of an omnidirectional pressure microphone and a velocity ribbon microphone (pressure-differential microphone) [12]. One implicit property of (10.17) is that for $0 \leq \alpha_1 \leq 1$ there is a maximum at $\theta = 0$ and a minimum at an angle between $\pi/2$ and π. For values of $\alpha_1 > 1/2$ the response has a minimum at 180°, although there is no zero in the response. An example of the response for this case is shown in Fig. 10.3(a). When $\alpha_1 = 1/2$, the parametric algebraic equation has a specific form which is called a cardioid. The cardioid pattern has a zero response at $\theta = 180°$. For values of $\alpha_1 < 1/2$ there is no longer a minimum at $\theta = 180°$, although there is a zero-response (null) at $90° < \theta < 180°$. Figure 10.3(b) shows a directivity response corresponding to this case. For the first-order system, the solitary null is located at

$$\theta_1 = \cos^{-1}(-\frac{a_0}{a_1}) = \cos^{-1}\left(\frac{-\alpha_1}{1 - \alpha_1}\right). \tag{10.18}$$

The directivities shown by Fig. 10.3 are actually a representation of a plane slice through the center line of the true three-dimensional directivity plot. The arrays discussed in this chapter are rotationally symmetric around their axes. Figure 10.4 shows a three-dimensional representation of the directivity pattern shown in Fig. 10.3(b).

The realization of a general first-order differential response is accomplished by adjusting the time delay between the two zero-order microphones that comprise the first-order system model. From (10.14) and (10.15), the value of τ determines the ratio of a_1/a_0. The value of τ is proportional to d/c, the propagation time for an acoustic wave to axially travel between the zero-order

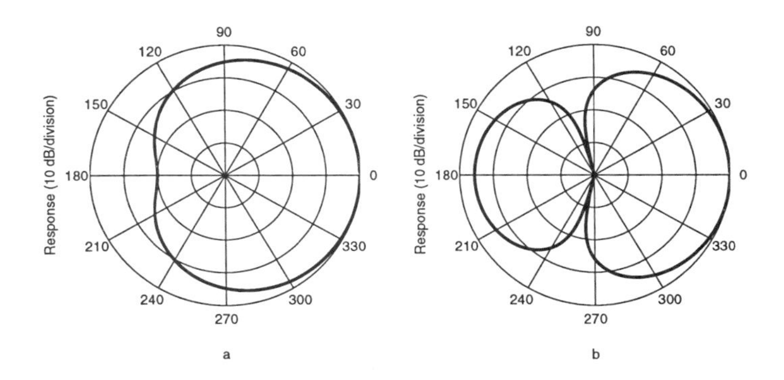

Figure 10.3 Directivity plots for first-order arrays (a) $\alpha_1 = 0.55$, (b) $\alpha_1 = 0.20$.

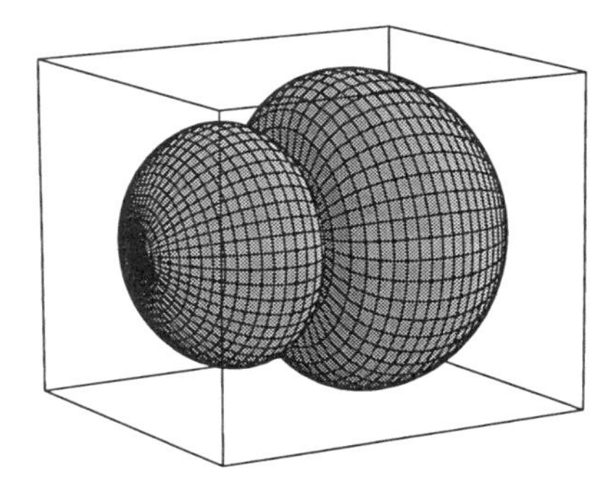

Figure 10.4 Three dimensional representation of directivity in Fig. 10.3(b). Note that the viewing angle is from the rear-half plane at an angle of approximately 225°. The viewing angle was chosen so that the rear-lobe of the array would not be obscured by the mainlobe.

microphones. This interelement propagation time is

$$\tau = \frac{d\, a_0}{c\, a_1} = \frac{d\alpha_1}{c(1-\alpha_1)}. \tag{10.19}$$

From (10.19) and (10.18), the pattern zero is at

$$\theta_1 = \cos^{-1}\left(-\frac{c\,\tau}{d}\right). \tag{10.20}$$

The nth order array can be written as the sum of the nth spatial derivative of the sound field plus lower-order terms. The nth-order array can also be written as

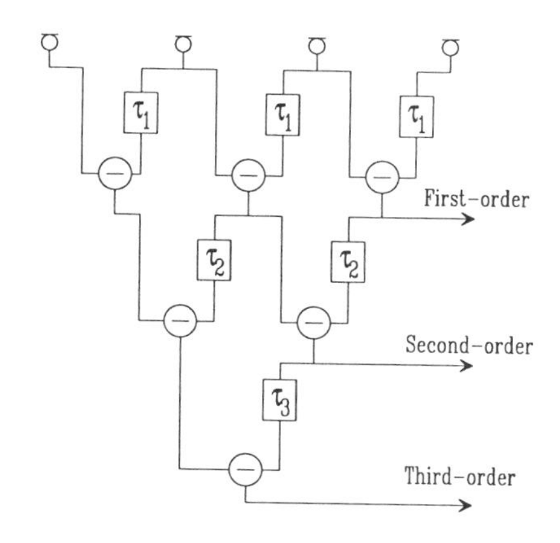

Figure 10.5 Construction of differential arrays as first-order differential combinations up to third-order.

the product of n first-order response terms as

$$E_n(\omega, \theta) = P_o \prod_{i=1}^{n} \left[1 - e^{-j\omega(\tau_i + d_i/c \cos\theta)}\right], \tag{10.21}$$

where the d_i relate to the microphone spacings, and the τ_i relate to chosen time delays. There is a design implementation advantage in expressing the array response in terms of the products of first-order terms since it is now simple to represent higher-order systems as cascaded systems of lower order. Figure 10.5 shows how a differential array can be constructed for orders up to three. The extension of the design technique to higher orders is straightforward. The values of τ_i can be determined by using the relationships developed in (10.14) and (10.19). The ordering of the τ_i is not important.

If again we assume that $kd_i \ll \pi$ and $\omega\tau_i \ll \pi$, then (10.21) can be approximated as

$$E_n(\omega, \theta) \approx P_o\omega^n \prod_{i=1}^{n} (\tau_i + d_i/c \cos\theta). \tag{10.22}$$

Equation (10.22) can be simplified by making the same substitution as was done in (10.14) and (10.15) for the arguments in the product term. If we set $\alpha_i = \tau_i/(\tau_i + d_i/c)$, then

$$E_n(\omega, \theta) \approx P_o\omega^n \prod_{i=1}^{n} [\alpha_i + (1 - \alpha_i) \cos\theta]. \tag{10.23}$$

If the product in (10.23) is expanded, a power series in $\cos\theta$ can be written for the response of the nth-order array to an incident plane-wave with frequency ω.

The result is

$$E_n(\omega, \theta) = P_o A\, \omega^n \left(a_0 + a_1 \cos\theta + a_2 \cos^2\theta + \ldots + a_n \cos^n\theta\right), \tag{10.24}$$

where the constant A is an overall gain factor and we have suppressed the explicit dependence of the directivity function E on the variables d_i and τ_i for compactness. The only frequency dependent term in (10.24) is the term ω^n. The frequency response of an nth-order differential array can therefore be easily compensated by a lowpass filter whose frequency response is proportional to ω^{-n}. By choosing the structure that places only a delay behind each element in a differential array, the coefficients in the power series in (10.24) are independent of frequency, resulting in an array whose beampattern is independent of frequency. To simplify the following exposition on the directional properties of differential arrays, we will assume that the amplitude factor can be neglected. Also, since the directional pattern described by the power series in $\cos\theta$ can have any general scaling, we will typically describe the normalized directional response as only a function of θ, such that

$$E_{N_n}(\theta) = a_0 + a_1 \cos\theta + a_2 \cos^2\theta + \ldots + a_n \cos^n\theta, \tag{10.25}$$

where the subscript N denotes a normalized response at $\theta = 0°$ [as in (10.17)] which implies

$$\sum_{i=0}^{n} a_i = 1. \tag{10.26}$$

In general therefore, the nth-order differential microphone has at most, n nulls (zeros). This follows directly from (10.25) and the fundamental theorem of algebra. Equation (10.25) can also be written in "canonic" form as the product of first-order terms

$$E_{N_n}(\theta) = \prod_{i=1}^{n} \left[\alpha_i + (1-\alpha_i)\cos\theta\right]. \tag{10.27}$$

Note that we have dropped the frequency dependent variable ω in (10.25) and (10.27) since we have argued that the frequency response for small interelement spacing is simply proportional to ω^n. The terms a_i in (10.25) can take on any desired value by adjusting the delays used in defining the desired differential microphone array. For the second-order array

$$E_{N_2}(\theta) = a_0 + a_1 \cos\theta + a_2 \cos^2\theta. \tag{10.28}$$

Equation (10.28) can also be factored into two first-order terms and written as

$$E_{N_2}(\theta) = [\alpha_1 + (1-\alpha_1)\cos\theta][\alpha_2 + (1-\alpha_2)\cos\theta], \tag{10.29}$$

where

$$\begin{aligned} a_0 &= \alpha_1 \alpha_2 \\ a_1 &= \alpha_1(1-\alpha_2) + \alpha_2(1-\alpha_1) \\ a_2 &= (1-\alpha_1)(1-\alpha_2), \end{aligned} \tag{10.30}$$

or

$$\begin{aligned} \alpha_1 &= a_0 + a_1/2 \pm \sqrt{(a_0 + a_1/2)^2 - a_0} \\ \alpha_2 &= a_0 + a_1/2 \mp \sqrt{(a_0 + a_1/2)^2 - a_0}. \end{aligned} \tag{10.31}$$

As shown, the general form of the second-order system is the sum of second-order, first-order and zero-order terms. If certain constraints are placed on the values of a_0 and a_1, it can be seen that there are two nulls (zeros) in the interval $0 \leq \theta < \pi$. The array response pattern is symmetric about $\theta = 0$. The appearance of four nulls in the interval $2\pi \leq \theta < 0$ are actually the same two nulls in the interval $0 \leq \theta < \pi$. These zeros can be explicitly found at the angles θ_1 and θ_2:

$$\theta_1 = \cos^{-1}\left(\frac{-\alpha_1}{1-\alpha_1}\right), \tag{10.32}$$

$$\theta_2 = \cos^{-1}\left(\frac{-\alpha_2}{1-\alpha_2}\right), \tag{10.33}$$

where now α_1 and α_2 can take on either positive or negative values. If the resulting beampattern is constrained to have a maximum at $\theta = 0°$, then the values of α_1 and α_2 can only take on certain values; we have ruled out designs that have a higher sensitivity at any angle other than $\theta = 0°$. The interesting thing to note is that negative values of α_1 or α_2 correspond to a null moving into the front half-plane. Negative values of α_1 for the first-order microphone can be shown to have a rear-lobe sensitivity that exceeds the sensitivity at 0°. Since (10.29) is the product of two first-order terms, emphasis of the rear-lobe, caused by a negative value of α_2, can be counteracted by the zero from the term containing α_1. As a result, a beam-pattern can be found for the second-order microphone that has maximum sensitivity at $\theta = 0°$ and a null in the front-half plane. This result also implies that the beamwidth of a second-order microphone with a negative value of α_2 is narrower than that of the second-order dipole.

It is straightforward to extend the previous results to the third-order case. For completeness, the equation governing the directional characteristics for the third-order array is

$$E_{N_3}(\theta) = a_0 + a_1 \cos\theta + a_2 \cos^2\theta + a_3 \cos^3\theta. \tag{10.34}$$

If certain constraints are placed on the coefficients in (10.34), it can be factored into three real roots:

$$E_{N_3}(\theta) = [\alpha_1 + (1-\alpha_1)\cos\theta][\alpha_2 + (1-\alpha_2)\cos\theta][\alpha_3 + (1-\alpha_3)\cos\theta]. \tag{10.35}$$

The third-order microphone has the possibility of three zeros that can be placed at desired locations in the directivity pattern. Solving this cubic equation yields expressions for α_1, α_2, and α_3 in terms of a_0, a_1, a_2, and a_3. However, these expressions are long and algebraically cumbersome and, as such, not repeated here.

3. ARRAY DIRECTIONAL GAIN

In order to "best" reject the noise in an acoustic field, we need to optimize the way we combine multiple microphones. Specifically we need to consider the directional gain, i.e. the gain of the microphone array in a noise field over that of a simple omnidirectional microphone. A common quantity used is the Directivity Factor Q, or equivalently, the Directivity Index DI $[10\log_{10}(Q)]$.

The directivity factor is defined as

$$Q(\omega,\theta_0,\phi_0) = \frac{|\,E(\omega,\theta_0,\phi_0)\,|^2}{\frac{1}{4\pi}\int_0^{2\pi}\int_0^{\pi} |\,E(\omega,\theta,\phi)\,|^2\, u(\omega,\theta,\phi)\sin\theta d\theta\, d\phi}, \tag{10.36}$$

where the angles θ and ϕ are the standard spherical coordinate angles, θ_0 and ϕ_0 are the angles at which the directivity factor is being measured, $E(\omega,\theta,\phi)$ is the pressure response of the array, and $u(\omega,\theta,\phi)$ is the distribution of the noise power, The function u is normalized such that

$$\frac{1}{4\pi}\int_0^{2\pi}\int_0^{\pi} u(\omega,\theta,\phi)\sin\theta d\theta\, d\phi = 1. \tag{10.37}$$

The directivity factor Q can be written as the ratio of two Hermitian quadratic forms [3] as

$$Q = \frac{\mathbf{w}^{\mathcal{H}}\mathbf{A}\mathbf{w}}{\mathbf{w}^{\mathcal{H}}\mathbf{B}\mathbf{w}}, \tag{10.38}$$

where

$$\mathbf{A} = \mathbf{S}_0\mathbf{S}_\mathbf{0}^{\mathcal{H}}, \tag{10.39}$$

$\mathbf{w}$ is the complex weighting applied to the microphones and $\mathcal{H}$ is the complex conjugate transpose. The elements of the matrix $\mathbf{B}$ are defined as

$$b_{mn} = \frac{1}{4\pi}\int_0^{2\pi}\int_0^{\pi} u(\omega,\theta,\phi)\exp\left[j\mathbf{k}\cdot(\mathbf{r}_m - \mathbf{r}_n)\right]\sin\theta\, d\phi\, d\theta, \tag{10.40}$$

and the elements of the vector $\mathbf{S}_0$ are defined as

$$s_{0n} = \exp(j\mathbf{k}_0 \cdot \mathbf{r}_n). \tag{10.41}$$

Note that for clarity we have left off the explicit functional dependencies of the above equations on the angular frequency ω. The solution for the maximum of Q, which is a Rayleigh quotient, is obtained by finding the maximum generalized eigenvector of the homogeneous equation

$$\mathbf{A}\mathbf{w} = \lambda_M \mathbf{B}\mathbf{w}. \tag{10.42}$$

The maximum eigenvalue of (10.42) is given by

$$\lambda_M = \mathbf{S}_0^{\mathcal{H}} \mathbf{B}^{-1} \mathbf{S}_0. \tag{10.43}$$

The corresponding eigenvector contains the weights for combining the elements to obtain the maximum directional gain

$$\mathbf{w}_{opt} = \mathbf{B}^{-1} \mathbf{S}_0. \tag{10.44}$$

In general, the optimal weights $\mathbf{w}_{opt}$ are a function of frequency, array geometry, element directivity and the spatial distribution of the noise field.

4. OPTIMAL ARRAYS FOR SPHERICALLY ISOTROPIC FIELDS

Acoustic reverberation in rooms has historically been modeled as spherically isotropic noise. A spherically isotropic noise field can be constructed by combining uncorrelated noises propagating in all directions with equal power. In room acoustics this noise field is referred to as a "diffuse" sound field and has been the model used for many investigations into the distribution of reverberant sound pressure fields. Since it is of interest to design microphone systems that optimally reject reverberant sound fields, the first optimization of array gain will assume a "diffuse" sound field.

4.1 MAXIMUM GAIN FOR OMNIDIRECTIONAL MICROPHONES

The literature on the maximization of the directional gain for an arbitrary array is quite extensive [17, 18, 13, 4, 16, 8]. Uzkov [17] showed that for uniformly spaced omnidirectional microphones that the directional gain reaches N^2 as the spacing between the elements goes to zero. The maximum value of directional gain is obtained when the array is steered to end-fire. Weston [18] has shown the same result by an alternative method. Parsons [13] has extended the proof to include nonuniformly spaced arrays that are much smaller than the acoustic wavelength, The proof given here also relies on the assumption that the elements are closely-spaced compared to the acoustic wavelength. The approach taken here is similar to that of Chu [4], Tai [16], and Harrington [8], who expanded the radiated/received field in terms of spherical wave functions. Chu did not look explicitly at the limiting case. Harrington did examine the limiting case of vanishing spacing, but his analysis involved approximations that are not necessary in the following analysis.

For a spherically isotropic field and omnidirectional microphones,

$$u(\omega, \theta, \phi) = 1. \tag{10.45}$$

In general, the directivity of N closely-spaced microphones can be expanded in terms of spherical wave functions. Now let us express the farfield pressure response $E(\theta, \phi)$ as a summation of orthogonal polynomials,

$$E(\theta, \phi) = \sum_{n=0}^{N-1} \sum_{m=0}^{n} h_{nm} P_n^m[\cos(\theta - \theta_z)] \cos m(\phi - \phi_z), \tag{10.46}$$

where we have limited the sum to be equal to the number of degrees of freedom in the N-element microphone case, where the P_n^m are the associated Legendre functions, and θ_z and ϕ_z are possible rotations of the coordinate system. Now define

$$G_{nm}(\theta, \phi) = P_n^m[\cos(\theta - \theta_z)] \cos\, m(\phi - \phi_z). \tag{10.47}$$

The normalization of the function G_{nm} is

$$\begin{aligned} N_{nm} &= \int_0^{2\pi} \cos^2 m\phi \int_{-1}^{1} \left[P_n^m(\eta)\right]^2 d\eta d\phi\,, \\ &= \frac{4\pi(n+m)!}{\varepsilon_m(2n+1)(n-m)!}, \end{aligned} \tag{10.48}$$

where ε_m is the Neumann factor, which equals 1 for $m = 0$ and 2 for $m \geq 1$. By using the orthogonal Legendre function expansion, we can write

$$Q(\theta_o, \phi_o) = \frac{\left[\sum_{n=0}^{N-1} \sum_{m=0}^{n} h_{nm} G_{nm}(\theta_o, \phi_o)\right]^2}{\frac{1}{4\pi} \sum_{n=0}^{N-1} \sum_{m=0}^{n} h_{nm}^2 N_{nm}}. \tag{10.49}$$

To find the maximum of (10.49), we set the derivative of Q with respect to h_{nm} for all n and m to zero. The resulting maximum occurs when

$$\begin{aligned} Q_{max} &= \sum_{n=0}^{N-1}\sum_{m=0}^{n} \frac{[G_{nm}(\theta_o, \phi_o)]^2}{\frac{1}{4\pi}N_{nm}} \\ &= \sum_{n=0}^{N-1}\sum_{m=0}^{n} \frac{\left[P_n^m(\cos(\theta_o-\theta_z))\cos m(\phi_o-\phi_z)\right]^2}{\frac{1}{4\pi}N_{nm}} \\ &\le \sum_{n=0}^{N-1}\sum_{m=0}^{n} \frac{\left[P_n^m(\cos(\theta_o-\theta_z))\right]^2}{\frac{1}{4\pi}N_{nm}}. \end{aligned} \tag{10.50}$$

The inequality in (10.50) infers that the maximum must occur when $\phi_o = \phi_z$. From the addition theorem of Legendre polynomials,

$$P_n(\cos\psi) = \sum_{m=0}^{n} \varepsilon_m \frac{(n-m)!}{(n+m)!} P_n^m(\cos\theta_o) P_n^m(\cos\theta_z)\cos(m(\phi_o-\phi_z)), \tag{10.51}$$

where ψ is the angle subtended by two points on a sphere and is expressed in terms of spherical coordinates as

$$\cos\psi = \cos\theta_o\cos\theta_z + \sin\theta_o\sin\theta_z\cos(\phi_o-\phi_z). \tag{10.52}$$

Equation (10.51) maximizes for all n, when $\psi = 0$. Therefore (10.50) maximizes when $\theta_o = \theta_z$. Since

$$P_n^m(1) = \begin{cases} 1 & m = 0 \\ 0 & \text{if } m > 0 \end{cases} \tag{10.53}$$

(10.50) can be reduced to

$$\begin{aligned} Q_{max} &= \sum_{n=0}^{N-1}(2n+1), \\ &= N^2. \end{aligned} \tag{10.54}$$

Thus, the maximum directivity factor Q for N closely-spaced omnidirectional microphones is N^2.

4.2 MAXIMUM DIRECTIVITY INDEX FOR DIFFERENTIAL MICROPHONES

As was shown in Section 2, there are an infinite number of possibilities for differential array designs. What we are interested in are the special cases that are optimal in some respect. For microphones that are axisymmetric, as is the case for all of the microphones that are covered, (10.36) can be written in a simpler form:

$$Q(\omega) = \frac{2}{\int_0^{\pi} \mid E_N(\omega, \theta, \phi) \mid^2 \sin\theta \, d\theta}, \tag{10.55}$$

where it has also been assumed that the directions θ_0 , ϕ_0 are in the direction of maximum sensitivity and that the array sensitivity function is normalized: $\mid E_N(\omega, \theta_0, \phi_0) \mid = 1$. If we now insert the formula from (10.25) and carry out the integration, we find the directivity factor

$$Q(a_0, \ldots, a_n) = \left[\sum_{i=0}^{n} \sum_{\substack{j=0 \\ i+j \text{ even}}}^{n} \frac{a_i a_j}{1+i+j} \right]^{-1} . \tag{10.56}$$

The directivity factor for a general n^{th}-order differential array (no normalization assumption) can be written as

$$Q(a_0, \ldots, a_n) = \left[\sum_{i=0}^{n} a_i \right]^2 \left[\sum_{i=0}^{n} \sum_{\substack{j=0 \\ i+j \text{ even}}}^{n} \frac{a_i a_j}{1+i+j} \right]^{-1} = \frac{\mathbf{a}^T \mathbf{B} \mathbf{a}}{\mathbf{a}^T \mathbf{H} \mathbf{a}}, \tag{10.57}$$

where $\mathbf{H}$ is a Hankel matrix given by

$$H_{i,j} = \begin{cases} \frac{1}{1+i+j} & \text{if i+j even} \\ 0 & \text{otherwise} \end{cases}$$

where

$$\mathbf{a}^T = \{a_0, a_1, \ldots, a_n\} \tag{10.58}$$

and

$$\mathbf{B} = \mathbf{b}\mathbf{b}^T, \tag{10.59}$$

where

$$\mathbf{b}^T = \overbrace{\{1, 1, \ldots, 1\}}^{n+1} . \tag{10.60}$$

Table 10.1 Table of maximum array gain Q, and corresponding eigenvector for differential arrays from first to fourth-order for spherically isotropic noise fields.

microphone order	maximum eigen- value	corresponding eigenvector
1	4	[1/4 3/4]
2	9	[-1/6 1/3 5/6]
3	16	[-3/32 -15/32 15/32 35/32]
4	25	[0.075 -0.300 -1.050 0.700 1.575]

From (10.57) we can see that the directivity factor Q is a Rayleigh quotient for two hermitian forms. The maximum of the Rayleigh quotient is reached at a value equal to the largest generalized eigenvalue of the equivalent generalized eigenvalue problem,

$$\mathbf{Bx} = \lambda \mathbf{Hx} \tag{10.61}$$

where, λ is the general eigenvalue and $\mathbf{x}$ is the corresponding general eigenvector. The eigenvector corresponding to the largest eigenvalue will contain the coefficients a_i which maximize the directivity factor Q. Since $\mathbf{B}$ equals a dyadic product there is only one eigenvector $\mathbf{x} = \mathbf{H}^{-1}\mathbf{b}$ with the eigenvalue $\mathbf{b}^T\mathbf{H}^{-1}\mathbf{b}$. Thus,

$$\max_{\mathbf{a}} Q \quad = \quad \lambda_m = \mathbf{b}^T\mathbf{H}^{-1}\mathbf{b}. \tag{10.62}$$

Table 10.1 gives the maximum array gain (largest eigenvalue), Q, for differential orders up to fourth-order. Note that the largest eigenvector has been scaled such that the microphone output is unity at $\theta = 0°$. The directivity index is a considerably useful measure in quantifying the directional properties of microphones and loudspeakers. The directivity index provides a rough estimate of the relative gain in signal-to-reverberation for a directional microphone in a reverberant environment. However, the directivity index is meaningless with respect to the performance of directional microphones in non-diffuse fields. A more detailed discussion on the relevance of the microphone directivity index with respect to room acoustics is given in Appendix A.

4.3 MAXIMIMUM FRONT-TO-BACK RATIO

Another possible measure of the "merit" of an array is the front-to-back rejection ratio, i.e., the gain of the microphone for signals propagating to the front of the microphone relative to signals propagating to the rear. One such quantity was suggested by Marshall and Harry [11] which will be referred to here as "F," for the front-to-back ratio. The ratio F is defined as

$$F(\omega) = \frac{\int_0^{2\pi} \int_0^{\pi/2} \mid E(\omega, \theta, \phi) \mid^2 \sin\theta d\theta \, d\phi}{\int_0^{2\pi} \int_{\pi/2}^{\pi} \mid E(\omega, \theta, \phi) \mid^2 \sin\theta d\theta \, d\phi}, \tag{10.63}$$

where the angles θ and ϕ are the spherical coordinate angles and $E(\omega,\theta,\phi)$ is the far-field pressure response. For axisymmetric microphones (10.63) can be written in a simpler form by uniform integration over ϕ,

$$F(\omega) = \frac{\int_0^{\pi/2} \mid E_N(\omega, \theta, \phi) \mid^2 \sin\theta \, d\theta}{\int_{\pi/2}^{\pi} \mid E_N(\omega, \theta, \phi) \mid^2 \sin\theta \, d\theta}. \tag{10.64}$$

Carrying out the integration of (10.64) and using the form of (10.25) yields,

$$F(a_0, \ldots, a_n) = \left[\sum_{i=0}^{n} \sum_{j=0}^{n} \frac{a_i a_j}{1+i+j}\right] \left[\sum_{i=0}^{n} \sum_{j=0}^{n} \frac{a_i a_j (-1)^{i+j}}{1+i+j}\right]^{-1}. \tag{10.65}$$

The supercardioid name is given to the differential system that has the maximum front-to-rear power ratio. In equation form, the front-to-rear power ratio can be written as

$$F(a_0, \ldots, a_n) = \left[\sum_{i=0}^{n} \sum_{j=0}^{n} \frac{a_i a_j}{1+i+j}\right] \left[\sum_{i=0}^{n} \sum_{j=0}^{n} \frac{(-1)^{i+j} a_i a_j}{1+i+j}\right]^{-1}$$

$$= \frac{\mathbf{a}^T \mathbf{B} \mathbf{a}}{\mathbf{a}^T \mathbf{H} \mathbf{a}}, \tag{10.66}$$

where $\mathbf{H}$ is a Hankel matrix given by

$$H_{i,j} = \frac{(-1)^{i+j}}{1+i+j} \tag{10.67}$$

and

$$\mathbf{a}^T = \{a_0, a_1, \ldots, a_n\}. \tag{10.68}$$

$\mathbf{B}$ is a special form of a Hankel matrix designated as a Hilbert matrix and is given by

$$B_{i,j} = \frac{1}{1+i+j}. \tag{10.69}$$

Table 10.2 Table of maximum F ratio and corresponding eigenvector for differential arrays from first to fourth-order for spherically isotropic noise fields.

microphone order	maximum eigen-value	corresponding eigenvector
1	$7+4\sqrt{3}$	$[\frac{1}{1+\sqrt{3}} \ \frac{1}{1+\sqrt{3}}]$
2	$127+48\sqrt{7}$	$[\frac{1}{2(3+\sqrt{7})} \ \frac{\sqrt{7}}{3+\sqrt{7}} \ \frac{5}{2(3+\sqrt{7})}]$
3	≈ 5875	$\approx$ [0.0184 0.2004 0.4750 0.3061]
4	≈ 151695	[0.0036 0.0670 0.2870 0.4318 0.2107]

From (10.66) we can see that, as was the case for the maximum directional gain, the front-to-back ratio can be represented as Rayleigh quotient of two hermitian forms. The maximum of the Rayleigh quotient is reached at a value equal to the largest eigenvalue of the equivalent generalized eigenvalue problem,

$$\mathbf{Bx} = \lambda \mathbf{Hx} \tag{10.70}$$

where λ is the eigenvalue and $\mathbf{x}$ is the corresponding eigenvector. Thus, as in the case of maximizing the directional gain Q, the maximization of the front-to-back ratio is a general eigenvalue problem with F as the largest eigenvalue. The matrices $\mathbf{H}$ and $\mathbf{B}$ are real Hankel matrices and are positive definite. The resulting eigenvalues are therefore positive real numbers and the eigenvectors are real. Table 10.2 summarizes the results for the maximum front-to-back ratios for differential arrays up to fourth-order.

As with the directivity index, the front-to-back ratio is a very useful measure in quantifying the directional properties of electroacoustic transducers.

The utility of the front-to-back ratio F measure in teleconferencing is clear if we consider the following scenario. In a typical teleconference, people sit along one side of a table facing a video screen that is generally acoustically reflective. The rear and sides of the room are usually absorptive. Maximizing the front-to-back rejection ratio will therefore minimize reflections from the front wall and video screen, as well as minimize the response to a wide distribution of loudspeakers used for transmitting the remote site audio to the room. The above example is somewhat contrived and the actual optimal microphone array will depend on the source and receiver locations and the room acoustics. For the particular case above, however, the ratio F yields a better measure of microphone performance than the directivity index.

4.4 MINIMUM PEAK DIRECTIONAL RESPONSE

Another approach that might be of interest, is to design differential arrays to have an absolute maximum sidelobe response. This specification would allow the designer to guarantee that the differential array response would not exceed a defined level, over a given angular range, where suppression of acoustic signals is desired.

The first suggestion of an equi-sidelobe *differential* array design was in a "comment" publication by V. I. Korenbaum [9], who only discussed a restricted class of nth-order microphones that have the following form:

$$E_{N_n}(\theta) = [\alpha_1 + (1 - \alpha_1) \cos \theta] \cos^{n-1} \theta. \qquad (10.71)$$

The restricted class defined by (10.71) essentially assumes that an nth-order differential microphone is the combination of an $(n-1)^{th}$-order dipole pattern and a general first-order pattern. The major reason for considering this restricted class is obvious; the algebra becomes very simple. Since we are dealing with systems of order less than or equal to three, we do not need to restrict ourselves to the class defined by (10.71).

A more general design of equi-sidelobe differential arrays can be obtained by using standard Dolph-Chebyshev design techniques [6]. With this method we can easily realize any order differential microphone that we desire. The roots of the Dolph-Chebyshev system are easily obtained. Knowledge of the roots simplifies formulation of the canonic equations that describe the nth order microphones as products of first-order differential elements.

We begin our analysis with the Chebyshev polynomials

$$T_n(x) = \begin{cases} \cos(n \cos^{-1} x), & -1 < x < 1 \\ \cosh(n \cosh^{-1} x), & 1 \leq |x|. \end{cases} \qquad (10.72)$$

The Chebyshev polynomial of order n has n real roots for arguments between -1 and 1, and grows proportional to x^n for arguments with a magnitude greater than 1. The design of the nth-order Chebyshev array requires a transformation of the variable x in (10.72). If we substitute $x = b + a \cos \theta$ in (10.72), then we can form a desired n^{th}-order directional response that follows the Chebyshev polynomial over any range. At $\theta = 0°$, $x = x_o \equiv a + b$, and the value of the Chebyshev polynomial is $T_n(x_o) \geq 1$. Setting this value to the desired mainlobe to sidelobe ratio, say S, we have

$$S = T_n(x_o) = \cosh(n \cosh^{-1} x_o). \qquad (10.73)$$

Equivalently,

$$x_o = a + b = \cosh\left(\frac{1}{n} \cosh^{-1} S\right). \qquad (10.74)$$

The sidelobe at $\theta = 180^\circ$ corresponds to $x = b - a = -1$. Therefore

$$\begin{aligned} a &= \frac{x_o + 1}{2} \\ b &= \frac{x_o - 1}{2}. \end{aligned} \tag{10.75}$$

Since the zeros of the Chebyshev polynomial are readily calculable, the null locations are easily found. From the definition of the Chebyshev polynomial given in (10.72), the zeros occur at

$$x_m = \cos\left[\frac{(2m-1)\pi}{2n}\right], \quad m = 1, \dots, n. \tag{10.76}$$

The nulls are therefore at angles

$$\theta_m = \cos^{-1}\left(\frac{2x_m - x_o + 1}{x_o + 1}\right). \tag{10.77}$$

4.5 BEAMWIDTH

Another useful measure of the array performance is the beamwidth. The beamwidth can be defined in many ways. It can refer to the angle enclosed between the zeros of a directional response, the 3 dB points, or the 6 dB points. We will use the 3 dB beamwidth definition in this chapter. For the first-order microphone response [(10.17)], the 3 dB beamwidth is simply

$$\theta_{B_1} = 2\cos^{-1}\left(\frac{-2a_0 + \sqrt{2}(a_0 + a_1)}{2a_1}\right). \tag{10.78}$$

For the second-order system, the algebra is somewhat more difficult but still straightforward. The result from (10.28) is

$$\theta_{B_2} = 2\cos^{-1}\left(\frac{-a_1 + \sqrt{a_1^2 + 2\sqrt{2}[a_2^2 + a_1 a_2 + (1 - \sqrt{2})a_0 a_2]}}{2a_2}\right), \tag{10.79}$$

where it is assumed that $a_2 \neq 0$. If $a_2 = 0$, the microphone degenerates into a first order array and the beamwidth can be calculated by (10.78).

Similarly, the beamwidth for a third-order array can be found although the algebraic form is extremely lengthy and, as such, has not been included here.

5. DESIGN EXAMPLES

For differential microphones with interelement spacing much less than the acoustic wavelength, the maximum directivity index is attained when all of the

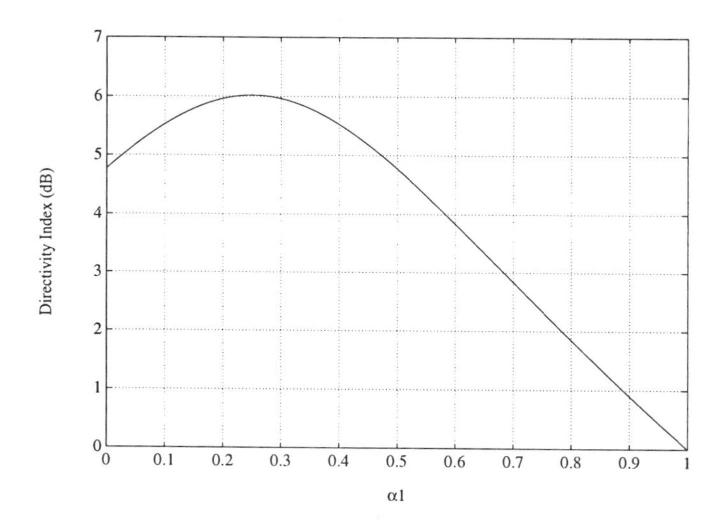

Figure 10.6 Directivity index of first-order microphone versus the first-order differential parameter α_1.

microphones are situated collinearly. For this case, the maximum directivity index is $20\ \log_{10}(n+1)$ where n is the order of the microphone [13]. For first, second, and third-order microphones, the maximum directivity indices are, 6.0, 9.5, and 12.0 dB respectively. Derivations of the specific results for $n \leq 3$ are given in the following sections.

As indicated in (10.25), there are an infinite number of possible designs for n^{th}-order differential arrays. Presently, the most common first-order microphones are: dipole, cardioid, hypercardioid, and supercardioid. The extension to higher orders is straightforward and is developed in later sections. Most of the arrays that are described in this chapter have directional characteristics that are optimal in some way; namely, the arrays are optimal with respect to one of the performance measures previously discussed: directivity index, front-to-back ratio, sidelobe threshold, and beamwidth. A summary of the results for first, second, and third-order microphones is given in Table 10.3.

5.1 FIRST-ORDER DESIGNS

Before we discuss actual first-order differential designs, we first examine the effects of the parameter α_1 on the directivity index DI, the front-to-back ratio F, and the beamwidth of the microphone. We have defined $\alpha_1 = a_0$ and $a_1 = 1 - \alpha_1$. Figure 10.6 shows the directivity index of a first-order system for values of α_1 between 0 and 1.

Table 10.3 Table of first-order differential, second-order differential, and third-order differential designs.

microphone type	DI (dB)	F (dB)	3dB Beamwidth	Null(s) (degrees)
		First-order designs		
dipole	4.8	0.0	90°	90
cardioid	4.8	8.5	131°	180
hypercardioid	6.0	8.5	105°	109
supercardioid	5.7	11.4	115°	125
		Second-order designs		
dipole	7.0	0.0	65°	90
cardioid	7.0	14.9	94°	180
OSW cardioid	8.8	14.9	76°	90, 180
hypercardioid	9.5	8.5	66°	73, 134
supercardioid	8.3	24.0	80°	104, 144
Korenbaum	8.9	17.6	76°	90, 146
−15 dB sidelobe	9.4	10.7	70°	78, 142
−30 dB sidelobe	8.1	18.5	84°	109, 152
min. rear peak	8.5	22.4	80°	98, 149
		Third-order designs		
dipole	8.5	0.0	54°	90
cardioid	8.5	21.0	78°	180
OSW cardioid	10.7	18.5	60°	90, 180
hypercardioid	12.0	11.2	48°	55, 100, 145
supercardioid	9.9	37.7	66°	97, 122, 153
−20 dB sidelobe	11.8	14.8	52°	62, 102, 153
−30 dB sidelobe	10.8	25.2	60°	79, 111, 156

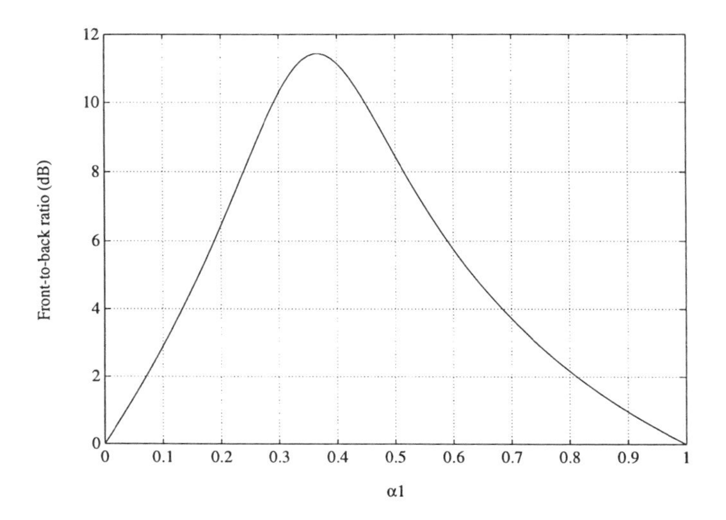

Figure 10.7 Front-to-back ratio of first-order microphone versus the first-order differential parameter α_1.

The first-order differential microphone that corresponds to the maximum in Fig. 10.6 is given the name *hypercardioid.* When $\alpha_1 = 0$, the first-order differential system is a dipole. At $\alpha_1 = 1$, the microphone is an omnidirectional microphone with 0 dB directivity index. Figure 10.7 shows the dependence of the front-to-back ratio F on α_1.

The maximum F value corresponds to the *supercardioid* design. Figure 10.8 shows the 3 dB beamwidth of the first-order differential microphone as a function of α_1.

When $\alpha_1 \approx 0.7$, the 3 dB down point is approximately at 180°. Higher values of α_1 correspond to designs that are increasingly omnidirectional. Figure 10.8 indicates that the first-order differential microphone with the smallest beamwidth is the dipole microphone with a 3 dB beamwidth of 90°.

5.1.1 Dipole. The dipole microphone is basically an acoustic particle-velocity microphone. The construction was described earlier; the dipole is normally a diaphragm that is open on both sides to the acoustic field. In (10.17), the dipole microphone corresponds the simple case where $a_0 = 0$, $a_1 = 1$,

$$E_{D_1}(\theta) = \cos\,\theta. \tag{10.80}$$

In Fig. 10.9(a), a polar plot of the magnitude of (10.80) shows the classic cosine pattern for the microphone.

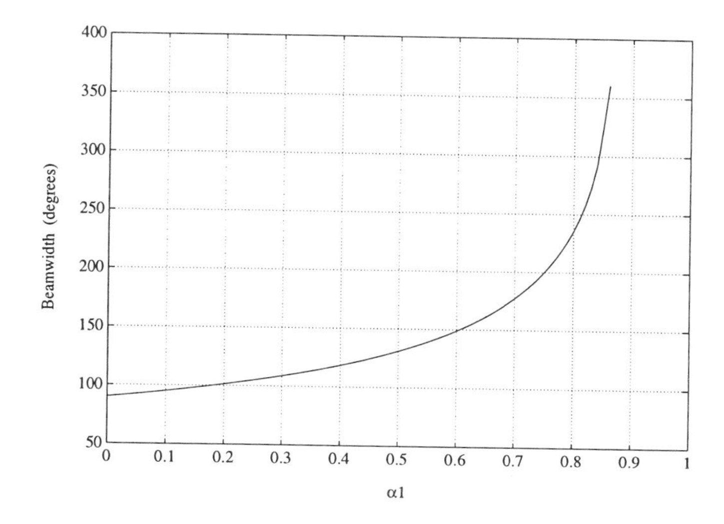

Figure 10.8 3 dB beamwidth of first-order microphone versus the first-order differential parameter α_1.

The fact that the output phase reverses in either direction is of no concern to us here. The directivity index is 4.8 dB and the 3 dB beamwidth for the dipole microphone is 90°. The zero in the response is at $\theta = 90°$. One potential problem, however, is that it is bidirectional; in other words, the pattern is symmetric about the axis tangential to the diaphragm or normal to the two zero-order microphone axis. As a result the front-to-back ratio is equal to 0 dB.

5.1.2 Cardioid. As shown earlier, all first-order patterns correspond to the "limaçon of Pascal" algebraic form. The special case of $\alpha_1 = 1/2$ is the cardioid pattern. The pattern is described by

$$E_{C_1}(\theta) = \frac{1 + \cos\theta}{2} \tag{10.81}$$

which is plotted in Fig. 10.9(b). Although the cardioid microphone is not optimal in directional gain or front-to-back ratio, it is the most commonly manufactured differential microphone. The cardioid directivity index is 4.8 dB, the same as that of the dipole microphone and the 3 dB beamwidth is 131°. The zero in the response is located at $\theta = 180°$. The front-to-back ratio is 8.5 dB.

5.1.3 Hypercardioid. The hypercardioid microphone has the distinction of having the highest directivity index of any first-order microphone. The

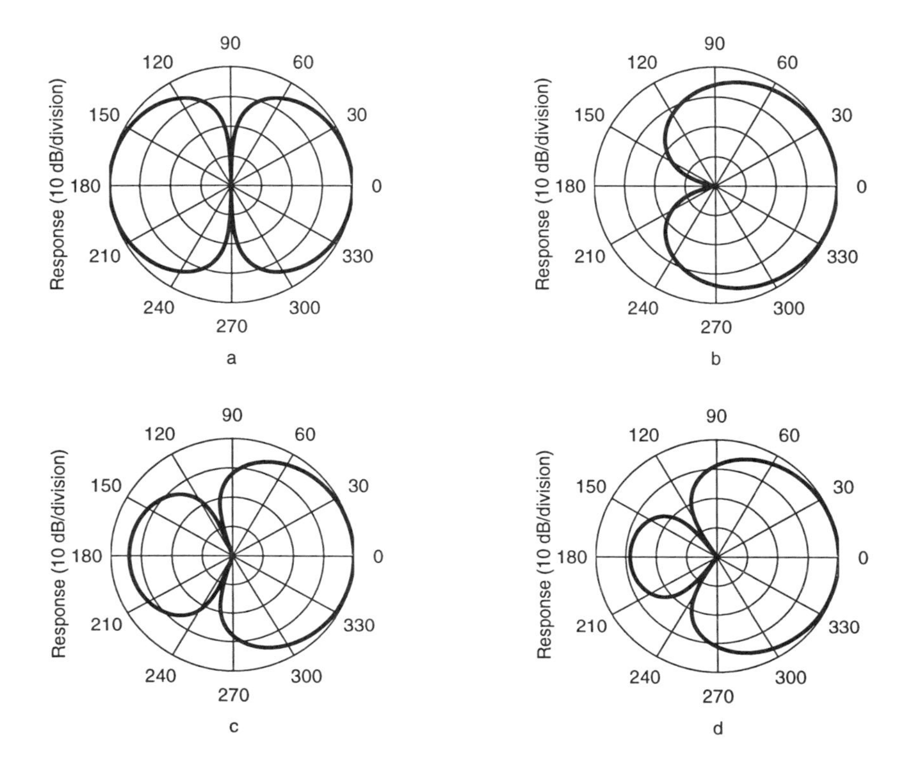

Figure 10.9 Various first-order directional responses, (a) dipole, (b) cardioid, (c) hypercardioid, (d) supercardioid.

derivation, which has apparently not been presented in the literature, is not difficult and was included in Section 4.2. The hypercardioid response can be written as

$$E_{HC_1}(\theta) = \frac{1 + 3\cos\theta}{4}. \tag{10.82}$$

Figure 10.9(c) is a polar plot of the absolute value of (10.82). The 3 dB beamwidth is equal to 105° and the zero is at 109°. The directivity index is 6 dB or 10 $\log_{10}(4)$, the maximum directivity index for a first-order system. The front-to-back ratio is equal to 8.5 dB.

5.1.4 Supercardioid. The name *supercardioid* is commonly used for the first-order differential design which maximizes the front-to-back received power; the term was probably coined by Shure engineers in the early 1940's, although the author could not find direct evidence of this. The first reference to the supercardioid design appears in a 1941 paper by Marshall and Harry [11].

The supercardioid is of interest since of all first-order designs it has the highest front-to-back power rejection for isotropic noise. The derivation of this result is contained in Section 4.3. The directional response can be written as

$$E_{SC_1}(\theta) = \frac{\sqrt{3} - 1 + (3 - \sqrt{3}) \cos \theta}{2}. \tag{10.83}$$

Figure 10.9(d) is a plot of the magnitude of (10.83). The directivity index for the supercardioid is 5.7 dB and the 3 dB beamwidth is 115°. The zero in the response is located at 125°. The front-to-back ratio is equal to 11.4 dB.

5.2 SECOND-ORDER DESIGNS

As with first-order systems, there are an unlimited number of second-order designs. Since second-order microphones are not readily available on the market today, there are no "common" designs. Two designs that have been suggested are the second-order cardioid and the second-order hypercardioid [12, 15]. Another group of proposed differential microphones is a restricted class of equi-sidelobe designs for arbitrary order n [9]. The following section presents some of these designs as well as a non-restricted equi-sidelobe design and a variety of second-order differential array designs based on common first-order microphones. The general second-order form as given in (10.28) has three parameters, a_0, a_1, and a_2. Equivalently, the second-order differential is the product of two first-order differential forms, as shown in (10.29).

The contours in Fig. 10.10 and Fig. 10.11 depict the dependence of DI and F on the parameters α_1 and α_2 from (10.29). Both figures are plotted for values of α_1 and α_2 between -1 and $+1$. Figure 10.10 has a DI maximum value of 9.5 dB and the interval between the contours is 0.5 dB. Figure 10.11 has a maximum value of 24.0 dB; the contours are in 1 dB steps.

5.2.1 Second-Order Dipole. By pattern multiplication, the second-order dipole directional response is the product of two first-order dipoles which have $\cos\theta$ response patterns given by

$$E_{D_2}(\theta) = \cos^2 \theta. \tag{10.84}$$

Figure 10.12(a) shows the polar magnitude response for this array. The directivity index is 7.0 dB, and by symmetry the front-to-back ratio is 0 dB. The 3 dB beamwidth is 65°.

5.2.2 Second-Order Cardioid. In general the term second-order cardioid implies that either first-order term in the second-order expression given in (10.29), can be a cardioid. However, we first adhere to the specification that a second-order cardioid corresponds to the case where both first-order terms are

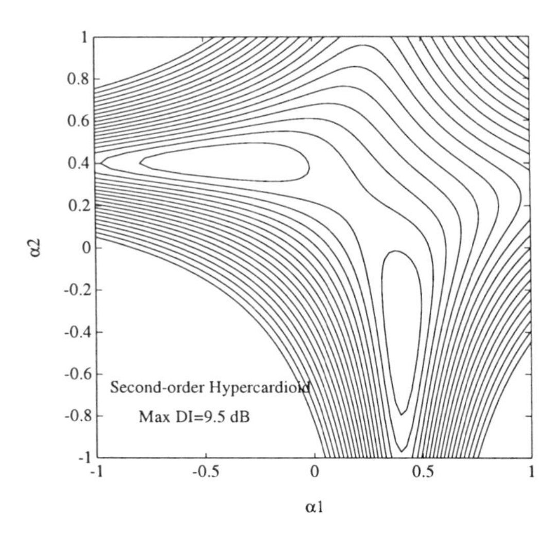

Figure 10.10 Contour plot of the directivity index DI in dB for second-order array versus α_1 and α_2. The contours are in 0.5 dB intervals.

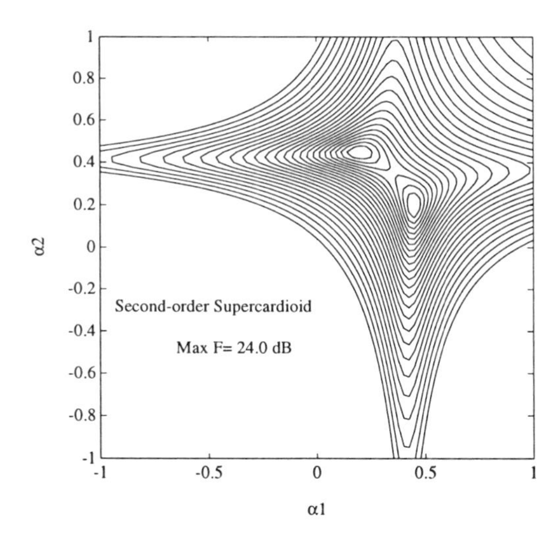

Figure 10.11 Contour plot of the front-to-back ratio in dB for second-order arrays versus α_1 and α_2. The contours are in 1 dB increments.

of the cardioid form. In equation form,

$$E_{C_2}(\theta) = \frac{(1 + \cos\theta)^2}{4}. \tag{10.85}$$

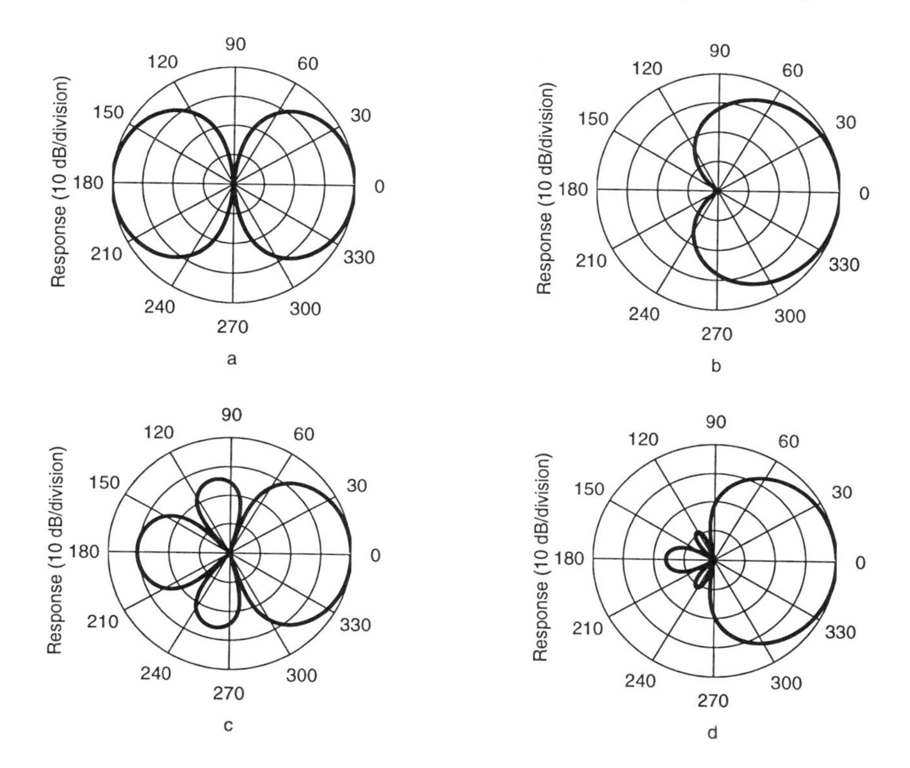

Figure 10.12 Various second-order directional responses, (a) dipole, (b) cardioid, (c) hypercardioid, (d) supercardioid.

In this case $\alpha_1 = \alpha_2 = 0.5$. The directivity index is 7.0 dB and the two zeros both fall at $\theta = 180°$. The front-to-back ratio is equal to 14.9 dB.

Returning to the generality mentioned above, we now consider a second-order cardioid formed from a cardioid and a general first-order response. second-order cardioid design can be easily seen The equation for this second-order differential cardioid is

$$E_{FC_2}(\alpha_1, \theta) = \frac{[\alpha_1 + (1 - \alpha_1) \cos \theta][1 + \cos \theta]}{2}. \qquad (10.86)$$

Olson [12] and Sessler and West [15] presented results for the specific case of α_1=0 in (10.86). Figure 10.13 is a plot of the magnitude of the response for this particular realization. The 3 dB beamwidth of the microphone is 76°. The directivity index is 8.8 dB, the nulls are at 90° and 180°, and the front-to-back ratio is equal to 14.9 dB.

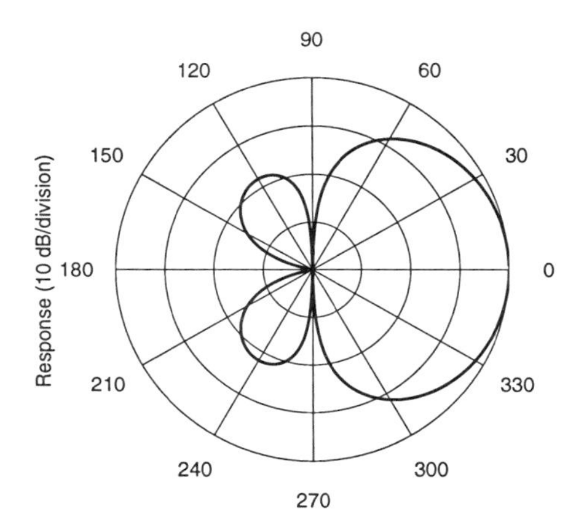

Figure 10.13 Second-order Olson-Sessler-West cardioid directional response.

5.2.3 Second-Order Hypercardioid. The second-order hypercardioid has the highest directivity index of a second-order system; its directivity index is 9.5 dB. The derivation of the directivity pattern and the parameters that determine the second-order hypercardioid are contained in Section 4.2. The results are:

$$\alpha_1 = \pm \frac{1}{\sqrt{6}} \approx \pm 0.41, \tag{10.87}$$

$$\alpha_2 = \mp \frac{1}{\sqrt{6}} \approx \mp 0.41. \tag{10.88}$$

These values correspond to the peaks in Fig. 10.10. The null locations for the second-order hypercardioid are at 73° and 134°. The front-to-back ratio is 8.5 dB, the same for the first-order differential cardioid and first-order differential hypercardioid. A polar response is given in Fig. 10.12(c).

5.2.4 Second-Order Supercardioid. The term second-order supercardioid designates an optimal design for the second-order differential microphone with respect to the front-to-back received power ratio. The derivation for the supercardioid microphone was given in Section 4.3. The results (repeated here) are:

$$\alpha_1 = \frac{\sqrt{7} - 2 \pm \sqrt{8 - 3\sqrt{7}}}{2} \approx 0.45,\ 0.20, \tag{10.89}$$

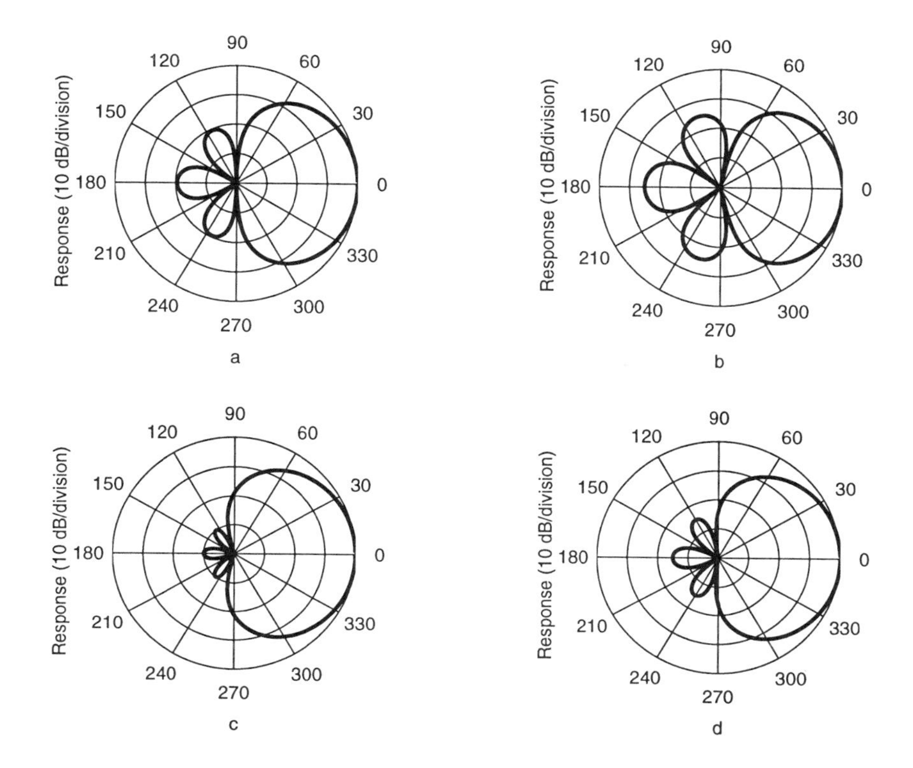

Figure 10.14 Various second-order equi-sidelobe designs, (a) Korenbaum design, (b) −15 dB sidelobes, (c) −30 dB sidelobes, (d) minimum rear half-plane peak response.

$$\alpha_2 = \frac{\sqrt{7} - 2 \mp \sqrt{8 - 3\sqrt{7}}}{2} \approx 0.20,\ 0.45. \tag{10.90}$$

These values correspond to the peaks in Figure 10.11. Figure 10.12(d) is a plot of the magnitude of the directional response. The directivity index is equal to 8.3 dB, the nulls are located at 104° and 144°, and the front-to-back ratio is 24.0 dB.

5.2.5 Equi-Sidelobe Second-Order Differential. Since a second-order differential microphone has two zeros in its response it is possible to design a second-order differential microphone such that the two lobes defined by these zeros are at the same level. Figure 10.14(a) shows the *only* second-order differential equi-sidelobe design possible using the form of (10.71). (The second-order differential cardioid as shown in Fig. 10.13 is not considered an equi-sidelobe design since the two lobes in the figure are actually the same lobe due to the symmetry of the polar pattern.) The directivity index of this *Koren-*

baum second-order differential array is 8.9 dB. The beamwidth is 76° and the front-to-back ratio is 17.6 dB.

We begin our analysis of second-order Chebyshev differential arrays begins by comparing terms of the Chebyshev polynomial and the second-order array response function. The Chebyshev polynomial of order 2 is

$$\begin{aligned} T_2(x) &= 2x^2 - 1 \\ &= 2b^2 - 1 + 4a\,b\,\cos\theta + 2a^2\cos^2\theta. \end{aligned} \tag{10.91}$$

Comparing like terms of (10.91) and (10.28) yields:

$$\begin{aligned} a_0 &= \frac{2b^2 - 1}{S} \\ a_1 &= \frac{4ab}{S} \\ a_2 &= \frac{2a^2}{S}, \end{aligned} \tag{10.92}$$

where S is again the sidelobe threshold.

By substituting the results of (10.75) into (10.92), we can determine the necessary coefficients for the desired equi-sidelobe second-order differential microphone:

$$\begin{aligned} a_0 &= \frac{x_o^2 - 2x_o - 1}{2S} \\ a_1 &= \frac{x_o^2 - 1}{S} \\ a_2 &= \frac{(x_o + 1)^2}{2S}, \end{aligned} \tag{10.93}$$

where

$$x_o = \cosh\left(\frac{1}{2}\cosh^{-1} S\right). \tag{10.94}$$

Thus for the second-order differential microphone,

$$\theta_{1,2} = \cos^{-1}\left(\frac{1 - x_o \pm \sqrt{2}}{1 + x_o}\right). \tag{10.95}$$

The zero locations given in (10.95) can be used along with (10.32) and (10.33) to determine the canonic first-order differential parameters α_1 and α_2. Figures 10.14(b) and 10.14(c) show the resulting second-order designs for -15 dB and -30 dB sidelobes respectively. The directivity indices for the two designs are respectively 9.4 dB and 8.1 dB. The null locations for the -15 dB sidelobes design are at 78° and 142°. By allowing a higher sidelobe level than the Korenbaum design (for $x_o = 1 + \sqrt{2}$, with $\theta_1 = 90°$), a higher directivity index can be achieved. In fact, the directivity index monotonically increases until the sidelobe levels exceed -13 dB; at this point the directivity index reaches its maximum at 9.5 dB, almost the maximum directivity index for a second-order differential microphone. For sidelobe levels less than -20.6 dB (Korenbaum design), both nulls are in the rear half-plane of the second-order microphone; the null locations for the -30 dB sidelobes design are at 109 and 152 degrees. Equi-sidelobe second-order directional patterns is always contain a lobe peak at $\theta = 180°$.

An interesting design that arises from the preceding development is a second-order differential microphone that minimizes the *peak* rear half-space response. This design corresponds to the case where the front-lobe response level at $\theta = 90°$ is equal to the equi-sidelobe level (for $x_o = 3$). Figure 10.14(d) is a directional plot of this realization. The canonic first-order differential parameters for this equi-sidelobe design are:

$$\begin{aligned} \alpha_1 &= \frac{5 \pm 2\sqrt{2}}{17} \\ \alpha_2 &= \frac{5 \mp 2\sqrt{2}}{17}. \end{aligned} \tag{10.96}$$

The directivity index is 8.5 dB. The nulls are located at 149° and 98°, and the front-to-back ratio is 22.4 dB.

Two other design possibilities can be obtained by determining the equi-sidelobe second-order design that maximizes either the directivity index DI or the front-to-back ratio F. Figure 10.15 is a plot of the directivity and front-to-back indices as a function of sidelobe level. As mentioned earlier, a -13 dB sidelobe level maximizes the directivity index at 9.5 dB. A sidelobe level of -27.5 dB maximizes the front-to-back ratio. Plots of these two designs are shown in Fig. 10.16. Of course, an arbitrary combination of DI and F could also be maximized for some given optimality criterion if desired.

5.2.6 Maximum Second-Order Differential DI and F Using Common First-Order Differential Microphones. Another typical approach to the design of second-order differential microphones involves the combinations of the outputs of two first-order differential microphones. Specifically, the combination is a subtraction of the first-order differential outputs after one is passed through a delay element. If the first-order differential microphone can be designed to have any desired canonic parameter α_1, then *any* second-order differential array can be designed. More commonly, however, the designer will have to work with off the shelf first-order differential microphones, such as, the

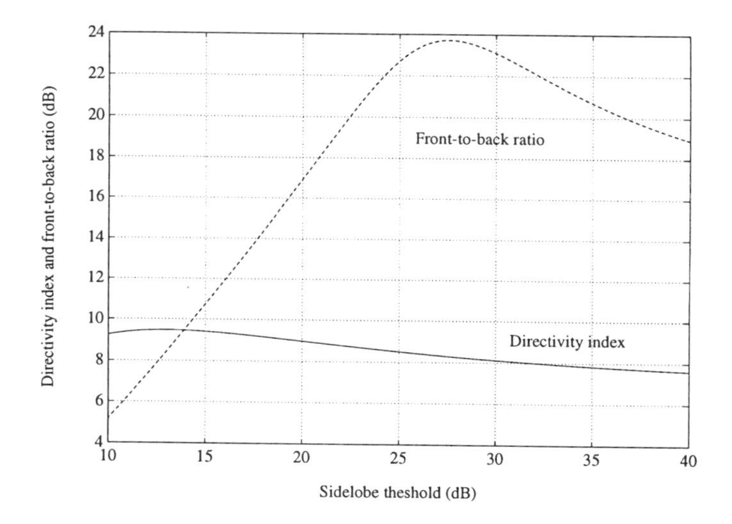

Figure 10.15 Directivity index (solid) and front-to-back ratio (dotted) for equi-sidelobe second-order array designs versus sidelobe level.

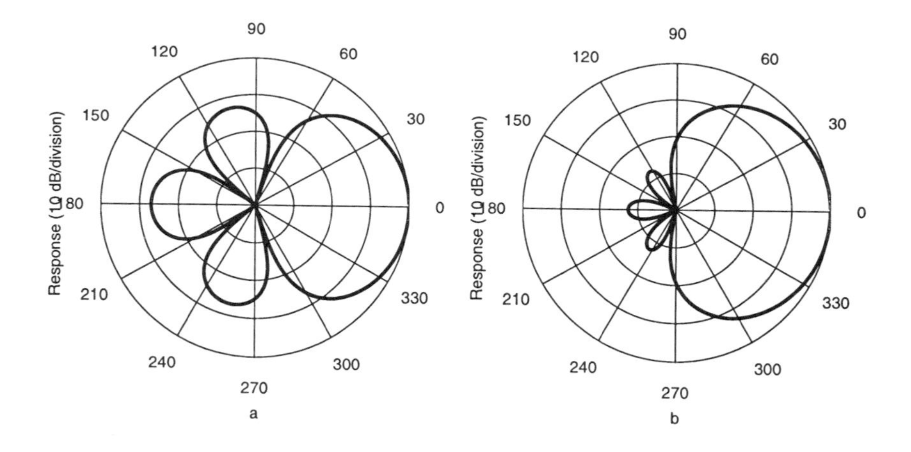

Figure 10.16 Directional responses for equi-sidelobe second-order differential arrays for, (a) maximum directivity index, and, (b) maximum front-to-back ratio.

standard first-order differential designs discussed in Section 3.1. If we constrain the second-order differential design to these typical first-order differential microphones, then we will not be able to reach the maximum directivity index and front-to-back ratio that is possible with the more general second-order differential array designed discussed previously. The following equations show how to

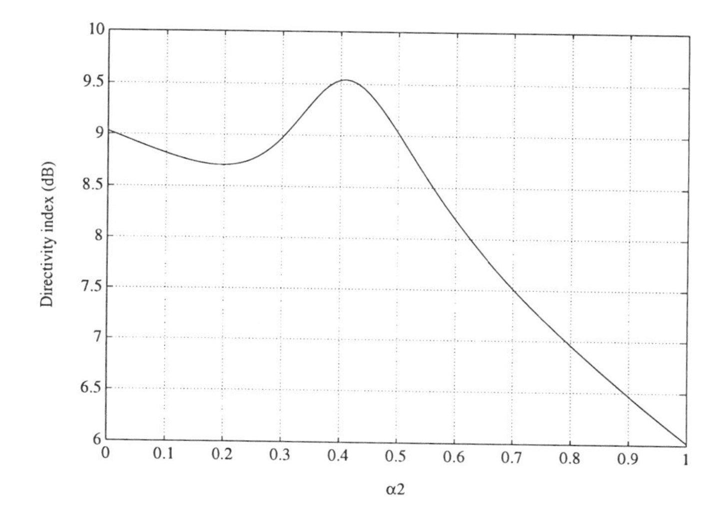

Figure 10.17 Maximum second-order differential directivity index DI for first-order differential microphones defined by (10.97).

implement an optimal design with respect to directivity index and front-to-back power rejection.

Given a first-order differential microphone with the directivity function,

$$E_{N_1}(\alpha_2, \theta) = \alpha_2 + (1 - \alpha_2) \cos \theta, \tag{10.97}$$

where α_2 is a constant, we would like to know how to combine two of these microphones so that the directivity index is maximized. The maximum is found by multiplying (10.97) by a general first-order response, integrating the square of this product from $\theta = 0$ to π, taking derivative with respect to α_1, and setting the resulting derivative to zero. The result is

$$\alpha_1 = \frac{3}{8} - \frac{5\alpha_2}{8(2 - 9\alpha_2 + 12\alpha_2^2)}. \tag{10.98}$$

A plot of the directivity index for $0 \leq \alpha_2 \leq 1$ is shown in Fig. 10.17. A maximum value of 9.5 dB occurs when $-\alpha_1 = \alpha_2 \approx 0.41$.

A similar calculation for the maximum front-to-back power response yields

$$\alpha_1 = \frac{\sqrt{8\alpha_2^4 + 8\alpha_2^3 + 8\alpha_2^2 - 12\alpha_2 + 3}\sqrt{12\alpha_2^4 + 6\alpha_2^3 + 17\alpha_2^2 - 20\alpha_2 + 5} - 8\alpha_2^3 - 12\alpha_2^2 + 13\alpha_2 - 3}{24\alpha_2^4 - 6\alpha_2 + 2}. \tag{10.99}$$

A plot of the front-to-back ratio for $0 \leq \alpha_2 \leq 1$ is shown in Fig. 10.18.

A maximum value of 24.0 dB occurs when $\alpha_2 \approx 0.45$ and $\alpha_1 \approx 0.20$, which are the values of the second-order supercardioid. By symmetry, the values of

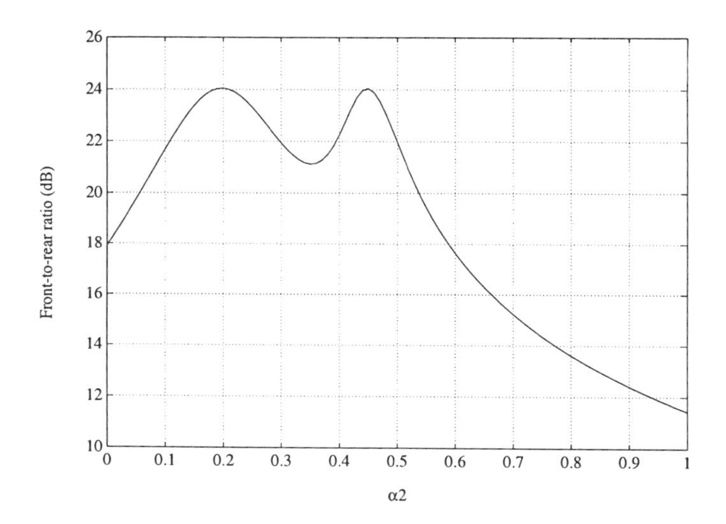

Figure 10.18 Maximum second-order differential front-to-back ratio for first-order differential microphones defined by (10.97).

α_1 and α_2 can obviously be interchanged. The double peak at $F = 24.0$ dB in Figure 10.18 is a direct result of this symmetry.

5.3 THIRD-ORDER DESIGNS

Very little can be found in the open literature on the construction and design of third-order differential microphones. The earliest paper in which an actual device was designed and constructed was authored by B. R. Beavers and R. Brown in 1970 [2]. This lack of any papers is not unreasonable given both the extreme precision that is necessary to realize a third-order array and the serious signal-to-noise problems. Recent advances in low noise microphones and electronics, however, support the feasibility of third-order microphone construction. With this in mind, the following section describes several possible design implementations. The mathematics that govern the directional response were given in (10.34) and (10.35).

5.3.1 Third-Order Differential Dipole. By the pattern multiplication the third-order dipole directional response is given by

$$E_{D_3}(\theta) = \cos^3 \theta. \tag{10.100}$$

Figure 10.19(a) shows the magnitude response for this array.

The directivity index is 8.5 dB, the front-to-back ratio is 0 dB and the 3 dB beamwidth is 54°.

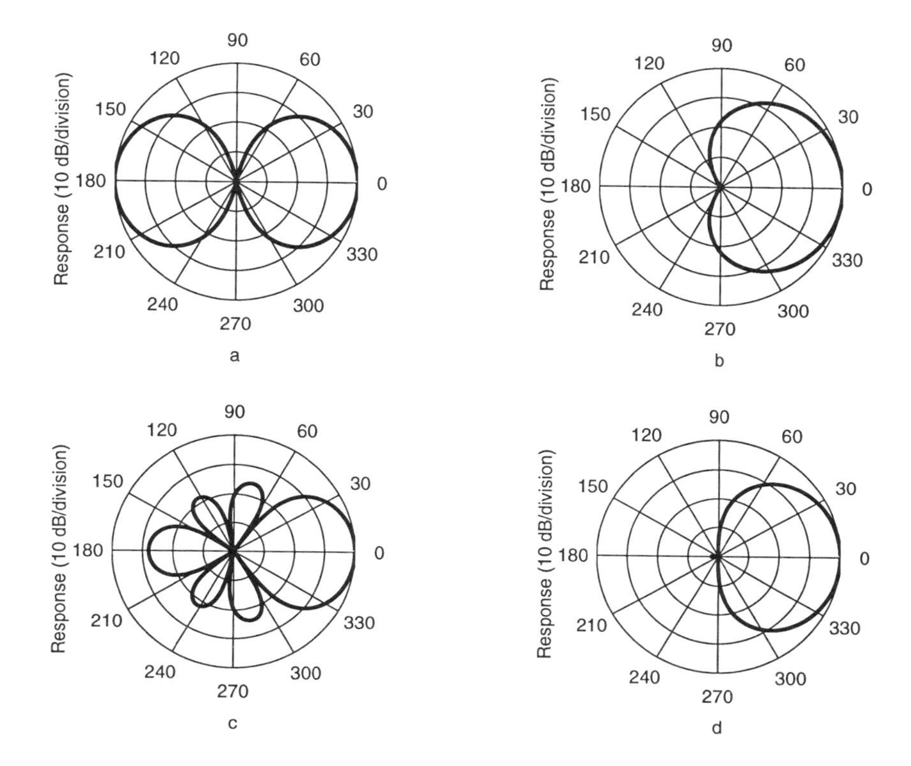

Figure 10.19 Various third-order directional responses, (a) dipole, (b) cardioid, (c) hypercardioid, (d) supercardioid.

5.3.2 Third-Order Differential Cardioid. In Section 3.2.2 we noted that the terminology of cardioids is ambiguous for second-order arrays. For the third-order, this ambiguity is even more pervasive. Nevertheless, we begin by suggesting the most obvious array possibility. If we form the cardioid by the straightforward pattern multiplication of three first-order differential cardioids we have

$$E_{c_3} = \frac{(1 + \cos\,\theta)^3}{8}. \tag{10.101}$$

Figure 10.19(b) shows the directional response for this array. The three nulls all fall at 180°. The directivity index is 8.5 dB and the front-to-back ratio is 21.0 dB.

Another possible design for the third-order cardioid corresponds to the pattern multiplication of a first-order differential cardioid with a second-order differential bidirectional pattern. The third-order differential OSW cardioid design implies that $\alpha_1 = 1$ and $\alpha_2 = \alpha_3 = 0$ in (10.35). Figure 10.20 shows

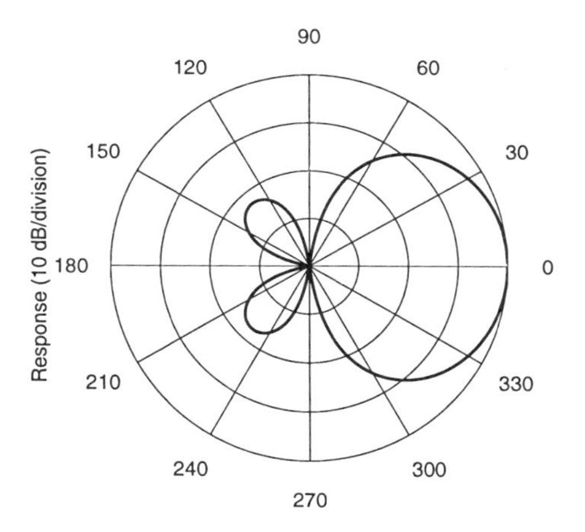

Figure 10.20 Third-order Olson-Sessler-West cardioid directional response.

the directional response of this third-order differential microphone. The directivity index is 10.7 dB and the front-to-back ratio is 18.5 dB. The nulls fall at $\theta = 180^\circ$ and at $\theta = 90^\circ$, the latter of which is a second-order zero.

5.3.3 Third-Order Differential Hypercardioid. The derivation of the third-order differential hypercardioid was given in Section 4.2. The results for the coefficients in (10.34) are:

$$\begin{aligned} a_0 &= -3/32 \\ a_1 &= -15/32 \\ a_2 &= 15/32 \\ a_3 &= 35/32. \end{aligned} \tag{10.102}$$

After solving for the roots of (10.34) with the coefficients given in (10.102), we can obtain the coefficients of the canonic representation given in (10.35), namely

$$\begin{aligned} \alpha_1 &= 1/2\left[\sqrt{5}\cos(\phi/3) - 1/2\right] \approx 0.45 \\ \alpha_2 &= 1/2\left[\sqrt{5}\cos(\phi/3 + 2\pi/3) - 1/2\right] \approx 0.15 \\ \alpha_3 &= 1/2\left[\sqrt{5}\cos(\phi/3 + 4\pi/3) - 1/2\right] \approx -1.35, \end{aligned} \tag{10.103}$$

where

$$\phi = \arccos\left(-2/\sqrt{5}\right). \tag{10.104}$$

Figure 10.19(c) shows the directional response of the third-order differential hypercardioid. The directivity index of the third-order differential hypercardioid is the maximum for all third-order designs: 12.0 dB. The front-to-back ratio is 11.2 dB and the three nulls are located at θ= 55°, 100°, and 145°.

5.3.4 Third-Order Differential Supercardioid. The third-order supercardioid is the third-order differential array with the maximum front-to-back power ratio. The derivation of this array was given in Section 4.3. The requisite coefficients a_i are:

$$\begin{aligned}
a_0 &= \frac{\sqrt{2}\sqrt{21-\sqrt{21}}-\sqrt{21}-1}{8} \approx 0.018 \\
a_1 &= \frac{21+9\sqrt{21}-\sqrt{2}(6+\sqrt{21})\sqrt{21-\sqrt{21}}}{8} \approx 0.200 \\
a_2 &= \frac{3[\sqrt{2}(4+\sqrt{21})\sqrt{21-\sqrt{21}}-25-5\sqrt{21}]}{8} \approx 0.475 \\
a_3 &= \frac{63+7\sqrt{21}-\sqrt{2}(7+2\sqrt{21})\sqrt{21-\sqrt{21}}}{8} \approx 0.306.
\end{aligned} \quad (10.105)$$

Figure 10.19(d) is a directivity plot of the resulting supercardioid microphone. The directivity index is 9.9 dB and the front-to-back ratio is 37.7 dB. The nulls are located at 97°, 122°, and 153°. The third-order supercardioid has almost no sensitivity to the rear half-plane. For situations where the user desires information from only one half-plane, the third-order supercardioid microphone performs optimally. Finding the roots of (10.34) with the coefficients given by (10.105) yields the parameters of the canonic expression given in (10.35). The results are:

$$\begin{aligned}
\alpha_1 &\approx .113 \\
\alpha_2 &\approx .473 \\
\alpha_3 &\approx .346.
\end{aligned} \quad (10.106)$$

5.3.5 Equi-Sidelobe Third-Order Differential. Finally, we discuss the design of equi-sidelobe third-order differential arrays. Like the design of equi-sidelobe second-order differential microphones, the design of third-order equi-sidelobe arrays relies on he use of Chebyshev polynomials and the Dolph-Chebyshev antenna synthesis technique. The basic technique was discussed earlier in Section 2.3. For the third-order microphone, n=3, and the Chebyshev polynomial is

$$T_3(x) = 4x^3 - 3x. \quad (10.107)$$

Using the transformation $x = b + a \cos\theta$ and comparing terms with (10.34) we have

$$\begin{aligned} a_0 &= \frac{b(4b^2-3)}{S} \\ a_1 &= \frac{3a(4b^2-1)}{S} \\ a_2 &= \frac{12a^2 b}{S} \\ a_3 &= \frac{4a^3}{S}. \end{aligned} \tag{10.108}$$

Combining (10.73), (10.74), (10.75), and (10.108) yields the coefficients describing the equi-sidelobe third-order differential. These results are:

$$\begin{aligned} a_0 &= \frac{x_o^3 - 3x_o^2 + 2}{2S} \\ a_1 &= \frac{3x_o(x_o+1)(x_o-2)}{2S} \\ a_2 &= \frac{3(x_o-1)(x_o+1)^2}{2S} \\ a_3 &= \frac{(x_o+1)^3}{2S}. \end{aligned} \tag{10.109}$$

Figures 10.21(a) and 10.21(b) show the resulting patterns for -20 dB and -30 dB sidelobe levels.

From (10.77), the nulls for the -20 dB sidelobe levels are at 62°, 102°, and 153°. The nulls for the -30 dB sidelobe design are at 79°, 111°, and 156°. The directivity indices are 11.8 dB and 10.8 dB, respectively. The front-to-back ratio for the -20 dB design is 14.8 dB and for the -30 dB design is 25.2 dB.

Finally, we examine the directivity index and the front-to-back ratio for the equi-sidelobe third-order array as a function of sidelobe level. Figure 10.22 shows these two quantities for equi-sidelobe levels from -10 dB to -60 dB.

The directivity reaches its maximum at 12.0 dB for a sidelobe level of -16 dB. The -16 dB equi-sidelobe design plotted in Fig. 10.23(a), approaches the optimal third-order differential hypercardioid of Fig. 10.19(c).

The front-to-back ratio reaches a maximum of 37.3 dB at a sidelobe level of -42.5 dB; the response is plotted in Fig. 10.23(b). For sidelobe levels less than -42.5 dB, the mainlobe moves into the rear half-plane. For sidelobe levels greater than -42.5 dB, the zero locations move towards $\theta = 0°$ and as a result the beamwidth decreases.

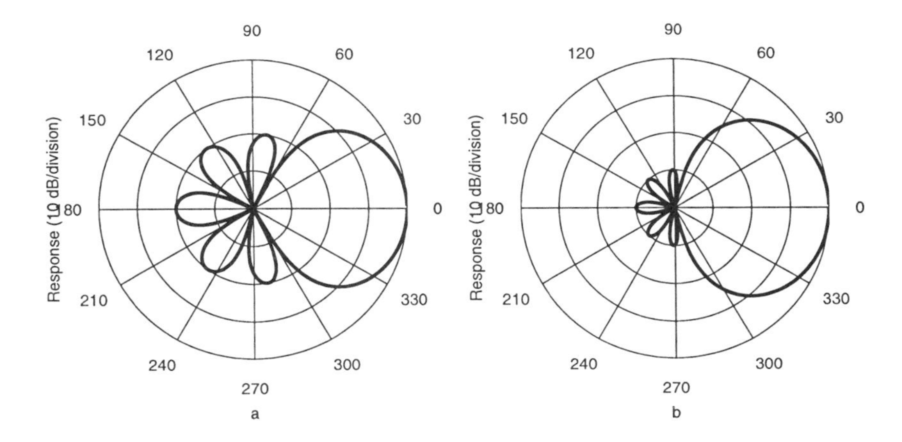

Figure 10.21 Equi-sidelobe third-order differential microphone for (a) −20 dB and (b) −30 dB sidelobes.

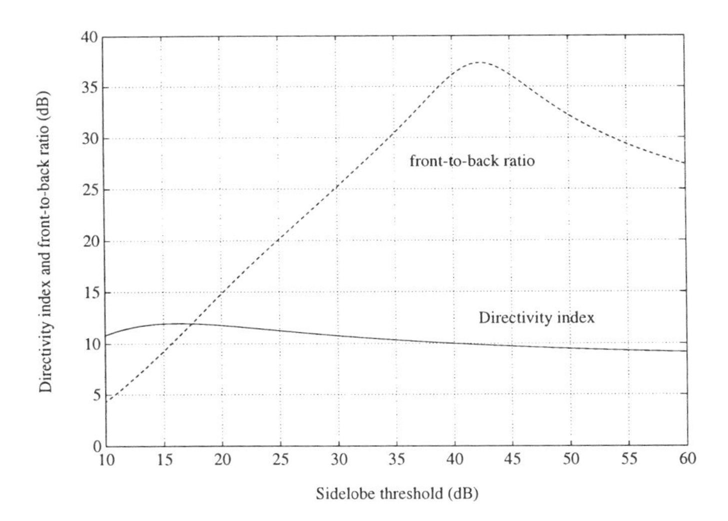

Figure 10.22 Directivity index and front-to-back ratio for equi-sidelobe third-order differential array designs versus sidelobe level.

5.4 HIGHER-ORDER DESIGNS

Due to sensitivity to electronic noise and microphone matching requirements, differential array designs higher that third-order are not practically realizable. These arrays will probably never be implemented on anything other than in a computer simulation. In fact, the design of higher-order supercardioid and

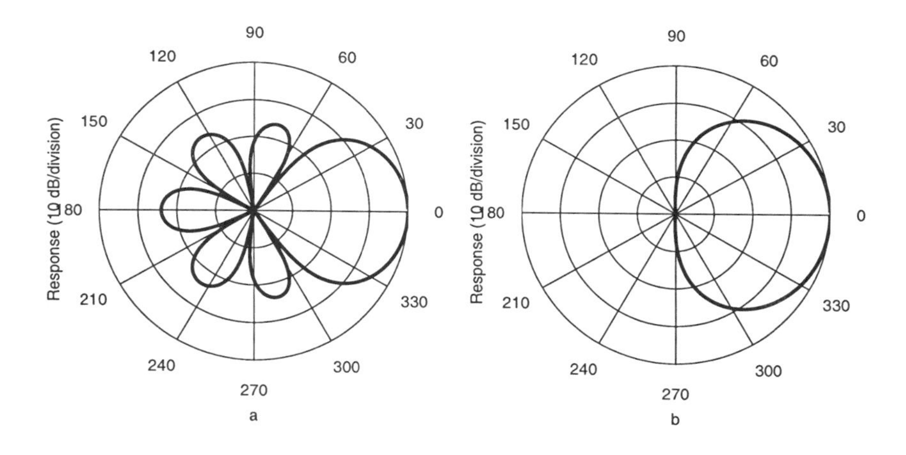

Figure 10.23 Directivity responses for equi-sidelobe third-order differential arrays for (a) maximum directivity index and (b) maximum front-to-back ratio.

hypercardioid differential arrays using the techniques discussed in Sections 4.2 and 4.3, can become difficult on present computers.

6. OPTIMAL ARRAYS FOR CYLINDRICALLY ISOTROPIC FIELDS

Although the standard to model for reverberant acoustic fields has been the "diffuse" field model (spherically isotropic noise), another noise field that is appropriate for room acoustics is cylindrical noise. In many rooms, where carpet and ceiling tiles are used, the main surfaces of absorption are the ceiling and the floor. As a result, a noise field model that has the noise propagating in the axial plane may be more appropriate. This type of noise field is better modeled as a cylindrically distributed noise field and the optimization of directional gain in this type of field is therefore of interest. The following sections deal with the design of differential arrays in a cylindrical noise field.

6.1 MAXIMUM GAIN FOR OMNIDIRECTIONAL MICROPHONES

For a cylindrically isotropic field we have plane waves arriving with equal probability from any angle ϕ and wave vector directions that lie in the ϕ plane. The cylindrical directivity factor in this field is defined as

$$Q_C(\omega, \phi_o) = \frac{| E(\omega, \phi_o) |^2}{\frac{1}{2\pi} \int_0^{2\pi} | E(\omega, \phi) |^2 \, u(\omega, \phi) d\phi}. \tag{10.110}$$

Again, the general weighting function u allows for the possibility of a nonuniform distribution of noise power and microphone directivity. For the case that we are presently interested in

$$u(\omega, \phi) = 1. \tag{10.111}$$

Following the development in the last section, we expand the directional response of N closely-spaced elements in a series of orthogonal functions. For cylindrical fields we can use the normal expansion in the ϕ dimension:

$$E(\phi) = \sum_{m=0}^{N-1} h_m \cos[m(\phi - \phi_z)]. \tag{10.112}$$

The normalization of these cosine functions is simply [1]:

$$\begin{aligned} N_m &= \int_0^{2\pi} \cos^2(m\phi)\, d\phi \\ &= \frac{2\pi}{\varepsilon_m}. \end{aligned} \tag{10.113}$$

The directivity factor can therefore be written as

$$Q_C(\phi_o) = \frac{\left[\sum_{m=0}^{N-1} h_m \cos(m(\phi_o - \phi_z))\right]^2}{\frac{1}{2\pi}\sum_{m=0}^{N-1} h_m^2 N_m}. \tag{10.114}$$

The maximum is found by equating the derivative of this equation for Q_C with respect to the h_m weights to zero. The result is

$$Q_{Cmax}(\phi_o) = \sum_{m=0}^{N-1} \frac{\cos^2[m(\phi_o - \phi_z))}{N_m}. \tag{10.115}$$

The equation for Q_C maximizes when $\phi_o = \phi_z$. Therefore

$$\begin{aligned} Q_{Cmax} &= \sum_{m=0}^{N-1} \varepsilon_m \\ &= 2N - 1. \end{aligned} \tag{10.116}$$

The above result indicates that the maximum directional gain of N closely-spaced omnidirectional microphones in a cylindrically correlated sound field is $2N - 1$. This result is apparently known in the microphone array processing community [5], but apparently there is no general proof that has been published. One obvious conclusion that can be drawn from the above result is that the rate of increase in directional gain as a function of the number of microphones is much slower in a cylindrically isotropic field than a spherically isotropic field. A plot comparing the maximum gain for microphone arrays up to $N = 10$ for both types of isotropic fields is shown in Fig. 10.24.

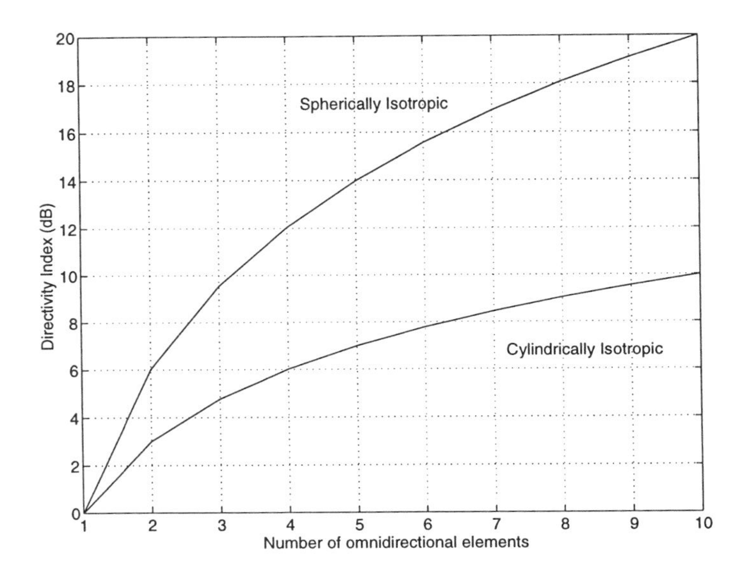

Figure 10.24 Maximum gain of an array of N omnidirectional microphones for spherical and cylindrical isotropic noise fields.

6.2 OPTIMAL WEIGHTS FOR MAXIMUM DIRECTIONAL GAIN

It was shown previously that the directivity pattern for a Nth-order differential array can in general be written as

$$E(\theta) = a_0 + a_1 \cos\theta + a_2 \cos^2\theta + ... + a_{N-1} \cos^{N-1}\theta, \tag{10.117}$$

where the coefficients a_i are general weighting scalars. The directivity factor, Q_C, from (10.38) is

$$Q_C(a_0, ...a_{N-1}) = \frac{\mathbf{a}^T\mathbf{B}\mathbf{a}}{\mathbf{a}^T\mathbf{H}\mathbf{a}}, \tag{10.118}$$

where the subscript C indicate a cylindrical field and $\mathbf{H}$ is a Hankel matrix given by

$$H_{i,j} = \begin{cases} \dfrac{(i+j-1)!!}{(i+j)!!} & \text{if i+j even ,} \\ 0 & \text{otherwise,} \end{cases}$$

where

$$\mathbf{a}^T = \{a_0, a_1, ..., a_n\}, \tag{10.119}$$

Table 10.4 Table of maximum eigenvalue and corresponding eigenvector for differential arrays from first to fourth-order, for cylindrically isotropic noise fields.

microphone order	maximum eigen-value	corresponding eigenvector
1	3	[1/3 1/2]
2	5	[-1/5 2/5 4/5]
3	7	[-1/7 -4/7 4/7 8/7]
4	9	[1/9 -4/9 -4/3 8/9 16/9]

$$\mathbf{B} = \mathbf{b}\mathbf{b}^T, \tag{10.120}$$

and

$$\mathbf{b}^T = \overbrace{\{1, 1, ..., 1\}}^{n+1}. \tag{10.121}$$

The double factorial function is defined as [1]: $(2n)!! = 2 \cdot 4 \ldots \cdot (2n)$ for n even and $(2n+1)!! = 1 \cdot 3 \cdot \ldots \cdot (2n+1)$ for n odd. As was stated previously, the maximum directivity factor Q_C is equal to the maximum eigenvalue of the equivalent generalized eigenvalue problem:

$$\mathbf{B}\mathbf{x} = \lambda \mathbf{H}\mathbf{x}\,. \tag{10.122}$$

where, λ is the general eigenvalue and $\mathbf{x}$ is the corresponding general eigenvector. The eigenvector corresponding to the largest eigenvalue will contain the coefficients a_i which maximize the directivity factor Q_C. Since $\mathbf{B}$ is a dyadic product there is only one eigenvector $\mathbf{x} = \mathbf{H}^{-1}\mathbf{b}$ with the eigenvalue $\mathbf{b}^T\mathbf{H}^{-1}\mathbf{b}$. Thus,

$$\max_{\mathbf{a}} Q_C = \lambda_m = \mathbf{b}^T\mathbf{H}^{-1}\mathbf{b}. \tag{10.123}$$

Table 10.4 gives the optimum values for array gain in isotropic noise for differential orders up to and including fourth-order. A plot of the highest array gain directivity patterns for differential orders up to fourth-order is given in Fig. 10.25.

6.3 SOLUTION FOR OPTIMAL WEIGHTS FOR MAXIMUM FRONT-TO-BACK RATIO FOR CYLINDRICAL NOISE

The front-to-back ratio F is defined as the ratio of the power of the output of the array to signals propagating from the front-half plane to the output power for

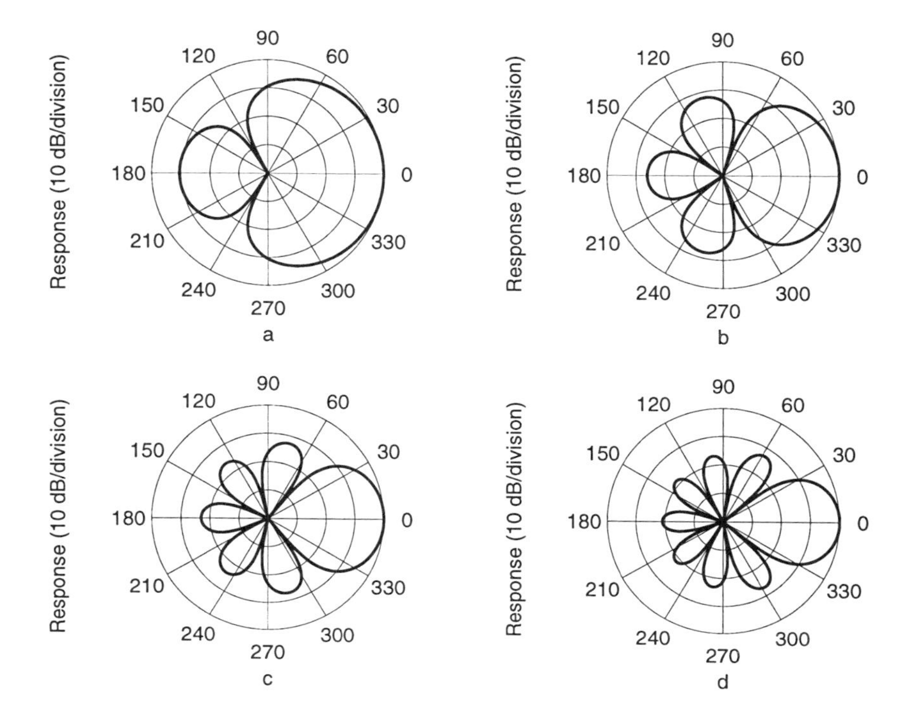

Figure 10.25 Optimum directivity patterns for differential arrays in a cylindrically isotropic noise field for (a) first, (b) second, (c) third, and (d) fourth-order

signals arriving from the rear-half plane. The ratio for cylindrically isotropic fields is mathematically defined as

$$F_C = \frac{\int_0^{\pi/2} \mid E(\phi) \mid^2 \, d\phi}{\int_{\pi/2}^{\pi} \mid E(\phi) \mid^2 \, d\phi} \,. \tag{10.124}$$

Using the differential pattern expansion from (10.117), we can write

$$F_C = \frac{\mathbf{a}^T \mathbf{B} \mathbf{a}}{\mathbf{a}^T \mathbf{H} \mathbf{a}} \,,$$

where

$$B_{ij} = \frac{\Gamma(\frac{1+i+j}{2})}{\Gamma(\frac{2+i+j}{2})}$$

and

$$H_{ij} = (-1)^{i+j} \frac{\Gamma(\frac{1+i+j}{2})}{\Gamma(\frac{2+i+j}{2})} \,, \tag{10.125}$$

Table 10.5 Table of maximum eigenvalue corresponding to the maximum front-to-back ratio and corresponding eigenvector for differential arrays from first to fourth-order, for cylindrically isotropic noise fields.

microphone order	maximum eigen-value	corresponding eigenvector
1	$7+4\sqrt{3}$	[$\sqrt{2}$-1 2-$\sqrt{2}$]
2	$\frac{9\pi^2+12\sqrt{22}\pi+88}{9\pi^2-88}$	$\approx$ [0.103 0.484 0.413]
3	$\approx$ 11556	$\approx$[0.002 0.217 0.475 0.286]
4	$\approx$ 336035	$\approx$ [0.00430 0.07429 0.29914 0.42521 0.19705]

where Γ is the Gamma function [1]. From (10.125) we can see that the front-to-back ratio is a Rayleigh quotient. The maximum of the Rayleigh quotient is reached at a value equal to the largest eigenvalue of the equivalent generalized eigenvalue problem

$$\mathbf{Bx} = \lambda \mathbf{Hx}, \tag{10.126}$$

where λ is the eigenvalue and $\mathbf{x}$ is the corresponding eigenvector. Thus, the maximization of the front-to-back ratio is again a general eigenvalue problem with F_C as the largest eigenvalue. As before, the corresponding eigenvector will contain the coefficients a_i, which maximize the front-to-back ratio F_C,

$$\max_{\mathbf{a}} F_C = \max_{k} \lambda_k = \frac{\mathbf{a}_k^T \mathbf{B} \mathbf{a}_k}{\mathbf{a}_k^T \mathbf{H} \mathbf{a}_k}, \tag{10.127}$$

where the subscript k refers to the kth eigenvalue or eigenvector. The matrices $\mathbf{H}$ and $\mathbf{B}$ are real Hankel matrices and are positive definite. The resulting eigenvalues are positive real numbers and the eigenvectors are real.

Table 10.5 contains the maximum front-to-back power ratios (given by the maximum eigenvalue). The maximum eigenvalue for the third and fourth-order cases result in very complex algebraic expressions. Therefore only the numeric results are given. A plot showing the highest front-to-back power ratio directivity patterns given by the optimum weightings (corresponding to the eigenvectors given in Table 10.5 are displayed in Fig. 10.26). One observation that can be made by comparing the cylindrical noise results with the spherical noise results is that the patterns are fairly similar although there are differences. The differences are due to the omission of the sine term in the denominator of

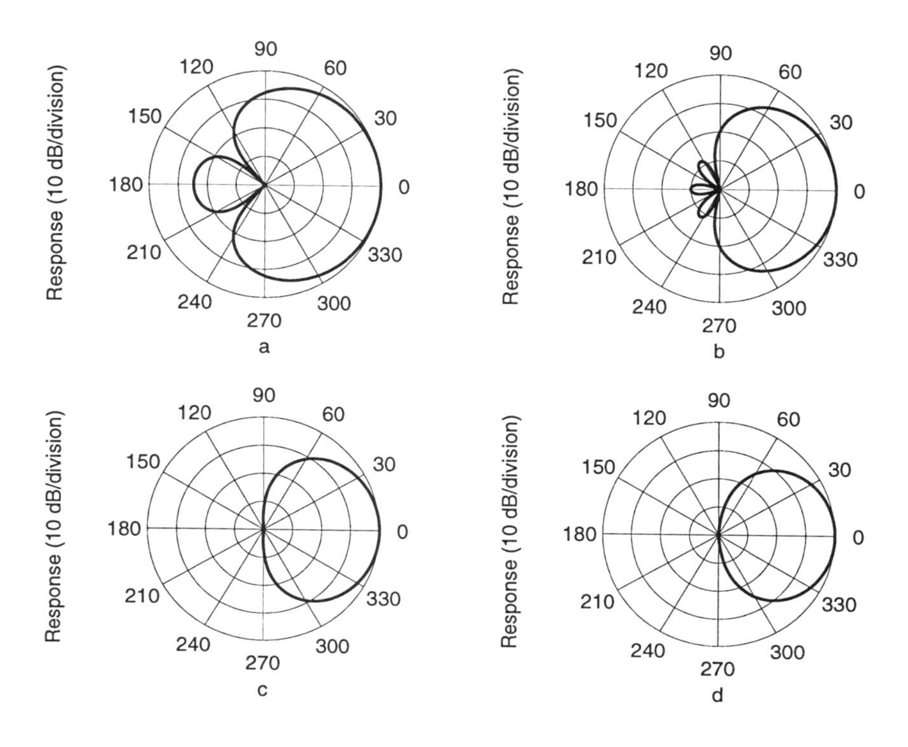

Figure 10.26 Directivity patterns for maximum front-to-back power ratio for differential arrays in a cylindrically isotropic noise field for (a) first, (b) second, (c) third, and (d) fourth-order

(10.36). The optimal patterns for cylindrically isotropic fields keep their sidelobes smaller in the rear-half of the microphone since this area is not weighted down by the sine term. Table 10.6 also summarizes the above results relative to the results for the spherically isotropic condition. Table 10.6 also contains the 3 dB beamwidth and the position of the pattern nulls. By knowing the null positions, these designs can be easily realized by combining first-order sections in a nulling tree architecture.

The results summarized in Table 10.6 also show that there is relatively small difference between the optimal designs of differential arrays for the spherically and cylindrically isotropic noise fields. Typically the differences between the optimum directional gains from either the cylindrical or spherical isotropy assumption results in less than a few tenths of a dB; most probably an insignificant amount. The most significant detail to notice is that the rate of increase in the directional gain versus differential array order is much smaller for cylindrically isotropic fields. This conclusion was also proven earlier and shown in Fig. 10.24.

Table 10.6 Table of maximum directional gain and front-to-back power ratio for differential arrays from first to fourth-order, for cylindrically and spherically isotropic noise fields.

Mic. order	DI_C (dB) cyl	F_C (dB) cyl	DI (dB) sph	F (dB) sph	Beam width degs	Null(s) degs
Maximum gain for cylindrical noise						
1st	4.8	10.9	5.9	11.1	112°	120
2nd	7.0	10.9	9.4	7.5	65°	72,144
3rd	8.5	13.9	11.8	10.3	46°	51,103,154
4th	9.5	13.9	13.7	8.9	36°	80,120,160
Maximum gain for spherical noise						
1st	4.6	7.4	6.0	8.5	105°	109
2nd	6.9	9.7	9.5	8.5	65°	73,134
3rd	8.3	12.4	12.0	11.2	48°	55,100,145
4th	9.4	13.8	14.0	11.2	38°	44,80,117,152
Maximum front-to-back ratio for cylindrical noise						
1st	4.6	12.8	5.4	10.9	120°	135
2nd	6.3	26.3	8.2	23.4	81°	106,153
3rd	7.2	40.6	9.8	37.0	66°	98,125,161
4th	7.8	55.3	10.9	51.1	57°	95,112,137,165
Maximum front-to-back ratio for spherical noise						
1st	4.8	12.0	5.7	11.4	115°	125
2nd	6.4	25.1	8.3	24.0	80°	104,144
3rd	7.2	39.2	9.9	37.7	65°	97,122,154
4th	7.8	53.6	11.0	51.8	57°	94,111,133,159

7. SENSITIVITY TO MICROPHONE MISMATCH AND NOISE

There is a significant amount of literature on the sensitivity of superdirectional microphone array design to interelement errors in position, amplitude, and phase [7, 10, 5]. Since the array designs discussed in this chapter have interelement spacings which are much less than the acoustic wavelength, differential arrays are indeed superdirectional arrays. Early work in superdirectional for *supergain* arrays involved over-steering a Dolph-Chebyshev array past endfire. When the effective interelement spacing becomes much less than the acoustic wavelength, the amplitude weighting of the elements oscillate between plus and minus, resulting in pattern differencing or differential operation. Curiously though, the papers in the field of superdirectional arrays never point out that at small spacings the array can be designed as a differential system as given by (10.25). The usual comment in the literature is that the design of superdirectional arrays requires amplitude weighting that is highly frequency dependent. For the application of the designs that we are discussing, namely differential systems where the wavelength is much larger than the array size, the amplitude weighting is constant with frequency as long as we do not consider the necessary time delay as part of the weighting coefficient. The only frequency correction necessary is the compensation of the output of the microphone for the ω^n high-pass characteristic of the nth order system.

One quantity which characterizes the sensitivity of the array to random amplitude and position errors is the sensitivity function introduced by Gilbert and Morgan [7]. The sensitivity function modified by adding a delay parameter τ is

$$K = \frac{\sum_{m=1}^{n} \mid b_m \mid^2}{\mid \sum_{m=1}^{n} b_m e^{-jk(r_m + c\tau_m)} \mid^2}, \tag{10.128}$$

where, r_m is the distance from the origin to microphone m, b_m are the amplitude shading coefficients of a linear array, and τ_m is the delay associated with microphone m. For the differential microphones discussed, the sensitivity function reduces to

$$K = \frac{n+1}{[\prod_{m=1}^{n} 2 \sin(k\,(d + c\tau_m)/2)]^2}, \tag{10.129}$$

where d is the microphone spacing. For values of $kd \ll 1$, (10.129) can be further reduced to

$$K \approx \frac{n+1}{[\prod_{m=1}^{n} k(d + c\tau_m)]^2}. \tag{10.130}$$

For the nth-order dipole case, (10.130) reduces to

$$K_D \approx \frac{n+1}{(kd)^{2n}}. \tag{10.131}$$

The ramifications of (10.131) are intuitively appealing. The equation merely indicates that the sensitivity to noise and array errors for n^{th}-order differential arrays is inversely proportional to the frequency response of the array.

The response of an array to perturbations of amplitude, phase, and position can be expressed as a function of a common error term δ^2. The validity to combining these terms into one quantity hinges on the assumption that these errors are small compared to the desired values. The reader is referred to the article by Gilbert and Morgan for specific details [7].

The error perturbation power response pattern is dependent on the error term δ^2, the actual desired beam pattern, and the sensitivity factor K. The response is given by

$$E_{N_{np}}(\theta) = E_{N_n}(\theta) + K\,\delta^2. \tag{10.132}$$

Typically δ is very small, and can be controlled by careful design. However, even with careful control, we can only hope for 1% tolerances in amplitude and position. Therefore, even under the best of circumstances we will have trouble realizing a differential array if the value of K approaches or exceeds 10,000 (40 dB). A plot of the value of K for various first-order microphones as a function of the dimensionless parameter kd, is shown in Figure 10.27(a). We note here that of all of the microphone designs discussed, the hypercardioid is the design that has the lowest K factor. This is in direct contradiction to most superdirectional array designs that can be found in the literature [5].

Typically, the higher the directional supergain, the higher the value of K. The reason for the apparent contradiction in Fig. 10.27(a) is that the overall gain of the hypercardioid is higher than the other microphones shown since it uses the highest delay. Other first-order designs with lower values of K are possible, but these do not exhibit the desired optimum directional patterns. Figure 10.27(b) shows the sensitivity function for first, second, and third-order dipoles. As is obvious, a higher differential array order results in a much larger sensitivity. Also, as the phase delay (d/c) between the elements is increased, the upper frequency limit of the usable bandwidth is reduced; in Figures 24a and 24b, it is clear that as the spacing or the frequency is reduced, the sensitivity increases exponentially.

Another problem that is directly related to the sensitivity factor K is the susceptibility of the differential system to microphone and electronic preamplifier noise. If the noise is approximately independent between microphones, the SNR *loss* will be proportional to the sensitivity factor K. At low frequencies

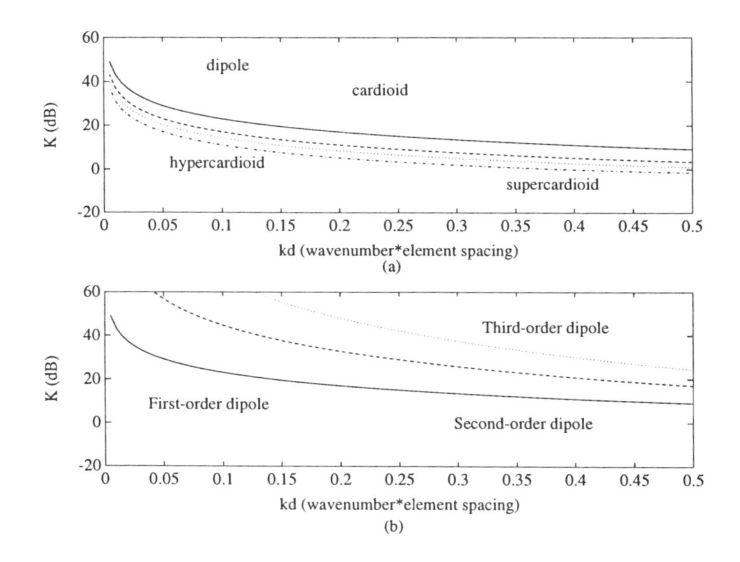

Figure 10.27 Sensitivity as a function of wavelength element-spacing product for, (a) various first-order differential microphones, and, (b) first, second, and third-order dipoles.

and small spacings, the signal to noise ratio can easily become less than 0 dB and that is of course not a very useful design.

As an example, we consider the case of a first-order dipole array with an effective dipole spacing of 1 cm. Assume that the self-noise of the microphone is equal to an equivalent sound pressure level of 35 dB re 20μPa, which is typical for available first-order differential microphones. Now, we place this first-order differential dipole at 1 meter from a source that generates 65 dB at 500 Hz at 1 meter (typical of speech levels). The resulting first-order differential microphone output SNR from Fig. 10.27(b), is only 9 dB. For a second-order array with equivalent spacing the SNR would be -12 dB, and for a third-order array, -33 dB. Although this example makes differential arrays higher than first-order look hopeless, there are design solutions that can improve the situation.

In the design of second-order arrays, West, Sessler, and Kubli [15] used baffles around first-order dipoles to increase the second-order differential signal-to-noise. The diffraction caused by the baffle effectively increases the dipole distance d, by a factor that is proportional to the baffle radius. The diffraction is angle and frequency dependent and, if used properly, can be exploited to offer superior performance to an equivalent dipole composed of two zero-order (omnidirectional) microphones. The use of the baffles discussed in reference [15] resulted in a an effective increase in the SNR by approximately 10 dB.

The benefit of the baffles used by West, Sessler, and Kubli, becomes clear by examining (10.131) and noting the aforementioned increase in the effective dipole distance.

Another possible technique to improve both the signal-to-noise ratio and reduce the sensitivity to microphone amplitude and phase mismatch, is to split the design into multiple arrays, each covering a specific frequency range. In the design of a differential array, the spacing must be kept small compared to the acoustic wavelength. Since the acoustic wavelength is inversely proportional to the frequency, the desired upper frequency cutoff for an array sets the array microphone spacing requirements. If we divide the differential array into frequency subbands, then the ratio of upper frequency to lower frequency cutoff can be reduced and the spacing for each subband can be made much larger than the spacing for a full-band differential array. The increase in signal-to-noise ratio is proportional to the relative increase in spacing allowed by the use of the subband approach. If the desired frequency range is equally divided into m subbands, the lowest subband SNR increase will be proportional to $20 \log_{10}(m)$. The increase in SNR for each increasing frequency subband will diminish until the highest subband which will have the same SNR as the full-band system. The subband solution does have some cost: the number of array elements must also increase. The increase is at least m, and by reuse of the array elements in the subband arrays, can be controlled to be less that nm, where n is the differential array order.

Finally, another approach to control microphone self-noise would be to construct many differential arrays that are very close in position to each other. By combining the outputs of many arrays with uncorrelated self-noise, the SNR can be effectively enhanced by $10 \log_{10}(u)$, where u is the number of individual differential arrays.

8. CONCLUSIONS

Very little work regarding the analytical development of optimal differential microphones can be found in the literature. The purpose of the work presented in this chapter is to provide a basis for the design of differential microphone array systems. Systems with differential orders greater than three require microphones and calibration with tolerances and noise levels not yet currently available. Also, higher order systems are somewhat impractical in that the relative gain in directivity is small $[O(\log_{10}(n))]$ as the order n of the microphone increases. Primarily though, differential microphone array designs are limited by the sensitivity to microphone mismatch and self-noise. One feasible design solution is to split the system into frequency bands and change the spacing as a function of frequency, thereby maintaining an approximately constant signal-to-noise ratio over the desired operating frequency range.

The results presented in this chapter have shown some novel optimal designs for second and third-order systems. Table 10.3 summarizes most of the results.

One possible critical issue that has not been discussed in this chapter is the sensitivity of these differential systems to non-propagating acoustic high-wavenumber noise. These forms of noise occur in turbulent flow and in the near-field of sources undergoing subsonic bending waves below the critical frequency. The problem of high-wavenumber noise can be counteracted by combining multiple differential arrays that are themselves part of a conventional delay-sum beamformer. The synergy here is that the conventional beamformer acts like a spatial lowpass filter, which can greatly reduce the high-wavenumber spatial reception by the differential elements. Another advantage of a hybrid combination of differential elements and a classical delay-sum beamformer is the increase in signal-to-noise ratio achieved by the addition of many independent elements. If the element noise is uncorrelated between differential elements in a classic uniformly weighted array, the gain in signal-to-noise will be $10\log_{10}(N)$, where N is the total number of array elements. The incorporation of differential sub-arrays as elements in a classic delay-sum beamformer has yet another advantage: the differential elements put a lower bound on the frequency dependent directivity index of the classic beamformer.

References

[1] M. Abramowitz and I. A. Stegun, *Handbook of Mathematical Functions*. Dover, New York, 1965.

[2] B. R. Beavers and R. Brown, "Third-order gradient microphone for speech reception," *J. Audio Eng. Soc.*, vol. 16, pp. 636-640, 1970.

[3] D. K. Cheng, "Optimization techniques for antenna arrays," *Proc. IEEE*, vol. 59, pp. 1664-1674, Dec. 1971.

[4] L. J. Chu, "Physical limitations of omni-directional antennas," *J. Applied Physics*, vol. 19, pp. 1163-1175, 1948.

[5] H. Cox, R. Zeskind, and T. Kooij, "Practical supergain," *IEEE Trans. Acoust., Speech, Signal Processing*, vol. ASSP-34, pp. 393-398, June 1986.

[6] C. L. Dolph, "A current distribution for broadside arrays which optimizes the relationship between beamwidth and sidelobe level," *Proc. IRE*, pp. 335-348, June 1946.

[7] E. N. Gilbert and S. P. Morgan, "Optimum design of directive antenna array subject to random variations," *Bell Syst. Tech. J.*, vol. 34, pp. 637-663, 1955.

[8] R. F. Harrington, "On the gain and beamwidth of directional antennas," *IRE Trans. Anten. and Prop.*, pp. 219-223, 1958.

[9] V. I. Korenbaum, "Comments on unidirectional, second-order gradient microphones," *J. Acoust. Soc. Am.*, 1992.

[10] Y. T. Lo, S. W. Lee, and Q. H. Lee, "Optimization of directivity and signal-to-noise ratio of an arbitrary antenna array," *Proc. IEEE*, vol. 54, pp. 1033-1045, 1966.

[11] R. N. Marshall and W. R. Harry, "A new microphone providing uniform directivity over an extended frequency range," *J. Acoust. Soc. Am.*, vol. 12, pp. 481-497, 1941.

[12] H. F. Olson, *Acoustical Engineering*. D. Van Nostrand Company, Inc., Princeton, NJ, 1957.

[13] A. T. Parsons, "Maximum directivity proof of three-dimensional arrays," *J. Acoust. Soc. Amer.*, vol. 82, pp. 179-182, 1987.

[14] M. R. Schroeder, "The effects of frequency and space averaging on the transmission responses of mulitmode media," *J. Acoust. Soc. Am.*, vol. 49, pp. 277-283, 1971.

[15] G. M. Sessler and J. E. West, "Second-order gradient unidirectional microphone utilizing an electret microphone," *J. Acoust. Soc. Am.*, vol. 58, pp. 273-278, July 1975.

[16] C. T. Tai, "The optimum directivity of uniformly spaced broadside arrays of dipoles," *IEEE Trans. Anten. and Prop.*, pp. 447-454, 1964.

[17] A. I. Uzkov, "An approach to the problem of optimum directive antenna design," *Compt. Rend. Dokl. Acad. Sci. USSR*, vol. 53, pp. 35-38, 1946.

[18] D. E. Weston, "Jacobi arrays and circle relationships," *J. Acoust. Soc. Amer.*, vol. 34, pp. 1182-1167, 1986.

Appendix: Directivity Factor and Room Acoustics

The relative sound pressure level (SPL) in a reverberant enclosure can be ideally modeled as the sum of the direct sound signal and a diffuse reverberant part. The ratio of these terms can be used to obtain a rough estimate of the signal-to-reverberation ratio. The relative sound-pressure level is

$$SPL_{rel} = 10 \log_{10} \left[\frac{Q}{4\pi r^2} + \frac{4}{R} \right], \tag{10.A.1}$$

where Q is the directivity factor, and R is the *room constant* equal to $S\bar{\alpha}/(1-\bar{\alpha})$, where S is equal to the room surface area and $\bar{\alpha}$ is equal to the average absorption coefficient. The term "critical distance" is given to the value of r for which the two terms of (10.A.1) are equal:

$$r_{\text{critical}} = \frac{1}{4}\sqrt{\frac{QR}{\pi}}. \tag{10.A.2}$$

If we use the simple reverberation equation of Sabine, $T_{60} \approx 0.16V/(S\bar{\alpha})$, where T_{60} is the -60 dB reverberation time and V is the volume in cubic meters, and if we assume that $\alpha \ll 1$ so that $R \approx S\bar{\alpha}$, then (10.A.2) can be written as

$$r_{\text{critical}} = \frac{1}{10}\sqrt{\frac{QV}{\pi T_{60}}}. \tag{10.A.3}$$

Equations (10.A.2) and (10.A.3) indicate that a higher value of the term Q corresponds to a larger critical distance; the higher values of Q correspond to an improved ratio of direct sound to reverberant sound at a given position. Equation (10.A.2) can also be interpreted in another way: if we define a new room constant R' such that $R' = QR$, then we have an equivalent situation of an omnidirectional transducer ($Q = 1$) in a more absorbent room, in other words a room with a larger $\bar{\alpha}$.

One caveat that should be mentioned here is that these expressions assume a *diffuse* sound field, where the term diffuse means that acoustic energy incident on a point in space, is equally probable from any direction. The diffuse assumption is valid only at frequencies where the room modal overlap is such that three modes fall into one modal bandwidth. The frequency where the modal density reaches this value depends on the room size and the absorption; this frequency is usually referred to as the *Schroeder cutoff frequency* [14]. It should be emphasized that the transition from non-diffuse to diffuse is of course not instantaneous. The original "Schroeder cutoff frequency" was formulated as twice the present value. The cutoff frequency is

$$f_c = C\sqrt{\frac{T_{60}}{V}}, \tag{10.A.4}$$

where

$$C \approx 2000\ (m/s)^{3/2}. \tag{10.A.5}$$

Below the *diffuse* frequency limit, the sound field is dominated by individual modes propagating in discrete directions. At low frequencies, the axial modes dominate and the use of the above analysis breaks down. Since axial modes have the longest mean-free-path, the decay rate for these modes is much slower than for tangential and oblique modes. Compounding the problem is the fact that wall absorption decreases at lower frequencies. To account for the effects individual modes, an optimal transducer design must necessarily depend on the position and orientation of the sensor in the room as well as the room acoustics.

Chapter 11

MICROPHONE ARRAYS FOR VIDEO CAMERA STEERING

Yiteng (Arden) Huang
Center for Signal and Image Processing (CSIP)
School of Electrical and Computer Engineering
Georgia Institute of Technology
arden@ece.gatech.edu

Jacob Benesty
Bell Laboratories, Lucent Technologies
jbenesty@bell-labs.com

Gary W. Elko
Bell Laboratories, Lucent Technologies
gwe@bell-labs.com

Abstract In this chapter, we consider the problem of passively estimating the acoustic source location by using microphone arrays for video camera steering in real reverberant environments. Within a two-stage framework for this problem, different algorithms for time delay estimation and source localization are developed. Their performance as well as computational complexity are analyzed and discussed. A successful real-time system is also presented.

Keywords: Time Delay Estimation, Acoustic Source Localization, Reverberation, Eigenvalue Decomposition, Least Squares, Real-Time Implementation

1. INTRODUCTION

In current video-conferencing environments, a set of human-controlled cameras have to be set up in different locations to provide video of the active talker as he or she contributes to the discussion. Usually, this tedious task needs full involvement of professional camera operators. Alternatively, acoustic and image object tracking techniques can be used to locate and track an active talker automatically in 3D space - determine his or her range, azimuth, as well as elevation. Thereafter, the need for one or more human camera operators can be eliminated and the practicability will be increased.

There are several possible choices for tracking an active talker. Broadly they can be dichotomized into the class of visual tracking and the class of acoustic tracking, depending on what particular information (visual or acoustic cues, respectively) is applied. Even though visual tracking techniques have been investigated for several decades and have had good success, acoustic source localization systems have some advantages that are not present in vision-based tracking systems. They receive acoustic signals omni-directionally and can act in the dark. Therefore they are able to detect and locate sound sources in the rear or sources that are hiding or occluded.

Humans, like most vertebrates, have two ears which form a microphone array, mounted on a *mobile* base (head). By continuously receiving and processing the propagating acoustic signals with such a binaural auditory system, we can accurately and instantaneously gather information about the environment, particularly about the spatial positions and trajectories of sound sources and about their states of activity. However, the extraordinary performance features demonstrated by our binaural auditory system form a big technical challenge for engineers mainly because of room reverberation. Microphone array processing is a rapidly emerging technique and, we believe, will play an important role in a practical solution.

Existing source localization methods can be loosely divided into three categories: steered beamformer-based, high-resolution spectral estimation-based, and time delay estimation-based locators [1]. With continued investigation over the last two decades, the time delay estimation-based locator has become the technique of choice, especially in recent digital systems. Here, we will concern ourselves strictly with the time delay estimation-based source localization techniques and the primary focus is on obtaining an optimal (in the sense of accuracy, robustness, and efficiency) source location estimator which can be implemented in real-time with a digital computer.

Time delay estimation-based localization systems determine the location of acoustic sources in a two-step process. In the first step, a set of time delay of arrivals (TDOAs) among different microphone pairs is calculated. For the second step, this set of TDOA information is then employed to estimate the acoustic source location with the knowledge of the microphone array geometry. Within such a system, errors are introduced in both steps. The errors incurred in the first step, such as those due to reverberation and quantization, will be propagated through the second step and reduce the overall accuracy of the whole system.

Extensive literature exists on the topic of time delay estimation. In Section 2, two different approaches suitable for real-time implementation will be developed. The first is the conventional approach which is based on cross-correlation calculation [2]. Even though the generalized cross-correlation (GCC) algorithms can be improved to achieve good performance in the presence of noise [3, 4], it has a fundamental drawback of inability to cope well with reverberation as shown clearly in [5]. Therefore, in Section 2.3, a more realistic solution is developed that tries to directly determine the relative delay between the direct paths of two estimated channel impulse responses. This approach, known as the adaptive eigenvalue decomposition algorithm (AEDA) [6, 7], performs reasonably well in a reverberant environment.

Locating point sources using measurements or estimates from passive, stationary sensor arrays has numerous applications in navigation, aerospace, and geophysics. Algorithms for radiative source localization have been studied for nearly 100 years, particularly for radar and underwater sonar systems. Many processing techniques have been proposed, with different complexity and restrictions. These methods, which are described in Section 3, include the maximum likelihood [8], triangulation [9], constrained least-squares, spherical intersection [10], spherical interpolation [11], and the one-step least-squares method [12].

Even though many algorithms for time delay estimation and subsequent source localization have been proposed, few real-time systems [9, 13, 14] have been presented. In Section 4, a successful real-time passive acoustic source localization system for video camera steering recently developed by the authors is presented. The system is based on the adaptive eigenvalue decomposition and one-step least-squares algorithms, and demonstrated the desired features of robustness, portability, and accuracy.

2. TIME DELAY ESTIMATION

Time delay estimation (TDE) is concerned with the computation of the relative time delay of arrival between different microphone sensors. It is a fundamental technique in microphone array signal processing and the first step of our passive acoustic source localization system. In general, there are two steps in developing a time delay estimation algorithm. The first is to choose an appropriate parametric model for the acoustic environment. In Section 2.1, two parametric acoustic models for TDE problems, namely ideal free-field and real reverberant models will be described in detail.

Once the model has been selected, the next step is to estimate the model parameters (here, the TDOAs) that provide *minimum* errors according to the received microphone signals. In this chapter, two classes of different approaches to TDE, suitable for real-time implementation, will be developed. The first, presented in Section 2.2, is a generalized cross-correlation (GCC) method that selects the time delay that maximizes the cross-correlation function between signals of two distinct microphones, as its estimate. As we will see, this approach assumes the ideal free-field model, and performs poorly in a reverberant environment. In Section 2.3, a solution is derived from directly estimating the acoustic channel impulse responses and demonstrates more immunity to room reverberation.

2.1 ACOUSTIC MODELS FOR THE TDE PROBLEM

2.1.1 Ideal Free-Field Model. In an anechoic open space as shown in Fig. 11.1(a), the given source signal $s(n)$ propagates radiatively and the signal acquired by the i-th ($i = 1, 2$) microphone can be expressed as follows:

$$x_i(n) = \alpha_i s(n - \tau_i) + b_i(n), \tag{11.1}$$

where α_i is an attenuation factor due to propagation loss, τ_i is the propagation time, and $b_i(n)$ is additive noise. The time delay of arrival between the two microphone signals 1 and 2 is defined as,

$$\tau_{12} = \tau_1 - \tau_2. \tag{11.2}$$

Further assume that $s(n)$, $b_1(n)$, and $b_2(n)$ are zero-mean, uncorrelated, stationary Gaussian random processes. In this case, a mathematically clear solution for τ_{12} can be obtained from the ideal model that is widely used for the classical TDE problem.

2.1.2 Real Reverberant Model. The ideal free-field model is simple and only few parameters need to be determined. But unfortunately, in a real acoustic environment as shown in Fig. 11.1(b), we must take into account the reverberation of the room and the ideal model no longer holds. Then, a more complicated but more realistic model for the microphone signals $x_i(n)$, $i = 1, 2$, can be expressed as follows:

$$x_i(n) = g_i * s(n) + b_i(n), \tag{11.3}$$

where $*$ denotes convolution and g_i is the acoustic impulse response of the channel between the source and the i-th microphone. Moreover, $b_1(n)$ and $b_2(n)$ might be correlated, which is the case when the noise is directional, e.g., from a ceiling fan or an overhead projector.

For the real reverberant model, we do not have an "ideal" solution to the TDE problem, as for the previous model, unless we can accurately (and blindly) determine the two impulse responses, which is a very challenging problem.

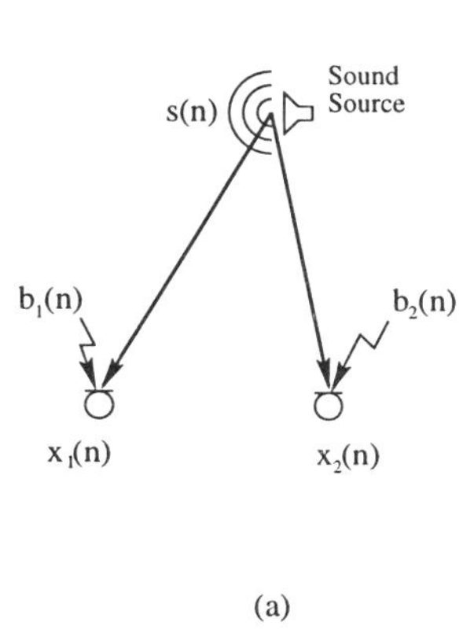

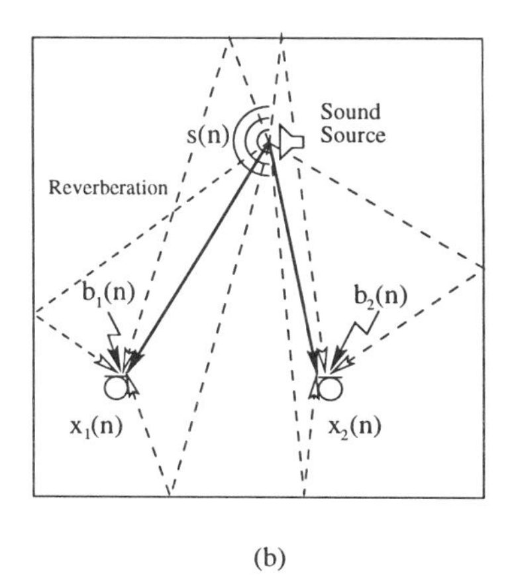

(a) (b)

Figure 11.1 Acoustic models for time delay estimation problems. (a) Ideal free-field model. (b) Real reverberant model.

2.2 THE GCC METHOD

In the GCC technique, which is based on the ideal free-field model, the time-delay estimate is obtained as the value of τ that maximizes the generalized cross-correlation function given by,

$$\begin{aligned} \psi_{x_1x_2}(\tau) &= \int_{-\infty}^{+\infty} \Phi(f) S_{x_1x_2}(f) \mathrm{e}^{j2\pi f\tau} df \\ &= \int_{-\infty}^{+\infty} \Psi_{x_1x_2}(f) \mathrm{e}^{j2\pi f\tau} df, \end{aligned} \tag{11.4}$$

where $S_{x_1x_2}(f) = E\left\{X_1(f)X_2^*(f)\right\}$ is the cross-spectrum, $\Phi(f)$ is a weighting function, and

$$\Psi_{x_1x_2}(f) = \Phi(f) S_{x_1x_2}(f) \tag{11.5}$$

is the generalized cross-spectrum. The GCC TDE may be expressed as:

$$\hat{\tau}_\phi = \arg\max_\tau \psi_{x_1x_2}(\tau). \tag{11.6}$$

The choice of $\Phi(f)$ is important in practice. The classical cross-correlation (CCC) method is obtained by taking $\Phi(f) = 1$. In the noiseless case, knowing that $X_i(f) = \alpha_i e^{-j2\pi f\tau_i} S(f),\ i = 1, 2$, we have:

$$\begin{aligned} \Psi_{x_1x_2}(f) &= \Psi_{\mathrm{cc}}(f) = E\{X_1(f)X_2^*(f)\} \\ &= \alpha_1\alpha_2 e^{-j2\pi f\tau_{12}} E\{|S(f)|^2\}. \end{aligned} \tag{11.7}$$

The fact that $\Psi_{\mathrm{cc}}(f)$ depends on the source signal can be problematic for TDE.

It is clear by examining (11.6) that the phase rather than the magnitude of cross-spectrum provides the TDOA information. Thereafter, the cross-correlation peak can be sharpened by pre-whitening the input signals, i.e. choosing $\Phi(f) = 1/|S_{x_1x_2}(f)|$, which leads to the so-called phase transform (PHAT) method [2], [15]. Also in the absence of noise, the cross spectrum,

$$\Psi_{x_1x_2}(f) = \Psi_{\text{pt}}(f) = e^{-j2\pi f \tau_{12}} \tag{11.8}$$

depends only on the τ_{12} and thus can, in general, achieve better performance than CCC, especially when the input signal-to-noise ratio (SNR) is low.

GCC is simple and easy to implement but will fail when the reverberation becomes significant because the fundamental model assumptions are violated.

2.3 ADAPTIVE EIGENVALUE DECOMPOSITION ALGORITHM

The adaptive eigenvalue decomposition algorithm (AEDA) focuses directly on the channel impulse responses for TDE and assumes that the system (room) is linear and time invariant. By following the real reverberant model (11.3) and the fact that (see Fig. 11.2):

$$x_1 * g_2 = s * g_1 * g_2 = x_2 * g_1, \tag{11.9}$$

in the noiseless case, we have the following relation at time n [16]:

$$\mathbf{x}^T(n)\mathbf{u} = \mathbf{x}_1^T(n)\mathbf{g}_2 - \mathbf{x}_2^T(n)\mathbf{g}_1 = 0 \tag{11.10}$$

where, $\mathbf{x}_i$ is the signal picked up by the i-th ($i = 1, 2$) microphone, ^{T} denotes transpose, and

$$\begin{aligned} \mathbf{x}_i(n) &= [x_i(n), x_i(n-1), \cdots, x_i(n-M+1)]^T, \\ \mathbf{g}_i &= [g_{i,0}, g_{i,1}, \cdots, g_{i,M-1}]^T, \quad i = 1, 2 \\ \mathbf{x}(n) &= \left[\mathbf{x}_1^T(n), \mathbf{x}_2^T(n)\right]^T, \\ \mathbf{u} &= \left[\mathbf{g}_2^T, -\mathbf{g}_1^T\right]^T, \end{aligned}$$

and M is the length of the impulse responses.

Multiplying (11.10) by $\mathbf{x}(n)$ and taking expectation yields,

$$\mathbf{R}(n)\mathbf{u} = \mathbf{0}, \tag{11.11}$$

where $\mathbf{R}(n) = E\{\mathbf{x}(n)\mathbf{x}^T(n)\}$ is the covariance matrix of the microphone signals $\mathbf{x}(n)$. This implies that the vector $\mathbf{u}$ (containing the two impulse responses) is in the null space of $\mathbf{R}(n)$. More specifically, $\mathbf{u}$ is the eigenvector of the covariance matrix corresponding to the eigenvalue equal to 0. If noise is present, $\mathbf{u}$ is estimated by minimizing $\mathbf{u}^T\mathbf{R}(n)\mathbf{u}$ subject to $\|\mathbf{u}\| = 1$. This is equivalent to finding $\mathbf{u}$ by computing the normalized eigenvector of $\mathbf{R}(n)$ corresponding to the smallest eigenvalue.

Before developing the adaptive eigenvalue decomposition algorithm to solve (11.11), it is worthwhile to expose the hidden factors that determine whether the acoustic channel impulse responses can be estimated, or if the acoustic channels are *identifiable*. From observation, $\mathbf{u}$ can be uniquely determined by solving (11.11) if and only if the covariance matrix $\mathbf{R}(n)$ is rank deficient by 1. Equivalently speaking, the information provided by the microphone signals is sufficient to uniquely identify the two acoustic channels if the following conditions hold [17]:

(I1) the polynomials $\mathbf{g}_i(m)$, $i = 1, 2$, $m = 0, ..., M-1$, are co-prime, or the channel transfer functions $G_i(z)$, $i = 1, 2$ do not share any common zeros,

(I2) the autocorrelation matrix of the source signal $s(n)$ is of full rank.

To avoid an ill-conditioned acoustic system, the two microphones can not be placed too close to each other (due to I1) and the analysis window should be long enough (due to I2). Both of the foregoing conditions are assumed in the AEDA and theoretically the unit-norm, non-zero solution to the normal equation $\mathbf{R}(n)\mathbf{u} = \mathbf{0}$ is uniquely equal to the desired acoustic channel impulse responses.

However, in practice, accurate estimation of the vector $\mathbf{u}$ is not trivial due to the nature of speech, the length of the impulse responses, the background noise, etc. However, for this application we only need to find an efficient way to detect the direct paths of the two impulse responses.

In order to efficiently estimate the eigenvector (here $\hat{\mathbf{u}}$) corresponding to the minimum eigenvalue of $\mathbf{R}(n)$, a constrained LMS algorithm [18] is often used. The error signal as illustrated in Fig. 11.2 is,

$$e(n) = \frac{\hat{\mathbf{u}}^T(n)\mathbf{x}(n)}{\|\hat{\mathbf{u}}(n)\|}, \tag{11.12}$$

and the constrained LMS algorithm may be expressed as,

$$\hat{\mathbf{u}}(n+1) = \hat{\mathbf{u}}(n) - \mu e(n)\nabla e(n), \tag{11.13}$$

where μ, the adaptation step, is a positive small constant, and

$$\nabla e(n) = \frac{1}{\|\hat{\mathbf{u}}(n)\|}\left[\mathbf{x}(n) - e(n)\frac{\hat{\mathbf{u}}(n)}{\|\hat{\mathbf{u}}(n)\|}\right]. \tag{11.14}$$

Substituting (11.12) and (11.14) in (11.13) and taking expectation after convergence gives:

$$\mathbf{R}\frac{\hat{\mathbf{u}}(\infty)}{\|\hat{\mathbf{u}}(\infty)\|} = E\{e^2(n)\}\frac{\hat{\mathbf{u}}(\infty)}{\|\hat{\mathbf{u}}(\infty)\|}, \tag{11.15}$$

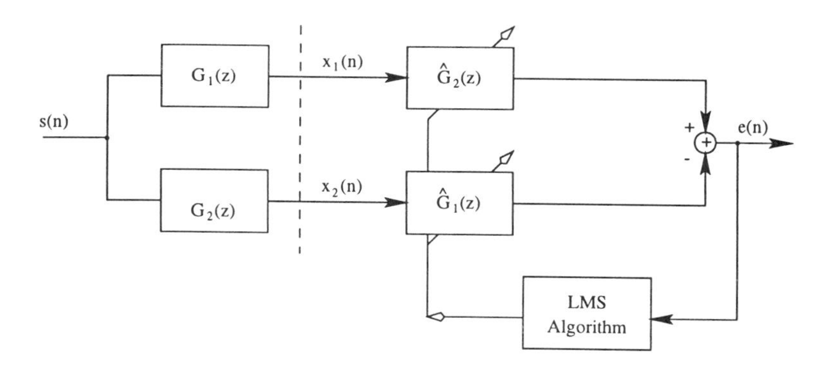

Figure 11.2 An adaptive filter for eigenvalue decomposition algorithm.

which is the desired result: $\hat{\mathbf{u}}$ converges in mean to the eigenvector of **R** corresponding to the smallest eigenvalue $E\{e^2(n)\}$. To avoid roundoff error propagation, normalization is imposed on the vector $\hat{\mathbf{u}}(n+1)$ after each update step. Finally the update equation is given by:

$$\hat{\mathbf{u}}(n+1) = \frac{\hat{\mathbf{u}}(n) - \mu e(n)\nabla e(n)}{\|\hat{\mathbf{u}}(n) - \mu e(n)\nabla e(n)\|}. \tag{11.16}$$

Note that if this normalization is used, then $\|\hat{\mathbf{u}}(n)\|$ (which appears in $e(n)$ and $\nabla e(n)$) can be removed, since we will always have $\|\hat{\mathbf{u}}(n)\| = 1$. If the smallest eigenvalue is equal to zero, which is the case here, the algorithm can be simplified as follows:

$$e(n) = \hat{\mathbf{u}}^T(n)\mathbf{x}(n), \tag{11.17}$$

$$\hat{\mathbf{u}}(n+1) = \frac{\hat{\mathbf{u}}(n) - \mu e(n)\mathbf{x}(n)}{\|\hat{\mathbf{u}}(n) - \mu e(n)\mathbf{x}(n)\|}. \tag{11.18}$$

Since the goal here is not to accurately estimate the two impulse responses g_1 and g_2 but rather the time delay, only the two direct paths are of interest. In order to take into account negative and positive relative delays, we initialize $\hat{u}_{M/2}(0) = 1$ which will be considered as an estimate of the direct path of g_2 and, during adaptation, keep it dominant in comparison with the other $M-1$ taps of the first half of $\hat{\mathbf{u}}(n)$ (containing an estimate of the impulse response g_2). A "mirror" effect will appear in the second half of $\hat{\mathbf{u}}(n)$ (containing an estimate of the impulse response $-g_1$): a negative peak will dominate which is an estimate of the direct path of $-g_1$. Thus the relative sample delay will be simply the difference between the indices corresponding to these two peaks.

To take advantage of the FFT, the filter coefficients are updated in the frequency domain in our real-time implementation using the unconstrained frequency-domain LMS algorithm [19]. The AEDA can be seen as a generalization of the LMS TDE proposed in [20].

3. SOURCE LOCALIZATION

In the TDE-based passive acoustic source localization system, the second step employs the set of TDOA estimates to calculate the location of the acoustic source. Since the equations describing source localization issues are highly nonlinear, the estimation and quantization errors introduced in the TDE step will be magnified more with some methods than others. In this section we first define the source localization problem. We then present the non-closed-form maximum likelihood locator, several closed-form locators, and their sensitivity to errors in TDOA. Detailed derivations of each localization method are developed and their performance features as well as computational complexities are analyzed.

3.1 SOURCE LOCALIZATION PROBLEM

The problem addressed here is the determination of the location of an acoustic source given the array geometry and relative TDOA measurements among different microphone pairs. The problem can be stated mathematically as follows:

A microphone array of $N+1$ microphones is located at,

$$\mathbf{r}_i \stackrel{\triangle}{=} (x_i, y_i, z_i)^T, \quad i = 0, ..., N \tag{11.19}$$

in Cartesian coordinates (see Fig. 11.3). The first microphone ($i = 0$) is regarded as the reference and placed at the origin of the coordinate system, i.e. $\mathbf{r}_0 = (0, 0, 0)$. The acoustic source is located at $\mathbf{r}_\mathrm{s} \stackrel{\triangle}{=} (x_\mathrm{s}, y_\mathrm{s}, z_\mathrm{s})^T$. The distances from the origin to the i-th microphone and the source are denoted by R_i and R_s, respectively.

$$R_i \stackrel{\triangle}{=} \|\mathbf{r}_i\| = \sqrt{x_i^2 + y_i^2 + z_i^2}, \quad i = 1, ..., N \tag{11.20}$$

$$R_\mathrm{s} \stackrel{\triangle}{=} \|\mathbf{r}_\mathrm{s}\| = \sqrt{x_\mathrm{s}^2 + y_\mathrm{s}^2 + z_\mathrm{s}^2}. \tag{11.21}$$

The distance between the source and the i-th microphone is denoted by,

$$D_i \stackrel{\triangle}{=} ||\mathbf{r}_i - \mathbf{r}_\mathrm{s}||_2 = \sqrt{(x_i - x_\mathrm{s})^2 + (y_i - y_\mathrm{s})^2 + (z_i - z_\mathrm{s})^2}\,. \tag{11.22}$$

The distance difference between microphones i and j from the source is given by,

$$d_{ij} \stackrel{\triangle}{=} D_i - D_j, \quad i, j = 0, ..., N. \tag{11.23}$$

The difference is usually termed as the *range difference* and is proportional to the time delay of arrival τ_{ij} with the speed of sound c,

$$d_{ij} = c \cdot \tau_{ij}, \tag{11.24}$$

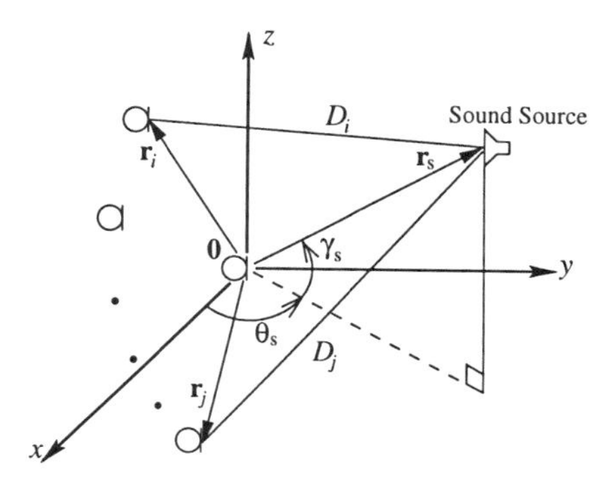

Figure 11.3 Spatial diagram illustrating notation defined in the source localization problem.

where, the speed of sound (in meters per second) can be estimated from the air temperature t_{air} (in degrees Celsius) according to the following approximate (first-order) formula,

$$c \approx 331 + 0.610 \times t_{\text{air}} \,. \qquad (11.25)$$

The localization problem is then to estimate $\mathbf{r}_s$ given the set of $\mathbf{r}_i$ and τ_{ij}. Note that there are $\frac{(N+1)N}{2}$ distinct TDOA estimates τ_{ij} which exclude the case $i = j$ and count the $\tau_{ij} = -\tau_{ji}$ pair only once. However, in the absence of *noise*, the space spanned by such TDOA estimates is of N dimensions. Any N linearly independent TDOAs determine all the rest. In a noisy environment, the TDOA redundancy can be utilized to improve the accuracy of the localization algorithms, at the expense of increasing the computational complexity of the corresponding source localization algorithm. For simplicity and also without loss of generality, we choose τ_{i0}, $i = 1, ..., N$ as the basis of such $\mathbb{R}^N$ space in this chapter.

3.2 IDEAL MAXIMUM LIKELIHOOD LOCATOR

Given a set of TDOA estimates $\boldsymbol{\tau} = (\tau_{10}, \tau_{20}, ..., \tau_{N0})^T$, the likelihood function of a source location is the joint probability of these TDOA estimates conditioned on the source location, i.e.

$$\mathcal{L}(\mathbf{r}_s) = \Pr(\boldsymbol{\tau}|\mathbf{r}_s) = \Pr(\tau_{10}, \tau_{20}, ..., \tau_{N0}|\mathbf{r}_s). \qquad (11.26)$$

The maximum likelihood solution is then given by,

$$\hat{\mathbf{r}}_{s,\text{ML}} = \arg\max_{\mathbf{r}_s} \Pr(\boldsymbol{\tau}|\mathbf{r}_s) = \arg\max_{\mathbf{r}_s} \Pr(\tau_{10}, \tau_{20}, ..., \tau_{N0}|\mathbf{r}_s). \qquad (11.27)$$

By assuming that the additive noise in TDOAs is zero mean and jointly Gaussian distributed, the joint probability density function of $\boldsymbol{\tau}$ conditioned on $\mathbf{r}_s$ is given by,

$$
\begin{aligned}
f(\boldsymbol{\tau}|\mathbf{r}_\mathrm{s}) &= \mathcal{N}(\boldsymbol{\tau};\ \boldsymbol{\tau}_\mathrm{m}(\mathbf{r}_\mathrm{s}),\ \boldsymbol{\Sigma}) \\
&= \frac{\exp\left\{-\frac{1}{2}[\boldsymbol{\tau} - \boldsymbol{\tau}_\mathrm{m}(\mathbf{r}_\mathrm{s})]^T\ \boldsymbol{\Sigma}^{-1}[\boldsymbol{\tau} - \boldsymbol{\tau}_\mathrm{m}(\mathbf{r}_\mathrm{s})]\right\}}{\sqrt{(2\pi)^N \det(\boldsymbol{\Sigma})}}, \quad (11.28)
\end{aligned}
$$

where, $\boldsymbol{\tau}_\mathrm{m}(\mathbf{r}_\mathrm{s})$ and $\boldsymbol{\Sigma}$ are respectively the mean vector and the covariance matrix of the estimated TDOAs, and det denotes determinant. In these parameters, only the mean vector $\boldsymbol{\tau}_\mathrm{m}(\mathbf{r}_\mathrm{s})$ is a function of the source location $\mathbf{r}_\mathrm{s}$. The covariance matrix depends only on the noise signal.

Since $\boldsymbol{\Sigma}$ is positive definite, the maximum likelihood solution is equivalent to that minimizing the error function defined as,

$$
\mathcal{E}(\mathbf{r}_\mathrm{s}) \triangleq [\boldsymbol{\tau} - \boldsymbol{\tau}_\mathrm{m}(\mathbf{r}_\mathrm{s})]^T\ \boldsymbol{\Sigma}^{-1}[\boldsymbol{\tau} - \boldsymbol{\tau}_\mathrm{m}(\mathbf{r}_\mathrm{s})]. \quad (11.29)
$$

Direct estimation of the minimizer is not practicable. If the noise is assumed uncorrelated, the covariance matrix is diagonal:

$$
\boldsymbol{\Sigma} = \mathrm{diag}(\sigma_1^2, \sigma_2^2, \ldots, \sigma_N^2) \quad (11.30)
$$

and the error function (11.29) can be rewritten as,

$$
\mathcal{E}(\mathbf{r}_\mathrm{s}) = \sum_{i=1}^{N} \frac{\left[\tau_{i0} - \tau_{i0,\mathrm{m}}(\mathbf{r}_\mathrm{s})\right]^2}{\sigma_i^2}. \quad (11.31)
$$

The steepest descent algorithm can be used to find $\hat{\mathbf{r}}_{\mathrm{s,ML}}$ iteratively by,

$$
\hat{\mathbf{r}}_\mathrm{s}(n+1) = \hat{\mathbf{r}}_\mathrm{s}(n) - \frac{1}{2}\mu \nabla \mathcal{E}(\hat{\mathbf{r}}_\mathrm{s}(n)), \quad (11.32)
$$

where μ is the step size.

The foregoing maximum likelihood locator is optimal in this problem only if the two assumptions made above hold. However, this is not the case in practice. The noise in the TDOA estimates comes from two sources: time delay estimation and quantization errors. The TDE error can be modeled as an additive Gaussian noise. For the quantization error, a uniformly distributed noise in $U[-T_\mathrm{s}/2, T_\mathrm{s}/2]$, where T_s is the sampling period, can certainly not be modeled as Gaussian noise. Moreover, in order to avoid local minima, we need to select a good initial guess of the source location, which is difficult to do in practice, and convergence of the iterative algorithm to the desired solution can not be guaranteed. Therefore, closed-form source localization methods have been studied.

3.3 TRIANGULATION LOCATOR

Of the closed-form source locators, triangulation is the most straightforward method. Consider the given set of TDOA estimates $\boldsymbol{\tau} = (\tau_{10}, \tau_{20}, ..., \tau_{N0})^T$ and convert these into a range difference vector $\mathbf{d} = (d_{10}, d_{20}, ..., d_{N0})^T$ by using (11.24). Each range difference d_{i0} in turn defines a hyperboloid by,

$$D_i - D_0 = \|\mathbf{r}_i - \mathbf{r}_s\| - \|\mathbf{r}_0 - \mathbf{r}_s\| = d_{i0} \, . \tag{11.33}$$

All points on such hyperboloid have the same range difference d_{i0} to the two microphones 0 and i. Since the acoustic source location is intended to be determined in $\mathbb{R}^3$ space, there are three unknowns and $N = 3$ (4 microphones) will make the problem possible to solve. Then the acoustic source lies on three hyperboloids and satisfies the set of hyperbolic equations:

$$\left\{ \begin{array}{lcl} \|\mathbf{r}_1 - \mathbf{r}_s\| - \|\mathbf{r}_0 - \mathbf{r}_s\| & = & d_{10} \\ \|\mathbf{r}_2 - \mathbf{r}_s\| - \|\mathbf{r}_0 - \mathbf{r}_s\| & = & d_{20} \\ \|\mathbf{r}_3 - \mathbf{r}_s\| - \|\mathbf{r}_0 - \mathbf{r}_s\| & = & d_{30} \end{array} \right. \tag{11.34}$$

This equation set is highly nonlinear. If microphones are arbitrarily arranged, a closed-form solution may not exist [21] and numerical methods must be used.

From a geometric point of view, the triangulation locator finds the source location by intersecting three hyperboloids. This approach is easy to implement. However, determinant (the number of unknowns is equal to that of equations) characteristics of the triangulation algorithm makes it very sensitive to noise in TDOAs. Small errors in TDOA can deviate the estimated source far from the true location.

3.4 THE SPHERICAL EQUATIONS

In order to avoid solving the set of hyperbolic equations (11.34) whose solution is very sensitive to noise, the source localization problem can be reorganized into a set of spherical equations. The acoustic source is potentially located on a group of spheres centered at the microphones.

Consider the distance from the i-th microphone to the acoustic source. From the definition of the range difference (11.23) and the fact that $D_0 = R_s$, we have:

$$D_i = R_s + d_{i0}. \tag{11.35}$$

From the Pythagorean theorem, D_i^2 can also be written as,

$$D_i^2 = \|\mathbf{r}_i - \mathbf{r}_s\|^2 = R_i^2 - 2\mathbf{r}_i^T \mathbf{r}_s + R_s^2 \, . \tag{11.36}$$

Substituting (11.35) into (11.36) yields,

$$(R_s + d_{i0})^2 = R_i^2 - 2\mathbf{r}_i^T \mathbf{r}_s + R_s^2 \, , \tag{11.37}$$

or equivalently,

$$\mathbf{r}_i^T \mathbf{r}_\mathrm{s} + d_{i0} R_\mathrm{s} = \frac{1}{2}(R_i^2 - d_{i0}^2), \quad i = 1, ..., N \,. \tag{11.38}$$

Putting the N equations together and writing them into a matrix form,

$$\mathbf{A}\,\theta = \mathbf{b}, \tag{11.39}$$

where,

$$\mathbf{A} \triangleq [\mathbf{S} \mid \mathbf{d}], \quad \mathbf{S} \triangleq \begin{bmatrix} x_1 & y_1 & z_1 \\ x_2 & y_2 & z_2 \\ & \vdots & \\ x_N & y_N & z_N \end{bmatrix},$$

$$\theta \triangleq \begin{bmatrix} x_\mathrm{s} \\ y_\mathrm{s} \\ z_\mathrm{s} \\ R_\mathrm{s} \end{bmatrix}, \quad \mathbf{b} \triangleq \frac{1}{2} \begin{bmatrix} R_1^2 - d_{10}^2 \\ R_2^2 - d_{20}^2 \\ \vdots \\ R_N^2 - d_{N0}^2 \end{bmatrix}.$$

By introducing one supplemental variable, the source range R_s, the source localization problem is *linearized.* The spherical equations are linear in $\mathbf{r}_\mathrm{s}$ given R_s and vice versa. This set of the spherical equations can be expanded into an over-determined system with ease. Therefore, the locators that try to solve (11.39) can take advantage of redundant microphone signals to achieve more accurate estimation performance without dramatically increasing their computational complexity.

3.5 CLS AND SPHERICAL INTERSECTION (SX) METHODS

In Section 3.4, we have reorganized the source localization problem into a linear system by introducing a supplemental variable. Therefore, the source location can be obtained by solving the spherical equations (11.39) subject to $\|\mathbf{r}_\mathrm{s}\| = R_\mathrm{s}$. This is in fact a constrained least-squares (CLS) problem. A Lagrangian can be written as,

$$J_c \triangleq (\mathbf{b} - \mathbf{A}\,\theta)^T(\mathbf{b} - \mathbf{A}\,\theta) + \lambda\left(\theta^T\,\Lambda\,\theta\right), \tag{11.40}$$

where $\Lambda = \mathrm{diag}[1, 1, 1, -1]$ is a diagonal matrix. Taking the gradient with respect to θ and equating the result to zero yields:

$$\theta_c = [\mathbf{A}^T\mathbf{A} + \lambda\,\Lambda]^{-1}\mathbf{A}^T\mathbf{b}, \tag{11.41}$$

from which the quadratic constraint $\theta^T \Lambda \theta = 0$ becomes,

$$\mathbf{b}^T\mathbf{A}[\mathbf{A}^T\mathbf{A} + \lambda\, \Lambda]^{-1}\, \Lambda[\mathbf{A}^T\mathbf{A} + \lambda\, \Lambda]^{-1}\mathbf{A}^T\mathbf{b} = 0. \tag{11.42}$$

This quadratic equation in the unknown λ can be solved only numerically using iterative methods which are computationally intensive in real-time implementation.

The SX locator solves the problem in two steps. It first finds the least-squares solution for $\mathbf{r}_s$ in terms of R_s,

$$\mathbf{r}_s = \mathbf{S}^\dagger(\mathbf{b} - R_s\mathbf{d}), \tag{11.43}$$

where,

$$\mathbf{S}^\dagger = \left(\mathbf{S}^T\mathbf{S}\right)^{-1}\mathbf{S}^T.$$

Then, substituting (11.43) into the constraint $R_s^2 = \mathbf{r}_s^T\mathbf{r}_s$, yields the quadratic equation,

$$R_s^2 = \left[\mathbf{S}^\dagger(\mathbf{b} - R_s\mathbf{d})\right]^T \left[\mathbf{S}^\dagger(\mathbf{b} - R_s\mathbf{d})\right]. \tag{11.44}$$

After expansion,

$$aR_s^2 + bR_s + c = 0, \tag{11.45}$$

where,

$$a = 1 - \|\mathbf{S}^\dagger\mathbf{d}\|^2,\ b = 2\mathbf{b}^T\mathbf{S}^{\dagger^T}\mathbf{S}^\dagger\mathbf{d},\ c = -\|\mathbf{S}^\dagger\mathbf{b}\|^2.$$

The positive real root is taken as an estimate of the source range $\hat{R}_s$ and is then substituted into (11.43) to calculate the SX estimate of the source location $\hat{\mathbf{r}}_{s,SX}$.

In the SX procedure, the solution of the quadratic equation (11.45) for the source depth R_s is required. This solution must be a positive value by all means. If no real positive root is available, the SX solution does not *exist*. On the other hand, if both of the roots are real and greater than 0, then the SX solution is not *unique*. In both cases, the SX locator fails to provide reliable estimate for source location. This is not desirable for real-time operation.

3.6 SPHERICAL INTERPOLATION (SI) LOCATOR

In order to overcome the drawback of spherical intersection, a spherical interpolation locator was proposed in [22] which attempts to relax the restriction $\hat{R}_s = \|\hat{\mathbf{r}}_s\|$ by estimating R_s in the least-squares sense.

To begin, substitute the least-squares solution (11.43) into the original spherical equations (11.39) to obtain,

$$R_s\mathbf{P}_{\mathbf{S}^\perp}\mathbf{d} = \mathbf{P}_{\mathbf{S}^\perp}\mathbf{b}, \tag{11.46}$$

where,

$$\mathbf{P}_{\mathbf{S}^{\perp}} \stackrel{\triangle}{=} \mathbf{I} - \mathbf{S}(\mathbf{S}^T\mathbf{S})^{-1}\mathbf{S}^T, \tag{11.47}$$

and $\mathbf{I}$ is an $N \times N$ identity matrix. The $\mathbf{P}_{\mathbf{S}^{\perp}}$ matrix projects a vector onto a space which is orthogonal to the column space of $\mathbf{S}$. It is symmetric ($\mathbf{P}_{\mathbf{S}^{\perp}} = \mathbf{P}_{\mathbf{S}^{\perp}}^T$) and idempotent ($\mathbf{P}_{\mathbf{S}^{\perp}} = \mathbf{P}_{\mathbf{S}^{\perp}}^2$). Then the least-squares solution to (11.46) is given by,

$$\hat{R}_{\mathrm{s,SI}} = \frac{\mathbf{d}^T\mathbf{P}_{\mathbf{S}^{\perp}}\mathbf{b}}{\mathbf{d}^T\mathbf{P}_{\mathbf{S}^{\perp}}\mathbf{d}}. \tag{11.48}$$

Substituting this solution into (11.43) yields the spherical interpolation estimate,

$$\hat{\mathbf{r}}_{\mathrm{s,SI}} = \left(\mathbf{S}^T\mathbf{S}\right)^{-1}\mathbf{S}^T\left[\mathbf{I}_N - \left(\frac{\mathbf{d}\mathbf{d}^T\mathbf{P}_{\mathbf{S}^{\perp}}}{\mathbf{d}^T\mathbf{P}_{\mathbf{S}^{\perp}}\mathbf{d}}\right)\right]\mathbf{b}. \tag{11.49}$$

In practice, the SI locator performs better, but is computationally more complex than the SX locator.

3.7 ONE STEP LEAST SQUARES (OSLS) LOCATOR

The SI method tries to solve the spherical equations (11.39) in two separate steps for the source range and its location, both calculated in the least-squares sense. Since the algorithm needs a significant amount of matrix multiplications to be performed, it is not computationally efficient especially when more microphones are applied (N gets larger). In this section, we will simplify the procedure by a one-step least-squares (OSLS) method and show that the OSLS method generates the same results as the SI method while dramatically decreasing the computational complexity, which is desirable for real-time implementations.

The least-squares solution of (11.39) for θ (the source location as well as its range) is given by:

$$\hat{\theta}_{\mathrm{OSLS}} = (\mathbf{A}^T\mathbf{A})^{-1}\mathbf{A}^T\mathbf{b}, \tag{11.50}$$

or written into block form as,

$$\hat{\theta}_{\mathrm{OSLS}} = \left[\begin{array}{c|c} \mathbf{S}^T\mathbf{S} & \mathbf{S}^T\mathbf{d} \\ \hline \mathbf{d}^T\mathbf{S} & \mathbf{d}^T\mathbf{d} \end{array}\right]^{-1}\left[\begin{array}{c} \mathbf{S}^T_{3\times N} \\ \hline \mathbf{d}^T \end{array}\right]\mathbf{b}. \tag{11.51}$$

First, write the matrix that appears in (11.51) as follows:

$$\left[\begin{array}{c|c} \mathbf{S}^T\mathbf{S} & \mathbf{S}^T\mathbf{d} \\ \hline \mathbf{d}^T\mathbf{S} & \mathbf{d}^T\mathbf{d} \end{array}\right]^{-1} = \left[\begin{array}{c|c} \mathbf{Q} & \mathbf{v} \\ \hline \mathbf{v}^T & k \end{array}\right], \tag{11.52}$$

where,

$$\begin{aligned}
\mathbf{v} &= -\left(\mathbf{S}^T\mathbf{S} - \frac{\mathbf{S}^T\mathbf{d}\mathbf{d}^T\mathbf{S}}{\mathbf{d}^T\mathbf{d}}\right)^{-1} \frac{\mathbf{S}^T\mathbf{d}}{\mathbf{d}^T\mathbf{d}}, \\
\mathbf{Q} &= \left(\mathbf{S}^T\mathbf{S}\right)^{-1}\left[\mathbf{I} - \left(\mathbf{S}^T\mathbf{d}\right)\mathbf{v}^T\right], \\
k &= \frac{1 - (\mathbf{d}^T\mathbf{S})\mathbf{v}}{\mathbf{d}^T\mathbf{d}}.
\end{aligned}$$

Define the projection matrix onto the **d**-orthogonal space:

$$\mathbf{P}_{\mathbf{d}^\perp} \triangleq \mathbf{I} - \frac{\mathbf{d}\mathbf{d}^T}{\mathbf{d}^T\mathbf{d}}, \tag{11.53}$$

and find,

$$\mathbf{v} = -\left(\mathbf{S}^T\mathbf{P}_{\mathbf{d}^\perp}\mathbf{S}\right)^{-1} \frac{\mathbf{S}^T\mathbf{d}}{\mathbf{d}^T\mathbf{d}}, \tag{11.54}$$

$$\mathbf{Q} = \left(\mathbf{S}^T\mathbf{P}_{\mathbf{d}^\perp}\mathbf{S}\right)^{-1}. \tag{11.55}$$

Substituting (11.52) with (11.54) and (11.55) into (11.51) yields the OSLS estimate,

$$\hat{\mathbf{r}}_{\mathrm{s,OSLS}} = \left(\mathbf{S}^T\mathbf{P}_{\mathbf{d}^\perp}\mathbf{S}\right)^{-1}\mathbf{S}^T\mathbf{P}_{\mathbf{d}^\perp}\mathbf{b}, \tag{11.56}$$

which is the minimizer of

$$J_{\mathrm{OSLS}}(\mathbf{r}_\mathrm{s}) = \|\mathbf{P}_{\mathbf{d}^\perp}\mathbf{b} - \mathbf{P}_{\mathbf{d}^\perp}\mathbf{S}\mathbf{r}_\mathrm{s}\|, \tag{11.57}$$

or the least-squares solution to the linear equations,

$$\mathbf{P}_{\mathbf{d}^\perp}\mathbf{S}\mathbf{r}_\mathrm{s} = \mathbf{P}_{\mathbf{d}^\perp}\mathbf{b}. \tag{11.58}$$

In fact, the OSLS algorithm tries to approximate the projection of the observation vector **b** with the projections of the column vectors of the microphone location matrix **S** onto the **d**-orthogonal space. The source location estimate is the coefficient vector associated with the *best* approximation. Clearly from (11.58), the OSLS algorithm is the generalization of the linear intersection method proposed in [23].

By using the Sherman-Morrison formula,

$$\left(\mathbf{A} + \mathbf{x}\mathbf{y}^T\right)^{-1} = \mathbf{A}^{-1} - \frac{\mathbf{A}^{-1}\mathbf{x}\mathbf{y}^T\mathbf{A}^{-1}}{1 + \mathbf{y}^T\mathbf{A}^{-1}\mathbf{x}}, \tag{11.59}$$

the item $\left(\mathbf{S}^T\mathbf{P}_{\mathbf{d}^\perp}\mathbf{S}\right)^{-1} = \left(\mathbf{S}^T\mathbf{S} - (\mathbf{S}^T\mathbf{d})(\mathbf{S}^T\mathbf{d})^T/(\mathbf{d}^T\mathbf{d})\right)^{-1}$ in (11.56) is expanded and finally it can be shown that the OSLS solution (11.56) is equivalent to the SI estimate (11.49), i.e. $\hat{\mathbf{r}}_{\mathrm{s,OSLS}} \equiv \hat{\mathbf{r}}_{\mathrm{s,SI}}$.

Now, we consider the computational complexity for the SI and OSLS methods. To calculate the inverse of an $M \times M$ matrix by using the Gauss-Jordan method without *pivoting*, the numbers of necessary scalar multiplications and additions are given by,

$$\mathrm{Mul_{mi}} = \frac{4}{3}M^3 - \frac{2}{3}M, \quad \mathrm{Add_{mi}} = \frac{4}{3}M^3 - \frac{3}{2}M^2 - \frac{1}{6}M. \tag{11.60}$$

To multiply matrix $\mathbf{X}_{p\times k}$ with matrix $\mathbf{Y}_{k\times q}$, the numbers of scalar multiplications and additions are,

$$\mathrm{Mul_{mm}} = pqk, \quad \mathrm{Add_{mm}} = pq(k-1). \tag{11.61}$$

For the SI locator (11.49), many matrix multiplications and one 3×3 matrix inverse need to be performed. Estimating the source location requires,

$$\mathrm{Mul_{SI}} = N^3 + 6N^2 + 22N + 35 \sim \mathrm{O}(N^3), \tag{11.62}$$
$$\mathrm{Add_{SI}} = N^3 + 4N^2 + 15N + 11 \sim \mathrm{O}(N^3), \tag{11.63}$$

multiplications and additions, respectively. Both are on the order of $\mathrm{O}(N^3)$. However, for the OSLS locator (11.50), only three matrix multiplications and one 4×4 matrix inverse are required. The numbers of scalar multiplications and additions are found as,

$$\mathrm{Mul_{OSLS}} = 36N + 84 \sim \mathrm{O}(N), \tag{11.64}$$
$$\mathrm{Add_{OSLS}} = 32N + 42 \sim \mathrm{O}(N), \tag{11.65}$$

which are on the order of only $\mathrm{O}(N)$. When $N \geq 4$ (5 microphones), the OSLS method is more computationally efficient than the SI method.

4. SYSTEM IMPLEMENTATION

A real-time passive acoustic source localization system for video camera steering has been developed recently by the authors at Bell Labs, Lucent Technologies. This system consists of a front-end 6-element microphone array with pre-amps, a Sonorus AUDI/O™ AD/24 8-channel A/D converter, a video camera, and a Pentium III™ 500 MHz PC equipped with Sonorus STUDI/O™ 16-channel digital audio interface board and video capture card. A block diagram of the system infrastructure is shown in Fig. 11.4.

In this system, the microphone array, shown in Fig. 11.5, was designed according to our simulation results. This array uses six Lucent Speech Tracker Directional™ hypercardioid microphones. The reference microphone 0 is located at the origin of the coordinate. Microphones 1, 2, and 4 are placed with equal distance of 50 centimeters from the origin on the $x-y$ (horizontal) plane. For symmetric performance to the horizontal plane, two more microphones, 3

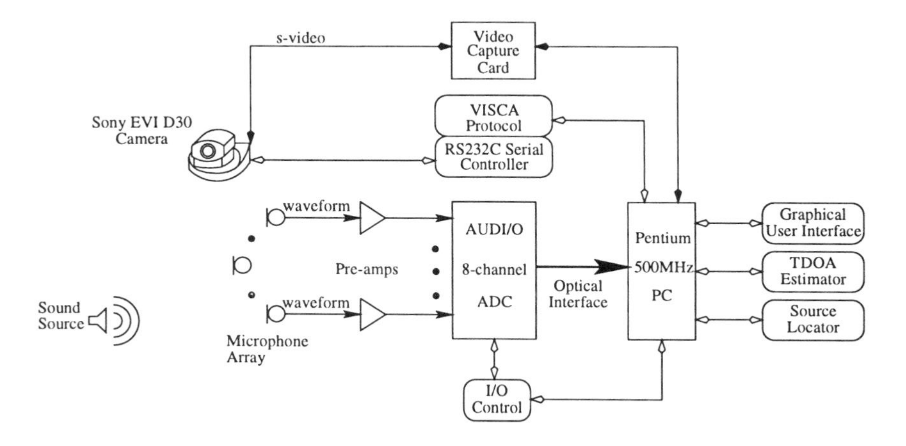

Figure 11.4 Schematic block diagram of the real-time system infrastructure.

and 5, are installed on z and $-z$ axis, both with 40 centimeters from the origin. In order to achieve desired portability, the microphone array is integrated with a Sony EVI-D30 video camera by using a frame. This camera, with pan, tilt, and zoom controls, allows for optimum coverage of a normal conference room. It has two high speed motors for pan and tilt. Pan, tilt, and zoom operation can be performed at the same time. Therefore, the camera is able to capture the full interaction of video-conference participants at remote locations.

The software for this system is running on Microsoft Windows98™. It consists of five major modules: acoustic signal I/O control, time delay estimator, source locator, graphical user interface, and video camera communications protocols. The acoustic signal is sampled at 8.82 kHz and quantized with 16 bits per sample. Five TDOAs were then estimated by the AEDA algorithm. The AEDA estimator uses a fixed step size which is $\mu = 0.01$ and the length of the adaptation vector $\mathbf{u}(n)$ is $L = 2M = 1024$. These TDOAs are supplied to the OSLS source locator. Benefiting from the efficient time delay estimator and source locator, the overall system has a relatively small computational requirement that can be easily met by the general-purpose Pentium III™ processor without any additional specific digital signal processor. The source location estimate is updated at 8 times/sec and used to drive the video camera. The PC communicates with the camera through two layers of protocol, namely RS232C serial control and the Video System Control Architecture (VISCA™) protocol. Finally, the video output is fed into the PC through video capture card and displayed on the system monitor.

This system has been tested by many people in two different rooms. One is a normal laboratory and the other is a conference room at Bell Labs. The laboratory has some computers and is much noisier than the conference room.

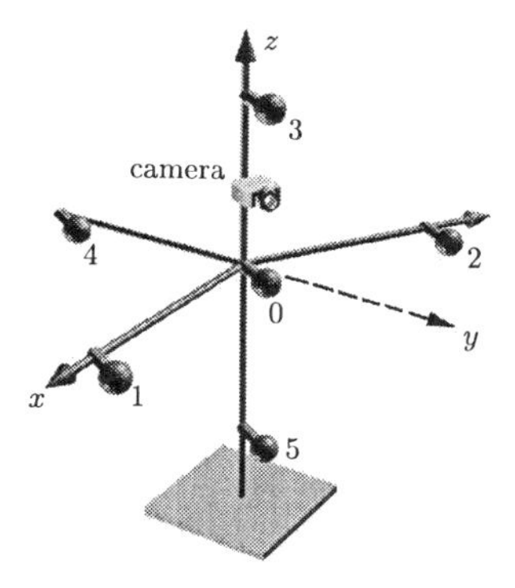

Figure 11.5 Three-dimensional microphone array for passive acoustic source localization.

The system can provide a quality video sequence in both environments and demonstrates robustness to noise. The speaker could be tracked with ease if he or she kept talking while moving. If the sound source is fixed at one arbitrary position, the system can drive the camera to focus on that position with high accuracy.

5. SUMMARY

In this chapter, we have looked at passive acoustic source localization techniques for video camera steering using a microphone array. The first part of the chapter was concerned with time delay estimation between distinct microphone sensor pairs. Beginning with the traditional generalized cross-correlation method, we found that this approach was of limited use since it did not take into account room reverberation and, as a result, performs poorly in a reverberant environment. We then derived the adaptive eigenvalue decomposition algorithm, which tried to directly estimate the speaker-microphone channel impulse response. This is a more realistic signal propagation channel model and the results demonstrated more immunity to noise and reverberation.

In the second part of the chapter, we considered the problem of determining the location of the sound source given the time delays of arrival and array geometry. A number of different approaches, including maximum likelihood estimation and several closed-form solutions, were derived. It was shown that the spherical equations could be used to localize the source. The least-squares solutions to the spherical equations were more robust to noise in the TDOAs than that to the hyperbolic equations. Even though all of the developed methods are somewhat limited, the one-step least-squares algorithm is favored over the others in terms of robustness, accuracy, and computational complexity.

In the last section of this chapter, we looked briefly at a real-time passive acoustic source localization system for video camera steering. The system applied the adaptive eigenvalue decomposition algorithm for time delay estimation and the one-step least-squares algorithm for source localization.

References

[1] M. S. Brandstein and H. F. Silverman, "A practical methodology for speech source localization with microphone arrays," *Comput., Speech, Language*, vol. 2, pp. 91-126, Nov. 1997.

[2] C. H. Knapp and G. C. Carter, "The generalized correlation method for estimation of time delay," *IEEE Trans. Acoust., Speech, Signal Processing*, vol. ASSP-24, pp. 320-327, Aug. 1976.

[3] M. S. Brandstein, "A pitch-based approach to time-delay estimation of reverberant speech," in *Proc. IEEE ASSP Workshop Appls. Signal Processing Audio Acoustics*, 1997.

[4] M. Bodden, "Modeling human sound-source localization and the cocktail-party-effect," *Acta Acoustica* 1, pp. 43-55, 1993.

[5] B. Champagne, S. Bédard, and A. Stéphenne, "Performance of time-delay estimation in the presence of room reverberation," *IEEE Trans. Speech Audio Processing*, vol. 4, pp. 148-152, Mar. 1996.

[6] Y. Huang, J. Benesty, and G. W. Elko, "Adaptive eigenvalue decomposition algorithm for realtime acoustic source localization system," in *Proc. IEEE ICASSP*, 1999, vol. 2, pp. 937-940.

[7] J. Benesty, "Adaptive eigenvalue decomposition algorithm for passive acoustic source localization," *J. Acoust. Soc. Am.*, vol. 107, pp. 384-391, Jan. 2000.

[8] M. Wax and T. Kailath, "Optimum localization of multiple sources by passive arrays," *IEEE Trans. Acoust., Speech, Signal Processing*, vol. ASSP-31, pp. 1210-1218, Oct. 1983.

[9] H. Wang and P. Chu, "Voice source localization for automatic camera pointing system in videoconferencing," in *Proc. IEEE ASSP Workshop Appls. Signal Processing Audio Acoustics*, 1997.

[10] H. C. Schau and A. Z. Robinson, "Passive source localization employing intersecting spherical surfaces from time-of-arrival differences," *IEEE Trans. Acoust., Speech, Signal Processing*, vol. ASSP-35, pp. 1223-1225, Aug. 1987.

[11] J. O. Smith and J. S. Abel, "Closed-form least-squares source location estimation from range-difference measurements," *IEEE Trans. Acoust., Speech, Signal Processing*, vol. ASSP-35, pp. 1661-1669, Dec. 1987.

[12] Y. Huang, J. Benesty, and G. W. Elko, "Passive acoustic source localization for video camera steering," Lucent Bell Laboratories technical memorandum, 1999.

[13] D. V. Rabinkin, R. J. Ranomeron, J. C. French, and J. L. Flanagan, "A DSP Implementation of Source Location Using Microphone Arrays," in *Proc. SPIE* 2846, 1996, pp. 88-99.

[14] C. Wang and M. S. Brandstein, "A hybrid real-time face tracking system," in *Proc. IEEE ICASSP*, 1998.

[15] M. Omologo and P. Svaizer, "Acoustic source location in noisy and reverberant environment using CSP analysis," in *Proc. IEEE ICASSP*, 1996, pp. 921-924.

[16] J. Benesty, F. Amand, A. Gilloire, and Y. Grenier, "Adaptive filtering algorithms for stereophonic acoustic echo cancellation," in *Proc. IEEE ICASSP*, 1995, pp. 3099-3102.

[17] G. Xu, H. Liu, L. Tong, and T. Kailath, "A least-squares approach to blind channel identification," *IEEE Trans. Signal Processing*, vol. 43, pp. 2982-2993, Dec. 1995.

[18] O. L. Frost, III, "An algorithm for linearly constrained adaptive array processing," *Proc. of the IEEE*, vol. 60, pp. 926-935, Aug. 1972.

[19] D. Mansour and A. H. Gray, JR., "Unconstrained frequency-domain adaptive filter," *IEEE Trans. Acoust., Speech, Signal Processing*, vol. ASSP-30, pp. 726-734, Oct. 1982.

[20] D. H. Youn, N. Ahmed, and G. C. Carter, "On using the LMS algorithm for time delay estimation," *IEEE Trans. Acoust., Speech, Signal Processing*, vol. ASSP-30, pp. 798-801, Oct. 1982.

[21] M. S. Brandstein, J. E. Adcock, and H. F. Silverman, "A closed-form method for finding source locations from microphone-array time-delay estimates," in *Proc. IEEE ICASSP*, 1995, pp. 3019-3022.

[22] J. S. Abel and J. O. Smith, "The spherical interpolation method for closed-form passive source localization using range difference measurements," in *Proc. IEEE ICASSP*, 1987, pp. 471-474.

[23] R. O. Schmidt, "A new approach to geometry of range difference location," *IEEE Trans. Aerosp. Electron.*, vol. AES-8, pp. 821-835, Nov. 1972.

Chapter 12

NONLINEAR, MODEL-BASED MICROPHONE ARRAY SPEECH ENHANCEMENT

Michael S. Brandstein
Division of Engineering and Applied Sciences
Harvard University
msb@hrl.harvard.edu

Scott M. Griebel
Division of Engineering and Applied Sciences
Harvard University
griebel@fas.harvard.edu

Abstract In this chapter we address the limitations of current approaches to using microphone arrays for speech acquisition and advocate the development of multi-channel techniques which employ non-traditional processing and an explicit model of the speech signal. The goal is to combine the advantages of spatial filtering achieved through beamforming with knowledge of the desired time-series attributes and intuitive nonlinear processing. We then offer a multi-channel algorithm which incorporates these principles. The enhanced speech is synthesized using a linear predictive filter. The excitation signal is computed from a nonlinear wavelet-domain process. It uses extrema clustering of the multi-channel speech data to discriminate portions of the linear prediction residual produced by the desired speech signal from those due to multipath effects and uncorrelated noise. The algorithm is shown to be capable of identifying and attenuating reverberant portions of the speech signal and reducing the effects of additive noise.

Keywords: Microphone Arrays, Multi-Channel, Speech Enhancement, Noise Reduction, Dereverberation, Wavelet, Nonlinear, Model-Based

1. INTRODUCTION

Speech has become an increasingly vital component in modern human-machine interfaces. Background noise, reverberation effects, and competing signals produce aesthetically undesirable effects and diminish the system's ability to convey information across the interface. Signal quality also plays a key role in the performance of related analysis systems. For instance, speech recognition and speaker identification techniques exhibit dramatic performance degradation when unpristine data is employed.

There is a direct relationship between the ease of use, effectiveness, and functionality of these interfaces and the capabilities of the speech acquisition system upon which it depends. Currently, most speech acquisition systems rely on the user to be physically close to a device to achieve reasonable sound quality. This close-talker condition significantly simplifies the acquisition problem by emphasizing the desired signal relative to background noise and other sources as well as by greatly reducing physical channel effects from the signal of interest. However, there are many situations in which it is not possible or desirable to have a talker or talkers physically linked to the acquisition device.

The ultimate goal of this work is to allow users the opportunity to roam unfettered in diverse environments while still providing a high quality speech signal and a robustness to background noise, interfering sources, and reverberation effects. Besides enhancing current human-machine interfaces (such as the speakerphone), advances in this technology are providing for a variety of novel approaches and new avenues for human-machine interaction. In recent years, the use of multiple sensors (microphone arrays) has received considerable attention as a means for dramatically improving the performance of traditional, single-microphone systems. This multi-sensor approach has opened up a wide range of additional topics of study [1, 2].

There is a great deal of potential for advancement in distant-talker speech acquisition research and a wealth of current and future technology dependent upon advances in this area. For single channel systems, recent research has focused on appropriate modeling of the speech signal, while multi-channel system research has emphasized improved spatial filtering. This work advocates an alternative approach to microphone array speech acquisition which incorporates nonlinear, model-based processing. By combining the advantages of multi-channel spatial filtering, non-traditional processing methods, and specific knowledge of the desired time-series attributes, this approach has the potential to significantly improve the quality of speech obtained in these challenging environments.

The next section provides a short summary of the existing single- and multi-channel approaches to distant-talker speech enhancement and discusses their limitations. We then introduce the principles of the nonlinear, model-based paradigm and detail a specific example of its use along with some results illustrating its merit. Finally, some conclusions and future issues for investigation are given.

2. SPEECH ENHANCEMENT METHODS

There is a rich history of work addressing the use of single channel methods for speech enhancement. Summaries of these techniques may be found in texts on the subject, such as [3, 4, 5]. While capable of improving perceived quality in restrictive environments (additive noise, no multipath, high to moderate signal-to-noise ratio (SNR), single source), these approaches do not perform well in the face of reverberant distortions, competing sources, and severe noise conditions. In recent years, more sophisticated speech models have been applied to the enhancement problem. In addition to utilizing the periodic features of the speech, as in the case of comb filtering, these systems exploit the signal's mixture of harmonic and stochastic components [6, 7, 8]. Such model-based techniques offer an improved performance, both in speech quality and intelligibility. Additionally, these methods, by virtue of their specific parameterization of the speech signal, offer some applicability to the more general acquisition problem. Currently, however, these model-based estimation schemes have been limited to single channel applications.

By employing spatial filtering in addition to temporal processing, multichannel systems (or microphone arrays) offer a distinct performance advantage over single-channel techniques in the presence of additive noise, interfering sources, and distortions due to multipath channel effects. (See [9] for a review of beamforming methods.) The simplest microphone array method is the Delay and Sum Beamformer which derives its output via an averaging of the time-synchronized microphone data. A variety of more sophisticated algorithms exists for adaptively 'steering' the array in the direction of the desired source and simultaneously adjusting the microphone 'weightings' to minimize the contributions of noise sources. These techniques usually assume the desired source is slowly varying and at a known location. While dynamic localization schemes, like those of [10], and weighting constraints may be incorporated into the adaptation procedure, these methods are very sensitive to steering errors which limit their noise source attenuation performance and frequently distort or cancel the desired signal. Furthermore, these algorithms are oriented solely toward noise reduction and have limited effectiveness at enhancing a desired signal corrupted by reverberations.

Another class of algorithms attempts to enhance the speech signal by employing any of the beamforming methods described above followed by a post-filter. The post-processor may take the form of multi-channel Wiener filtering [11, 12, 13, 14], spectral subtraction [15, 16], or a combination of the two [17]. These algorithms are limited in their assumption that the additive noise is uncorrelated. Once again the emphasis here is additive noise reduction. The performance (and appropriateness) of these post-filters quickly degrades in the presence of multi-path distortions and coherent noise.

A variant approach is based upon attempting to undo the effects of multipath propagation. The channel responses themselves are in general not minimum phase and are thus non-invertible. By beamforming to the direct path and the major images, it is possible to use the multipath reflections constructively to increase SNR's well beyond those achieved with a single beamformer. The result is a matched filtering process [18] which is effective in enhancing the quality of reverberant speech and attenuating noise sources. Unfortunately, this technique has a number of practical shortcomings. The matched filter is derived from the source location-dependent room response and as such is difficult to estimate dynamically. The channel responses obtained in this manner do not address the issue of non-stationary or unknown source locations or changing acoustic environments. These problems are addressed by Affes and Grenier [19] in attempting to adaptively estimate the channel responses and incorporate the results into an adaptive beamforming process.

In general, beamforming research has dealt with algorithms to attenuate undesired sources and noise, track moving sources, and deconvolve channel effects. These approaches, while effective to some degree, are fundamentally limited by the nature of the distant talker environment. Array design methods are overly sensitive to variations in their assumptions regarding source locations and radiation patterns, and are inflexible to the complex and time-varying nature of the enclosure's acoustic field. Motion as little as a few centimeters or a talker turning his or her head is frequently sufficient to compromise the optimal behavior of these schemes in practical scenarios [21, 20]. Similarly, matched filter processing, while shown to be capable of tracking source motion to a limited degree, requires significant temporal averaging and is not adaptable at rates sufficient to effectively capture the motions of a realistic talker.

3. NONLINEAR, MODEL-BASED PROCESSING

Single-channel techniques exploit various features of the speech signal; multi-channel methods focus primarily on improving the quality of the spatial filtering process. While advances will continue to be made, these methods pursued independently have provided limited success. It is unlikely that such schemes in their present forms will achieve the stated goal of this research, namely to acquire a high-quality speech signal from an unconstrained talker in a hands-free environment surrounded by interfering sources (the "cocktail party" problem). In addition to the spatial filtering benefits of microphone arrays, an effective solution to this challenging problem will require the application of knowledge regarding the desired signal content. Furthermore, existing approaches to this problem have relied almost entirely on traditional linear filtering and least-squares methods. As will be shown, it is possible to utilize some highly nonlinear, but still intuitive, techniques to suppress the effects of reverberation.

In [20] we proposed an alternative to these methods by explicitly incorporating the Dual Excitation Speech Model [7] into the beamforming process. Our work in [22] extended this idea using a multi-channel version of the Multi-Pulse Linear Predictive Coding (MPLPC) model [23, 24] and a nonlinear event-based processing method to discriminate impulses in the received signals due to channel effects from those present in the desired speech. These approaches were shown to suppress the deleterious effects of both reverberations and additive noise without explicitly identifying the channel and to be adaptive on a frame by frame basis. However, this earlier work was restricted to a small class of speech signals and limited to individual analysis frames. Here we take these concepts a step further by employing the wavelet domain for a multi-resolution signal analysis and reconstruction of the LPC residual. This allows the nonlinear, model-based processing paradigm to be applied for effective speech enhancement under general conditions. The proposed algorithm is now explained in detail.

4. A MULTI-CHANNEL SPEECH ENHANCEMENT ALGORITHM

The reverberant speech signal, $x_i[n]$, observed at the ith microphone ($i = 1, 2, \ldots, I$) can be modeled as:

$$x_i[n] = h_i[n] * s[n] + u_i[n] \tag{12.1}$$

where $s[n]$ is the clean speech utterance, $u_i[n]$ is noise, and $h_i[n]$ is the room impulse response between the speech source and the ith microphone. Under all but ideal circumstances, the room impulse response is a complicated function of the enclosure environment. The noise term, $u_i[n]$, is assumed to be uncorrelated, both temporally and spatially.

A very general model for speech production approximates the vocal tract as a time-varying all-pole filter [4]. In the case of voiced speech, the filter excitation is modeled as a quasi-periodic impulse train where the average width between consecutive impulses is the pitch period. For unvoiced speech, the excitation signal is approximated by random noise. The proposed algorithm relies on the assumption that the detrimental effects of additive noise and reverberations introduce only zeros into the overall system and will primarily affect only the nature of the speech excitation sequence, not the all-pole filter. It is also assumed that the noise and errant impulses contributed to the excitation sequences are relatively uncorrelated across the individual channels, while the excitation impulses due to the original speech are invariant to the environmental effects. Essentially, the approach will be to identify the clean speech excitation signal from a set of corrupted excitations. The enhanced speech is then reconstructed by using the enhanced excitation signal as input to an estimate of the all-pole filter representing the vocal system.

The idea of enhancing the excitation signal was explored in [22] using the LPC residual derived from I microphone channels. The proposed algorithm represents a more effective method for estimating and then reconstructing the excitation signal by employing a class of wavelets to decompose the LPC residuals. In [25], quadratic spline wavelets were shown to be effective at detecting singularities in a signal. Because these quadratic spline wavelets are the derivatives of smoothing functions, significant wavelet transform (WVT) coefficients, or wavelet extrema, prove to be appropriate indicators of discontinuities in a signal. The set of extrema at each wavelet scale is an effective representation of the impulses in the excitation signals. By locating the extrema which are well clustered across all channels, we attempt to capture the underlying impulsive structure (for voiced speech) of the original non-reverberant LPC residual signal. The estimated 'clean' multi-scale extrema can be used to reconstruct the desired excitation signal with minimal approximation error [25].

The overall procedure is outlined in Fig. 12.1. The received signals, $x_i[n]$, are assumed to have been time-aligned either through *a priori* knowledge of the source and microphone geometry or as the result of applying a relative time-delay estimation procedure to the original channel data. The microphone signals are converted to LPC residuals and transformed to the wavelet domain. For each wavelet scale the multi-channel extrema are clustered and scaled by a local coherence measure. The resulting extrema at each scale are used to generate a set of synthetic wavelets which produce the enhanced excitation signal. This synthesized excitation (presumably with less of the detrimental reverberation and noise effects) is then fed into the estimated LPC filter. Finally, a global coherence measure is used to modulate the LPC filter output in an effort to reduce larger time-scale reverberation effects. The algorithm stages will now be discussed in detail. For the illustrative plots that follow in the remainder of this section we employ 400 msec reverberated speech with additive white Gaussian noise with SNR of 10 dB. Details of the simulation procedure are provided in the following section.

4.1 ALGORITHM DETAILS

4.1.1 Inverse LPC Filtering. A joint LPC filter is derived from N-point windows of the I channels of speech. The linear prediction formalism assumes the speech source signal can be modeled as:

$$s[n] = \sum_{p=1}^{P} a_p s[n-p] + e[n] \tag{12.2}$$

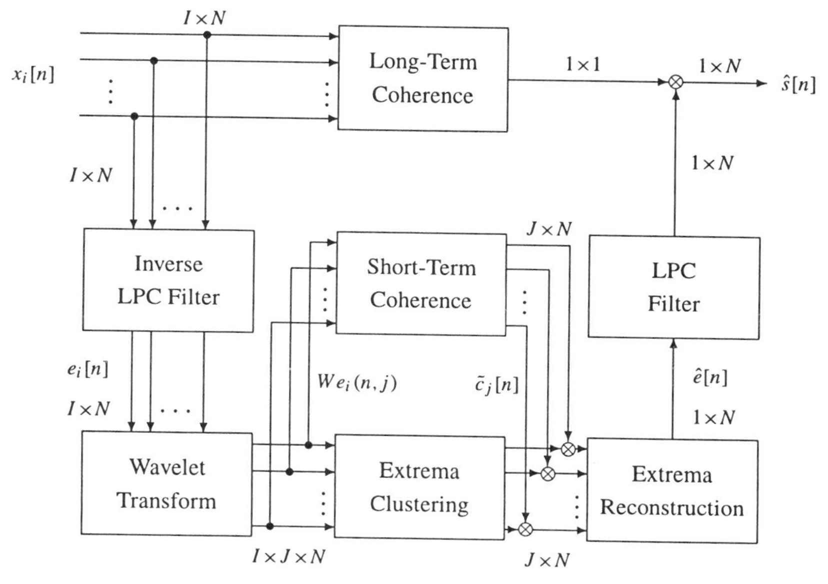

Figure 12.1 Outline of the proposed algorithm.

where $e[n]$ is the residual or excitation signal discussed previously. The LPC coefficients, $\{a_1, a_2, \ldots, a_P\}$, are estimated by minimizing the average energy of the residual signal. For this multi-channel setting the criterion is expressed as:

$$\frac{1}{I}\sum_{i=1}^{I}\sum_{n=1}^{N} e_i^2[n] = \frac{1}{I}\sum_{i=1}^{I}\sum_{n=1}^{N}\left\{s[n] - \sum_{p=1}^{P} a_p s[n-p]\right\}^2. \tag{12.3}$$

The minimum energy solution is achieved by solving a system of linear equations as in the single-channel case, but involves a multi-channel autocorrelation function, given by:

$$r[k] = \frac{1}{I}\sum_{i=1}^{I}\sum_{n=k}^{N} x_i[n]x_i[n-k]. \tag{12.4}$$

The resulting LPC coefficients can be used to generate a set of residual signals, $e_i[n]$, via inverse filtering. The goal of the following wavelet extrema reconstruction will be to use the multiple residual signals to create a single enhanced residual signal, $\hat{e}[n]$. This residual will then be used as input to the LPC filter to generate an enhanced speech signal, $\hat{s}[n]$.

For the work presented in this chapter, the analysis frame is 32 msec ($N = 256$ samples at 8kHz sampling rate), the number of channels is $I = 8$, and the LPC model order is $P = 13$.

4.1.2 Wavelet Transform. The first J scales of the wavelet transform (WVT) are calculated for each of the I residual signals. Here the WVT is calculated for the first $J = 5$ scales. The WVT of the residual signal of the ith channel, $e_i[n]$, is denoted by $We_i(n, j)$ where j is the scale parameter and n is the temporal index within that scale. We use the non-decimated quadratic spline wavelet discussed in [25]. Because the wavelet is the derivative of a smoothing function, the resulting WVT coefficients will contain local extrema at large changes in the signal. Therefore, impulses in the residual signal of voiced speech segments will be manifested as large WVT coefficients at different scales. This same wavelet has been used on non-reverberant speech for pitch detection [26].

Another advantage of working in the wavelet domain is the ability to approximately reconstruct a signal from the local extrema of its WVT coefficients at each scale [27]. To summarize this process, suppose the wavelet transform of a signal, $f[n]$, with local extrema at $n = n_{j,k}$ in scale j is to be approximated by $\hat{f}[n]$. By minimizing the norm of $\hat{f}[n]$ and imposing the constraint $Wf(n_{j,k}, j) = W\hat{f}(n_{j,k}, j)$, it is possible to construct $\hat{f}[n]$ via a conjugate gradient approach [27]. Figure 12.2 shows a clean voiced speech segment and its wavelet extrema reconstruction after 5 and 25 iterations of the conjugate gradient method. Note that this method tends to approximate a signal to within a very high frequency error term, which, for the speech signals we examined, was inaudible.

In the following stages, this algorithm is used to reconstruct an enhanced residual signal, where wavelet extrema at different scales correspond to impulses in the voiced excitation. While the significance of the wavelet extrema in the unvoiced case is not as clearly delineated, this approach is effective in reconstructing an appropriate noise-like residual signal.

4.1.3 Extrema Clustering. Based upon the observation that residual impulses due to the original speech tend to be clustered across channels and scales while those due to reverberation effects are relatively uncorrelated in both amplitude and time, a clustering technique is now applied to estimate the 'clean' extrema. In [22] we presented a nonlinear clustering criterion for this application, while in [28] we employed a Gaussian smoothing approach. However, the local extrema of WVT coefficients of the delay-and-sum beamformed speech have been found to perform sufficiently well in this context. Because of its computational advantages and near comparable performance, we will employ this delay and sum strategy for extrema clustering in what follows. The left-hand plot in Fig. 12.3 illustrates a single scale's WVT coefficients for 8 channels and the resulting clustered extrema. In general, for voiced speech, extrema due to glottal excitations tend to be more closely clustered than those due to additional impulses from the reverberant channel responses. This causes

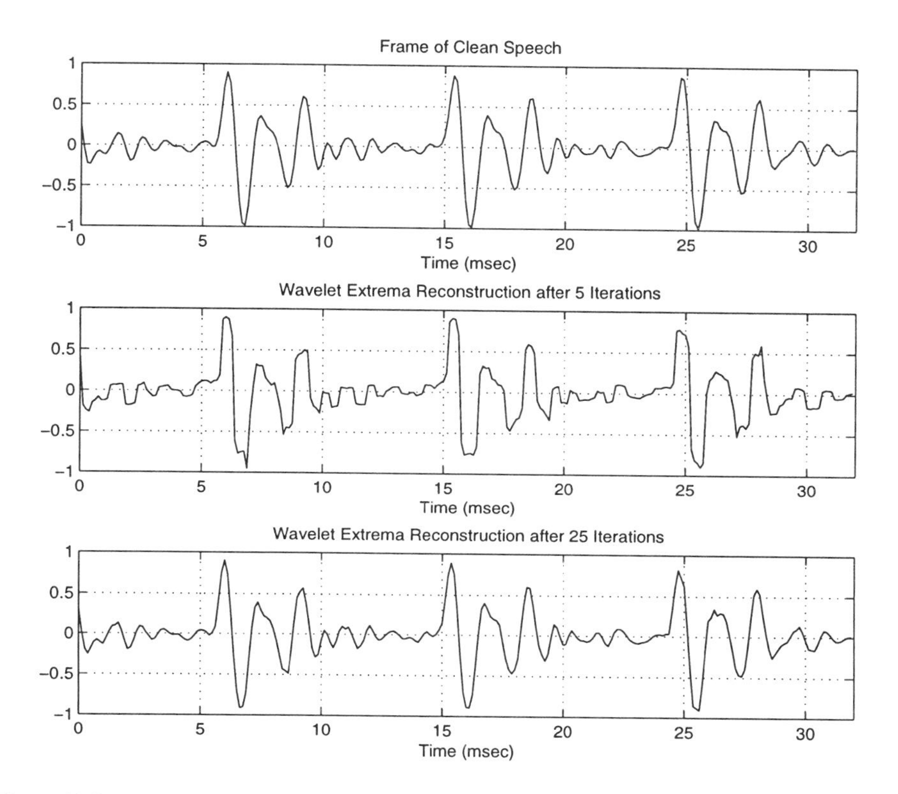

Figure 12.2 Clean speech and wavelet extrema reconstructions after 5 and 25 iterations.

the less clustered peaks to have a relatively small amplitude. The clustering operation maps the multi-channel WVT extrema data to a single WVT extrema representation suitable for input to the WVT extrema reconstruction algorithm described previously.

4.1.4 Short-Term Coherence Scaling. We now apply the first of two coherence-based scaling functions. The purpose of the amplitude modulation scheme detailed here is to further de-emphasize the clustered peaks due to room reverberations beyond what is achieved by the averaging process associated with the extrema clustering.

The following short-term coherence window is multiplied by the wavelet extrema at each scale j:

$$c_j[n] = \frac{2}{I-1} \frac{\sum_{i=1}^{I} \sum_{k=i+1}^{I} \sum_{l=-L}^{L} We_i(n+l, j) We_k(n+l, j)}{\sum_{i=1}^{I} \sum_{l=-L}^{L} We_i(n+l, j)^2} \tag{12.5}$$

where $2L + 1 \ll N$ is the length of a small subwindow within the analysis frame, on the order of 3 to 5 msec. With this window, any clustered extrema in

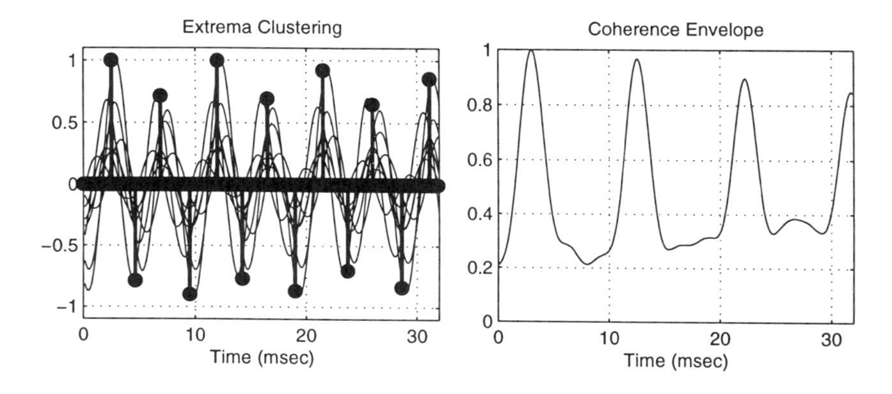

Figure 12.3 Clustering results and coherence envelope.

areas of low coherence will be de-emphasized, while highly coherent extrema will be relatively unaffected. The coherence function as given above can assume values between -1 and 1. A thresholding scheme is incorporated to maintain nonnegative scaling terms and to normalize the coherence values within the subwindow. For each scale, the modified coherence window is calculated from:

$$\tilde{c}_j[n] = \begin{cases} c_j[n]/\max_n c_j[n] & c_j[n] \geq T \\ T/\max_n c_j[n] & c_j[n] < T \end{cases} \tag{12.6}$$

where $0 \leq T \leq 1$. A reasonable value of T was determined experimentally to be 0.1. With $T > 0$, the thresholding ensures that the coherence window does not completely eliminate any wavelet extrema, which could result in excessive smoothing of the resulting speech signal. The right-hand plot in Fig. 12.3 shows a typical coherence envelope used for weighting the clustered WVT extrema.

4.1.5 Extrema Reconstruction. After multiplying the clustered peaks for each wavelet scale by the appropriate scale-dependent coherence envelope, the enhanced LPC residual is reconstructed using the conjugate gradient algorithm discussed previously. Listening tests show that the variations in amplitude produced by the extrema synthesis process across consecutive periods of voiced speech can produce audible distortions in the resulting speech signal. To lessen this effect, a quasi-pitch-synchronous smoothing approach is employed at the coarser wavelet scales to ensure that the peaks from consecutive voiced periods are slowly varying. This procedure involves averaging WVT extrema with their corresponding extrema in neighboring pitch periods and implicitly assumes a pitch estimate is available. In [29] a variety of multi-channel pitch estimators are evaluated in the environmental conditions considered here. Several are shown to be effective for reverberation levels of 0 to 1000 ms.

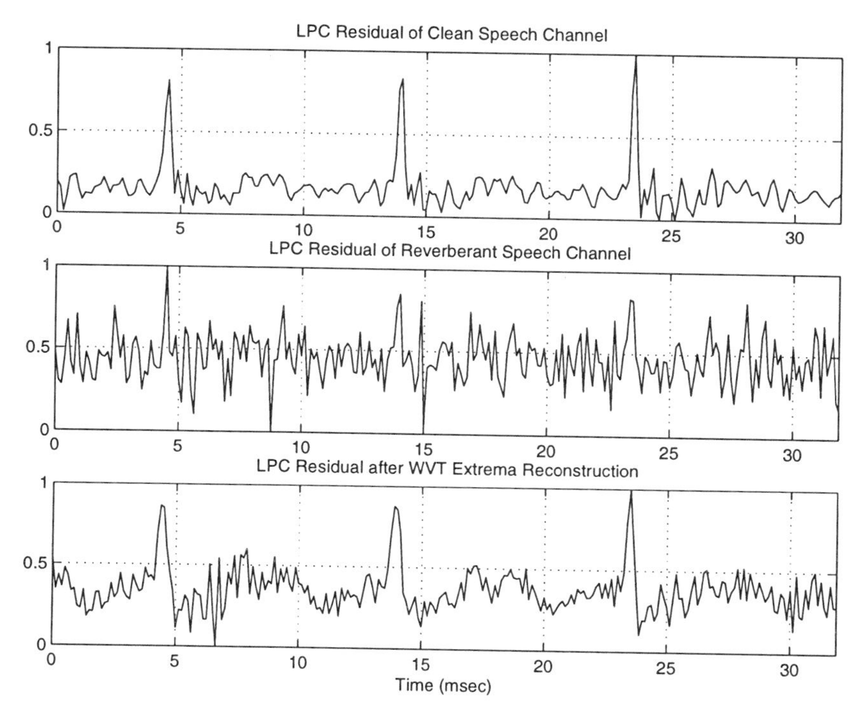

Figure 12.4 LPC residual of clean speech, after beamforming, and after wavelet clustering technique.

Figure 12.4 shows an example of the LPC residual for clean and reverberant speech frames followed by the residual signal resulting from the WVT extrema reconstruction algorithm.

4.1.6 LPC Filter. For each analysis frame, the reconstructed residual signal, $\hat{e}[n]$, is applied to the estimated LPC filter to generate the enhanced speech signal. Figures 12.5 and 12.6 offer the results for two different frames of speech. Figure 12.5 displays a single analysis frame corresponding to a little over one period of a voiced speech segment. The four plots compare the reconstructed speech signal to the original speech utterance, a single reverberant channel, and a simple delay-and-sum beamforming of the I reverberant speech channels. It is apparent that the primary effect of the multipath distortion is to inflate the signal energy between 987 ms and 997 ms. The beamformer is ineffective at attenuating these distortions. The proposed algorithm is capable of detecting the excitation energy due to the clean speech and produces a synthesized signal very similar to the original. Figure 12.6 demonstrates a voiced speech segment from early in a vowel segment. At this point, the speech is more

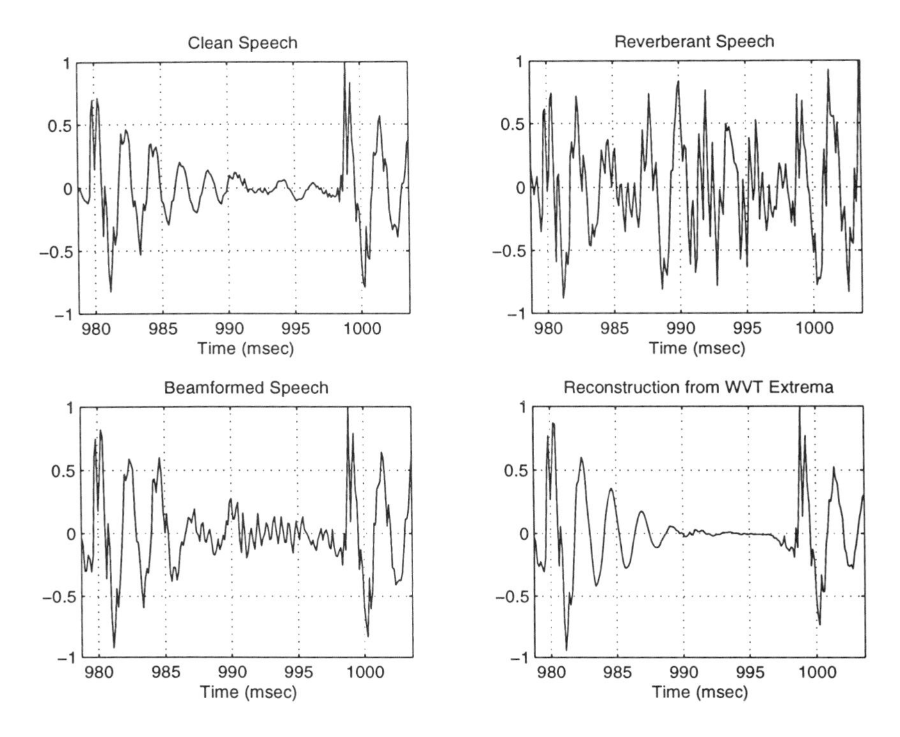

Figure 12.5 Comparison 1: clean, reverberant, beamformed, and WVT extrema reconstructed speech.

affected by additive noise than reverberations, and the enhancement procedure diminishes its presence.

4.1.7 Long-Term Coherence Scaling. The final step in the enhancement procedure is to scale the reconstructed speech frames by a measure that is based on both the long-term coherence of the speech channels and the energy of the beamformed signal. This second weighting procedure serves two purposes. First, it adjusts the energy of each frame to provide a relatively smoothly varying output. The short-term coherence window can vary substantially from frame to frame. The energy of the resulting speech will vary accordingly, potentially resulting in audible distortions. The long-term coherence window incorporates an energy statistic to address this issue. Second, it allows for the de-emphasis of lengthy signal portions which are due primarily to reverberation effects. This is especially the case at the end of voiced sounds. The long-term coherence window is able to detect this condition and is used to adjust the corresponding regions of the reconstructed signal.

The calculation of this coarse scaling term, $g[n]$, is as follows:

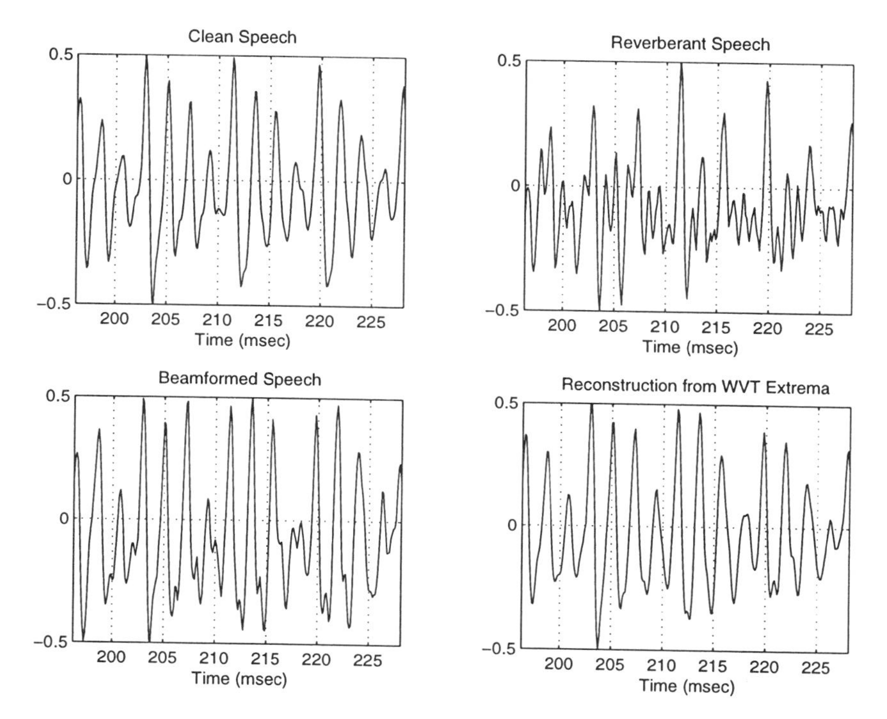

Figure 12.6 Comparison 2: clean, reverberant, beamformed, and WVT extrema reconstructed speech.

1. Calculate the coherence of observed speech signals using:

$$c[n] = \frac{2}{I-1} \frac{\sum_{i=1}^{I} \sum_{k=i+1}^{I} \sum_{l=-L}^{L} x_i[n+l] x_k[n+l]}{\sum_{i=1}^{I} \sum_{l=-L}^{L} x_i[n+l]^2} \tag{12.7}$$

2. Calculate the short term energy, $b[n]$, of the beamformed signal.

3. If the energy of the beamformed signal decreases rapidly, then:

$$g[n] = b[n]. * c[n] / \max_{k \in [0,N]} c[n-k] \tag{12.8}$$

otherwise, $g[n] = b[n]$.

Figure 12.7 shows one channel of reverberant speech and the resulting $g[n]$. Each reconstructed speech frame is multiplied by the average value of $\sqrt{g[n]}$ over the length of the frame. The individual frames are then combined using an overlap-add procedure. The resulting enhanced speech signal has an energy profile which follows $g[n]$.

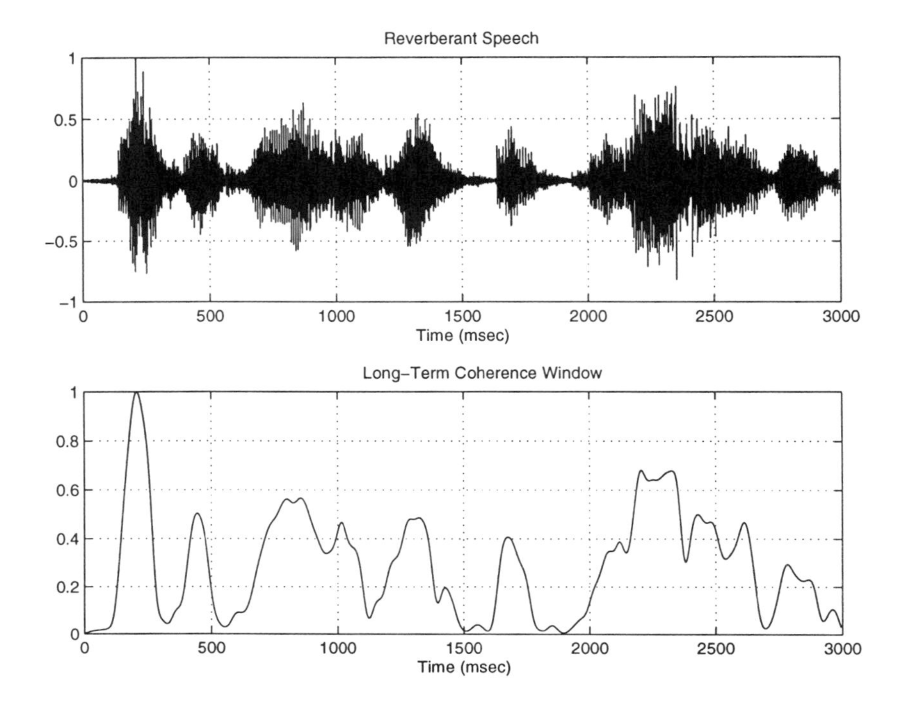

Figure 12.7 Long-term coherence window.

4.2 SIMULATIONS

Reverberant speech segments were generated using the image model of [30] with intra-sample interpolation [31] and up to sixth order reflections. Room reverberation times, T, ranged from 100 msec to 1 sec. The corresponding reflection ratio, β, used by the image model was calculated via Eyring's formula [32]:

$$\beta = \exp\{-13.82/[c(L_x^{-1} + L_y^{-1} + L_z^{-1})T]\}$$

where L_x, L_y, and L_z are the room dimensions and c is the speed of sound in air ($\approx$ 342 m/sec). The room was a 4 m $\times$ 4 m $\times$ 3 m rectangular enclosure assumed to have plane reflective surfaces and uniform, frequency-independent reflection coefficients. A total of eight microphones were used: one pair on 3 of the walls and another pair on the ceiling, as shown in Fig. 12.8. The source was located near the wall containing no microphones. The microphones and sources were all assumed to have cardioid patterns.

Figures 12.9 and 12.10 illustrate a series of three second speech segments which have been simulated in the reverberant enclosure described above. Each figure plots the original speech, the signal received at a single microphone,

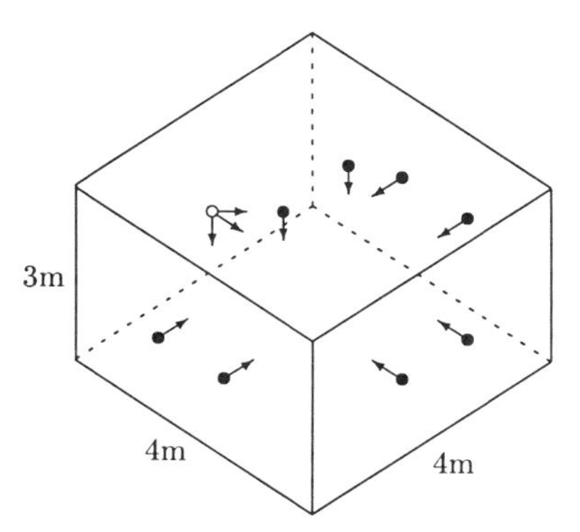

Figure 12.8 Room setup - • represents microphones, ∘ represents the speech source.

the product of delay-and-sum beamforming, and the result achieved using the proposed enhancement algorithm. Figure 12.9 shows the results derived for 400 msec reverberant speech while Fig. 12.10 shows the case for 400 msec reverberant speech plus additive noise. In this latter case, uncorrelated white Gaussian noise was added to each channel so that the SNR relative to the original non-reverberant utterance was 10 dB. The superior performance of the proposed algorithm relative to delay-and-sum beamforming at removing both long and short term reverberation effects and additive noise is apparent. Listening tests confirm the efficacy of the procedure. The synthesized speech reduces the audible reverberation and noise effects which are clearly present in both the degraded and delay-and-sum beamformed speech.

To quantify the improvements of the proposed algorithm, the average Bark Spectral Distortion (BSD) [33] was calculated over all frames of speech. The BSD is a speech quality assessment method which employs a series of perceptually motivated transformations to the signal spectra prior to computing a spectral distance. The result is an objective distortion measure which correlates well with perceptual quality ratings. Figure 12.11 shows the Bark spectral distance results for reverberation times from 100 msec to 1 sec with additive white noise (SNR = 10 dB). The proposed algorithm shows an improvement over delay-and-sum beamforming at all reverberation levels. These figures are consistent will informal listening tests.

5. CONCLUSION

By focusing on residual onset data and incorporating specific knowledge of the nature of the speech signal, the proposed wavelet-domain algorithm is capable of discriminating impulses of the excitation residual generated by the desired speech signal from those brought about by multipath echoes and uncor-

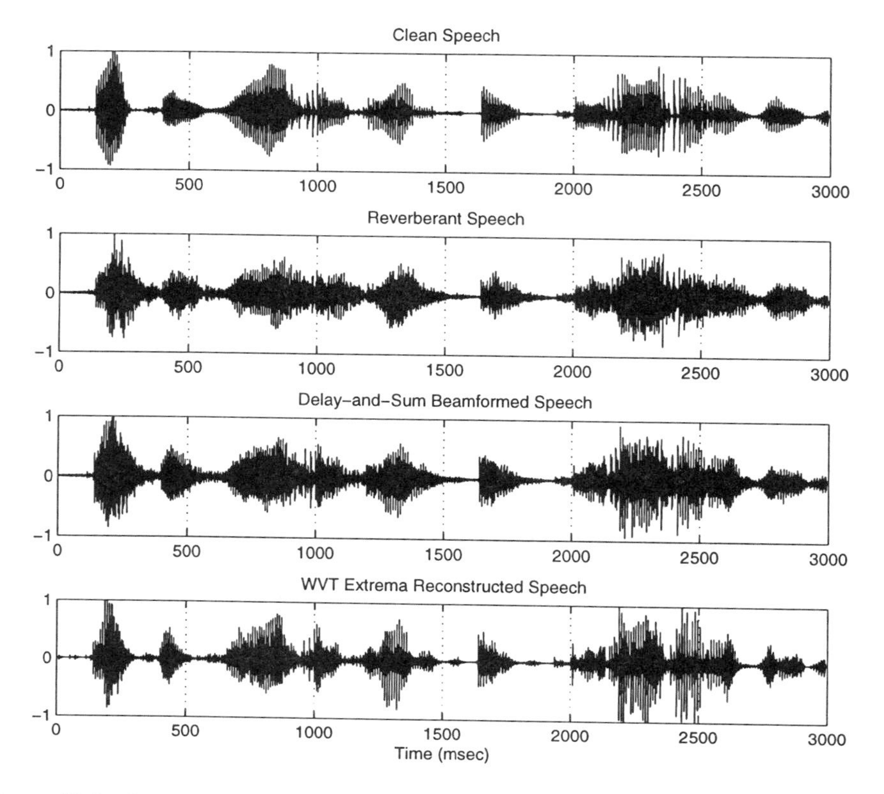

Figure 12.9 Comparison of clean, reverberant, beamformed, and the proposed algorithm (reverberation-only case).

related noise. The process achieves a significant attenuation of environmental reverberations and additive noise without explicitly requiring estimation of the channel or noise characteristics. This illustrates the effectiveness and practicality of the advocated fusion of nonlinear and model-based processing for microphone array applications.

These principles may be extended to address the case of interfering sources and unknown source locations. Essentially, the impulse events may be associated with one or more source locations by evaluating their relative delays across the channels given knowledge of the microphone placements. Once this association is performed, the individual speech signals may be independently reconstructed through the synthesis procedure outlined above. This approach to speaker isolation represents a distinct contrast to the spatial filtering paradigm which relies on very specific knowledge and assumptions regarding talkers' locations and radiation patterns to generate appropriate channel weightings. The procedure, by virtue of its underlying speech model and exploitation of

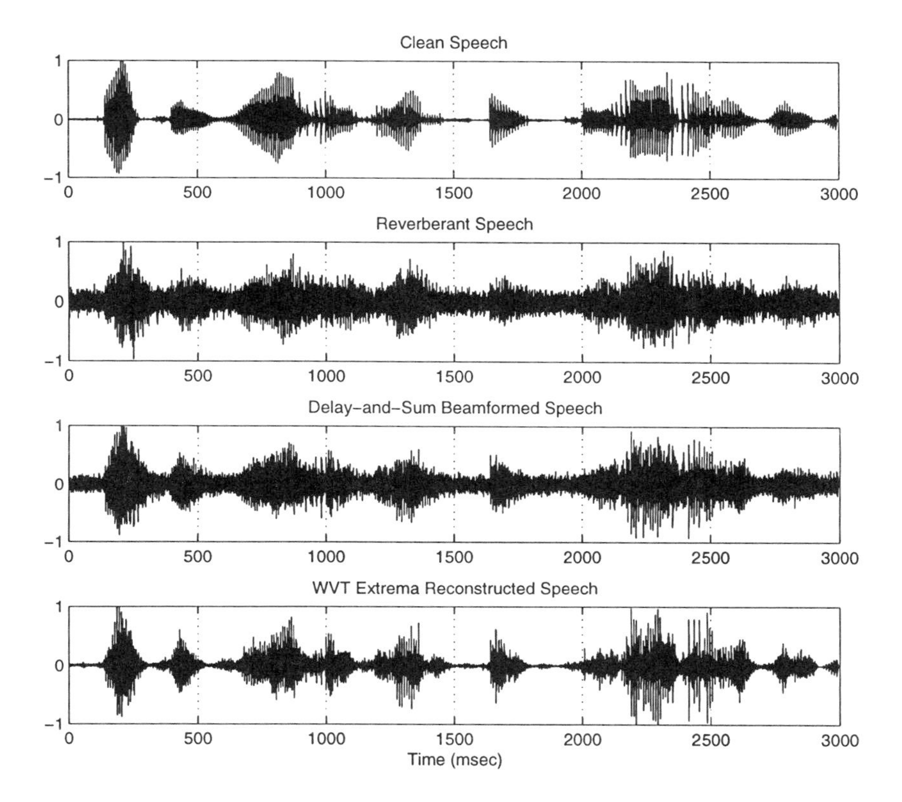

Figure 12.10 Comparison of clean, reverberant, beamformed, and the proposed algorithm (reverberation plus noise case).

impulse data alone, has the potential to be much less sensitive to environmental uncertainties (e.g. imperfect source localization, non-ideal radiator effects, and unknown channels).

References

[1] J. L. Flanagan and H. F. Silverman, "Material for international workshop on microphone-array systems: Theory and practice," LEMS Technical Report 113, LEMS, Division of Engineering, Brown University, Providence, RI 02912, Oct. 1992.

[2] J. L. Flanagan and H. F. Silverman, "Material for international workshop on microphone-array systems: Theory and practice," Technical report, CAIP, Rutgers University, Piscataway, NJ 08855, Oct. 1994.

[3] J. Lim, editor, *Speech Enhancement*. New Jersey: Prentice-Hall, 1983.

[4] J. Deller, J. Proakis, and J. Hansen, *Discrete-Time Processing of Speech Signals*. New Jersey: Prentice Hall, first edition, 1987.

[5] S. Furui and M. Sondhi, editors, *Advances in Speech Signal Processing*. New York: Marcel Dekker, first edition, 1992.

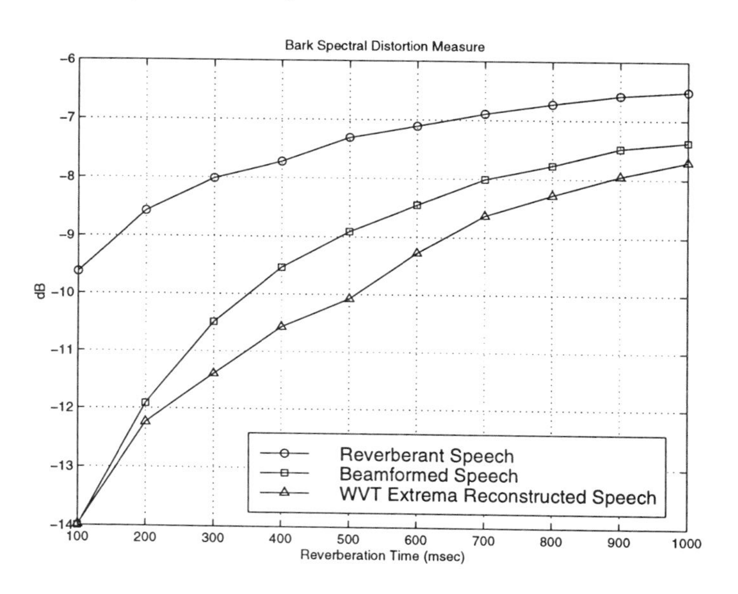

Figure 12.11 Bark spectral distortion results.

[6] R. McAulay and T. Quatieri, "Speech analysis/synthesis based on a sinusoidal representation," *IEEE Trans. Acoust., Speech, Signal Processing*, vol. ASSP-34, pp. 744-754, Aug. 1986.

[7] J. Hardwick, *The Dual Excitation Speech Model*, PhD thesis, Massachusetts Institute of Technology, Cambridge, MA, June 1992.

[8] J. Laroche, Y. Stylianou, and E. Moulines, "HNS: Speech modification based on a harmonic + noise model," in *Proc. IEEE ICASSP*, 1993, pp. II-550–II-553.

[9] D. Johnson and D. Dudgeon, *Array Signal Processing- Concepts and Techniques*. New Jersey: Prentice Hall, first edition, 1993.

[10] M. Brandstein and H. Silverman, "A practical methodology for speech source localization with microphone arrays," *Computer, Speech, and Language*, vol. 11, pp. 91-126, Apr. 1997.

[11] R. Zelinski, "A microphone array with adaptive post-filtering for noise reduction in reverberant rooms," in *Proc. IEEE ICASSP*, 1988, pp. 2578-2580.

[12] K. Simmer and A. Wasiljeff, "Adaptive microphone arrays for noise suppression in the frequency domain," in *Second Cost 229 Workshop on Adaptive Algorithms in Communications*, Bordeaux, France, Sept. 1992, pp. 185-194.

[13] Z. Yang, K. Simmer, and A. Wasiljeff, "Improved performance of multi-microphone speech enhancement systems," in *Proceedings of the* 14th *GRETSI Symposium*, 1993, pp. 479-482.

[14] C. Marro, Y. Mahieux, and K. Simmer, "Analysis of noise reduction and dereveberation techniques based on microphone arrays with postfiltering," *IEEE Trans. Speech Audio Proc.*, vol. 6, pp. 240-259, May 1998.

[15] S. Gierl, "Noise reduction for speech input systems using an adaptive microphone array," in *Proceedings* 22nd *ISATA*, 1990, pp. 517-524.

[16] M. Dahl, I. Claesson, and S. Nordebo, "Simultaneous echo cancellation and car noise suppression employing a microphone array," in *Proc. IEEE ICASSP*, 1997, pp. 239-242.

[17] J. Meyer and K. Simmer, "Multi-channel speech enhancement in a car environment using wiener filtering and spectral subtraction," in *Proc. IEEE ICASSP*, 1997, pp. 1167-1170.

[18] J. Flanagan, A. Surendran, and E. Jan, "Spatially selective sound capture for speech and audio processing," *Speech Communication*, vol. 13, pp. 207-222, 1993.

[19] S. Affes and Y. Grenier, "A signal subspace tracking algorithm for microphone array processing of speech," *IEEE Trans. Speech Audio Proc.*, vol. 5, pp. 425-437, Sept. 1997.

[20] M. Brandstein, "On the use of explicit speech modeling in microphone array applications," in *Proc. IEEE ICASSP*, 1998, pp. 3613-3616.

[21] B. Radlovic, R. Williamson, and R. Kennedy, "On the poor robustness of sound equalization in reverberant environments," in *Proc. IEEE ICASSP*, 1999, pp. 881-884.

[22] M. Brandstein, "An event-based method for microphone array speech enhancement," in *Proc. IEEE ICASSP*, 1999, pp. 953-956.

[23] B. S. Atal and J. R. Remde, "A new model of lpc excitation for producing natural-sounding speech at low bit rates," in *Proc. IEEE ICASSP*, 1982, pp. 614-617.

[24] S. Singhal and B. S. Atal, "Improving performance of multi-pulse lpc coders at low bit rates," in *Proc. IEEE ICASSP*, 1984, pp. I-131–I-134.

[25] S. Mallat and S. Zhong, "Characterization of signals from multiscale edges," *IEEE Trans. on Pattern Analysis and Machine Intelligence*, vol. 14, pp. 710-732, July 1992.

[26] S. Kadambe and G. Faye Boudreaux-Bartels, "Applications of the wavelet transform for pitch detection of speech signals," *IEEE Trans. Information Theory*, vol. 38, pp. 917-924, Mar. 1992.

[27] S. Mallat, *A Wavelet Tour of Signal Processing*. Boston: Academic Press, 1998.

[28] S. Griebel and M. Brandstein, "Wavelet transform extrema clustering for multi-channel speech dereverbearation," in *IEEE Workshop on Acoustic Echo and Noise Control*, Pocono Manor, Pennsylvania, Sept. 1999, pp. 52-55.

[29] S. M. Griebel, "Multi-channel wavelet techniques for reverberant speech analysis and enhancement," Technical Report 5, HIMMEL, Harvard University, Cambridge, MA, Feb. 1999.

[30] J. B. Allen and D. A. Berkley, "Image method for efficiently simulating small room acoustics," *J. Acoust. Soc. Am.*, vol. 65, pp. 943-950, Apr. 1979.

[31] P. M. Peterson, "Simulating the response of multiple microphones to a single acoustic source in a reverberant room," *J. Acoust. Soc. Amer.*, vol. 80, pp. 1527-1529, Nov. 1986.

[32] H. Kuttruff, *Room Acoustics*. London: Elsevier, third edition, 1991.

[33] S. Wang, A. Sekey, and A. Gersho, "An objective measure for predicting subjective quality of speech coders," *IEEE J. Selected Areas in Communications*, vol. 10, pp. 819-829, June 1992.

V

VIRTUAL SOUND

Chapter 13

3D AUDIO AND VIRTUAL ACOUSTICAL ENVIRONMENT SYNTHESIS

Jiashu Chen
Lucent Technologies
jiashuchen@lucent.com

Abstract This chapter discusses the problem and solutions of synthesizing 3D (three dimensional) audio and complex virtual acoustic environments where heavy computation load is required for 3D positioning of multiple sound sources and their respective reflections. Spatial feature extraction and a regularization model for measured head-related transfer functions (HRTFs) are reviewed. For some functional representations of the HRTF, fast algorithms are derived. These fast algorithms can reduce the computational burden by an order of magnitude compared with conventional methods. Several computing architectures are illustrated for the implementation of virtual acoustic environment synthesis.

Keywords: Spatial Hearing, 3D Audio, Virtual Acoustic Environment Synthesis (VAES), HRTF Modeling, Fast Algorithms

1. INTRODUCTION

Basic findings in sound localization research prompted scientists and engineers to electronically synthesize three dimensional sound. Commercial stereo technology uses multiple channel (microphone) recordings and digital or analog means to master them such that the mixed two channel sounds can represent the original with high fidelity. The sound is mastered by the recording engineer with the objective of maximizing pleasing effects; the authenticity of the spatial attributes of the acoustic scene is not the main focus. In addition, stereo is designed mainly for loudspeaker delivery in relatively uncontrolled environments. Typically, the best stereo sound effects can only be perceived by the listener in a space between the two loudspeakers. For headphone presentation the sound field is collapsed inside the head and between the two ears. Therefore, a stereo system is not considered to be a complete 3D system even though *spatialness* is one important consideration in 3D design. True 3D sound/binaural recordings were almost simultaneously considered and researched when stereo systems were initially developed. Due to variety of reasons, in particular, technical limitations (at the time), the current stereo system was chosen as a standard.

Thanks to the rapid advancement of psychoacoustics, digital signal processing, and computers, we now largely understand the science of HRTFs and have enough processing power to both measure and process sound through them in real-time. Starting from late 80's, commercial systems that can render 3D sound emerged onto the market. They originated from NASA, and U.S. Department of Defense supported research prototypes. Compared with stereo systems, 3D audio systems can interactively position sound sources in 3D space. A direct application of this technology is computer gaming. New games with 3D sound positioning capability are gaining popularity in the marketplace. The second application of 3D sound technology is to enhance the current stereo system. Combined with the so-called "crosstalk canceler," 3D audio technology extends sound field rendering beyond the physical span of the loudspeakers. That is, a pair of frontally positioned loudspeakers can reproduce sounds that appear to be very lateralized or even appear to be from the back of the user. The third application of 3D audio is to convert multichannel audio signals into a 2-channel presentation. For example, Dolby™ Prologic system requires 5 channels of audio processing and 5 speakers to reproduce a 3D audio space, while 3D-audio technology can approximate this performance with only a two-loudspeaker presentation.

It is interesting to note that an important technique of creating the spatial perception of sound is to artificially add echoes to the anechoic source signals. In this regard, echoes are not *always* a nuisance.

The chapter is organized as follows. Section 2 describes background work related to 3D audio. Previous studies of HRTF modeling are summarized. It is also shown how 3D audio can be synthesized with direct implementation of identified acoustic cues. Section 3 describes a specific modeling technique for measured HRTFs that leads to a functional representation of HRTFs and fast algorithms for virtual acoustic environment synthesis. Section 4 introduces some advanced methods in 3D audio synthesis. Computing architectures that implement SFER (spatial feature extraction and regularization) models are also introduced in Section 4. Section 5 discusses some specific issues for VAES (virtual acoustic environment synthesis) and Section 6 contains the conclusions.

2. SOUND LOCALIZATION CUES AND SYNTHETIC 3D AUDIO

The ability of localizing sounds in a three dimensional space is important to human listeners in terms of their awareness of aural happenings in their

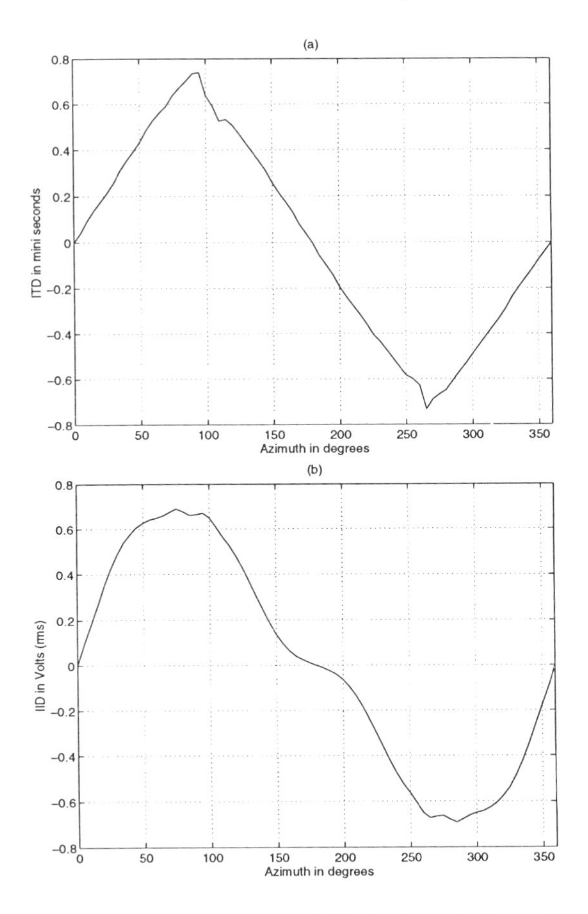

Figure 13.1 Interaural difference of a B&K HATS in horizontal plane. (a) ITD. (b) IID.

surroundings as well as in their every day social contact. For humans and most mammals, this ability is based on the fact that they have two ears.

2.1 INTERAURAL CUES FOR SOUND LOCALIZATION

Sound emitted from a source located away from the median plane of the head arrives at two ears with an interaural time difference (ITD) and an interaural intensity difference (IID). It has been recognized for more than a century that the ITD and IID are the primary cues for sound localization. Specifically, ITD is primarily responsible for providing cues for sound with low frequency (below 1.0 kHz) since the ITD creates distinguishable phase difference between two ears at low frequencies. On the other hand, due to the head shadowing effect, IID is created at the high frequency range (above 2.0 kHz) that helps listeners to

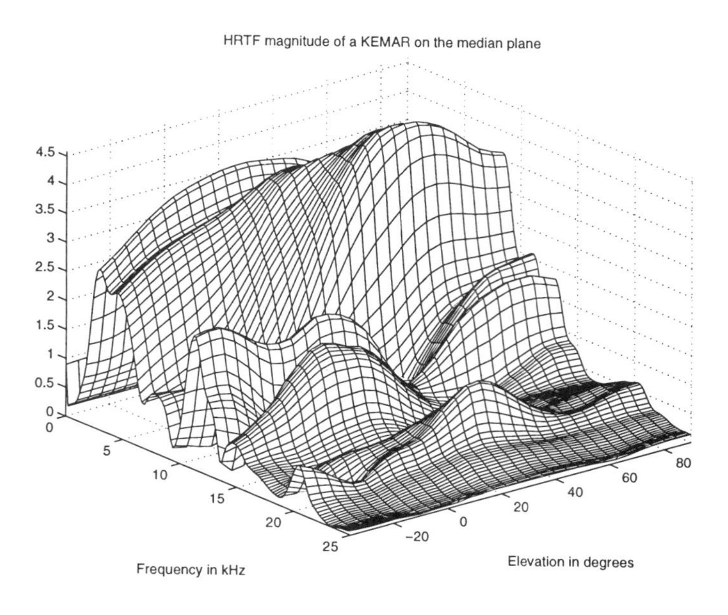

Figure 13.2 HRTF variations in median plane for a KEMAR manikin.

localize sound with high frequency components. This observation has led to the so-called "duplex theory" [24]. The duplex theory explained many experimental results and life listening experiences and thus had been widely used by scientists in conducting spatial hearing research. Figure 13.1 (a) depicts the ITD of a dummy head on horizontal plane and (b) the IID. It can be seen that the magnitudes of both ITD and IID are minimized when source is at front and back positions (azimuth equals to 0 or 180 degrees) and are maximized at both left and right sides (azimuth equals to 90 and 270 degrees). Note that the front direction is defined as the origin of both azimuth and elevation. Clockwise the azimuth increases from zero to 360 degrees. While the 90-degree elevation is the upward direction and the -90 degree direction is the downward.

2.2 HEAD-RELATED TRANSFER FUNCTION (HRTF)

Regardless of the success of the duplex theory in explaining much of the binaural experimental data, it does not adequately address why a listener can resolve a sound source moving up and down in the median plane where both ITD and IID are diminished. It also fails to explain why a listener can localize sound with only one ear, although the localization accuracy is poor. Only recently, a third class of cues has been identified as being responsible for median plane and monaural sound localization [5, 6]. Many experiments have been conducted to reveal that before sound arrives at the ear-drums it has been

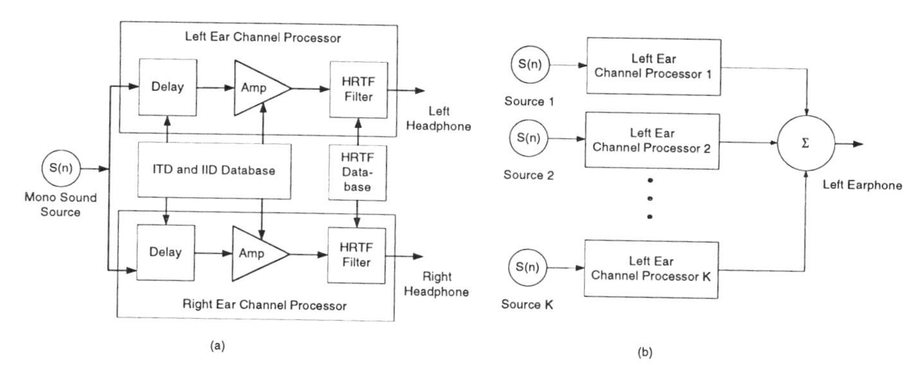

Figure 13.3 Simple implementation of 3D sound. (a) A single sound source is 3D positioned. (b) Multiple sound sources are virtualized.

modified by a listener's external ear (pinnae), head, and torso. In other words, the incoming sound is "transformed" by an acoustic filter. This acoustic filter is now commonly referred to as the head-related transfer function (HRTF). Moreover, the manner and degree of typical modifications of the sound due to the HRTF is dependent upon the incident angle of the sound source in a sort of systematic fashion. Figure 13.2 displays a set of HRTFs as a function of both frequency and elevation in the median plane. It has been speculated that the systematic changes of the notches' and peaks' positions in the frequency domain with respect to elevation change provide localization cues [12, 15, 21].

It has also been discovered that a synthetic headphone presentation of sound processed only by ITD and IID is typically heard as having been originated from inside the listeners head, although the sound source can be lateralized. It is the lack of filtering by an HRTF that causes this "internalized" perceived sound image. This can easily be experienced by listening to a CD using a headphone set versus a speaker array. Therefore, another function of the HRTF is to provide sound image *externalization.*

2.3 SYNTHETIC 3D AUDIO

Figure 13.3 depicts a 3D audio system that employs delay lines, amplifiers, and digital filters to simulate the three major acoustic cues, ITD, IID, and HRTF filtering, respectively. Panel (a) illustrates a single source system where a mono source is first delayed and then individually amplified in both left and right channels. The difference of delays and amplification between left and right channel is controlled by the ITD and IID database. The generated two-channel signals are then fed into a pair of HRTF filters from an HRTF database. Note that the ITD, IID, and HRTF data all coherently correspond to the target location of the virtual source.

In many applications, multiple sound sources are present. Multiple source 3D audio systems are required in rendering spatially distributed and independent multiple sound sources, e. g., to synthesize a virtual concert with singer, guitar, and drum set standing at different positions on the stage. Another example of a multiple source application is the simulation of room acoustics, where not only the direct source but also the reflections (echoes) from the walls are considered. The topic of room acoustic modeling is beyond the scope of this chapter but a simple way of doing it is to introduce "images" of the source to represent the wall reflections [1]. In this case, the direct source and its multiple images have to be rendered simultaneously. Panel (b) of Fig. 13.3 depicts such a system. It is implemented by simply stacking single source systems together. The outputs of each single source system are summed together to form the overall output. Note that only the left ear channel is shown in the figure and the right channel is identical, hence omitted.

The major advantage of using the methods of Fig. 13.3 is their simplicity. Computationally, delay and amplification are inexpensive. The major burden is to execute the HRTF filtering which involves convolution of the HRTF filter coefficients (most commonly in the format of a finite impulse response (FIR) filter) and the signal. If only one source is considered, a pair of HRTFs are required. Assume the HRTFs are represented by 64-tap FIR filters, $64 \times 2 = 128$ multiplications per sample, are required to implement this filter. If the sample rate is 22.05 kHz, this generates a $22050 \times 128 = 2.8$ MIPS computing load. For most off-the-shelf DSP hardware this is an easy task. When a multiple source application is considered, however, the computing load increases proportional to the number of sources simulated. For example, to render a 3D audio image in a room with reasonable spatial impression, the reflections of the walls must be produced. Each "image" is also subject to HRTF filtering as they usually come from different directions. If only the first order reflections are considered, there will be 6 additional sources to be simulated. This will increase the computing load by a factor of 7, which requires 20 MIPS of computing resources. If the secondary reflections are considered, then 36 more images have to be simulated. This immediately exhausts the computing power of current commercially available single chip DSPs, without mentioning that there are potentially many independent sources (and their respective images) to be simulated in most applications. This problem has been addressed in some experimental systems with multiple DSPs, which naturally results in higher system cost.

2.4 MODELING THE MEASURED HRTFS

In the direct implementation of a 3D audio system, the use of measured HRTFs in 3D audio synthesis presents a serious limitation. First, a very large number of HRTFs have to be available to synthesize the moving sound objects with continuity. Second, as we will see in the last section, using many HRTFs to synthesize many sound sources can be computationally prohibitive. These limitations thus have stimulated interests in functional representations of the HRTFs. That is, one seeks a mathematical model or equation that represents the HRTFs as a function of frequency and direction.

The early work of HRTF modeling started with Batteau [2, 3]. He conjectured that the external ear could be modeled as a three-channel two-delay and sum acoustic coupler. One delay varies with the sound source elevation and the other with the source azimuth. Two authors [29, 26] used this model in localization research to synthesize crude eardrum signals. From today's standard, this model seems to be an oversimplification of the HRTF cues and is not able to address more complicated experimental data identified in later research [28]. A recent effort is that of Genuit [10] where, a filter-bank of 16 time-delay channels is adopted to represent acoustic contributions of external ear including the pinna, head, shoulders, and torso. He used classic physical acoustics to establish the relationship between the filter parameters and external ear geometry. This is an attractive approach because it avoids the need for HRTF measurements. However, thorough acoustic and behavioral validation of this model is not yet reported.

More recently, a functional model of the external ear derived from measured HRTFs was proposed by the author [8]. In this study, the external ear is modeled as a multisensor broadband beamformer, with the sensor geometry and beamformer weight set chosen to represent the physical characteristics of the external ear. This model provides an explicit mathematical relationship between the HRTF and source location. Hence, it interpolates HRTF's in arbitrary directions. However, this model only had limited success in modeling a large sector of the space due to its non-orthogonal nature.

Low-dimension and orthogonal representation for measured HRTFs have been generated [14, 13, 16] by applying principal component analysis (PCA) to the logarithms of the HRTFs' magnitudes after directionally independent frequency dependence is removed. It is important that, in their study, the perceptual validity of this method was established by comparing human listener's judgements of direction based on HRTFs as measured empirically with judgements based on HRTFs as reconstructed from the PCA. Due to the non-functional and non-linear nature of this model it does not result in a method that would address the two major issues in 3D audio synthesis mentioned above. Two additional references [7, 22] provide recent reviews for HRTF modeling work.

3. SPATIAL FEATURE EXTRACTION AND REGULARIZATION (SFER) MODEL FOR HRTFS

To address the issues presented in the previous modeling work, another functional model was proposed by the author and colleagues at the University of Wisconsin [7, 9]. In this model, the HRTF's are expressed as weighted combinations of a set of complex valued eigentransfer functions (EFs). The EFs form an orthogonal set of frequency-dependent functions; the weights applied to each EF are functions only of spatial location and are thus termed spatial characteristic functions (SCFs). Estimates of the EFs are obtained from a discrete Karhunen-Loève expansion procedure applied to a sample covariance matrix constructed from measured HRTFs represented on a linear scale. Samples of the SCFs at the measurement locations are obtained by projecting each of the EFs onto the measured HRTFs. A functional representation or mathematical equation for each SCF is obtained by fitting the SCF samples with a thin-plate (two-dimensional) spline using procedures of regularization theory [23, 25]. It hence adopts the descriptive terminology "spatial feature extraction and regularization" model. An extensive acoustic validation of the SFER model establishes its effectiveness. It was shown in this study that over a large angular sector the relative approximation and interpolation mean squared errors are less than one percent.

3.1 SFER MODEL FOR HEAD-RELATED IMPULSE RESPONSE

While the previous SFER model was build upon the frequency domain HRTF data, in practice, HRTF filtering is almost always implemented in time domain. A time domain SFER model can be derived from the so-called head-related impulse response (HRIR) [17] which is the discrete inverse Fourier transform of the HRTF. Expressing the HRIR as an N-by-1 vector, we have the HRIR sample covariance matrix

$$\mathbf{C} = \sum_{i=1}^{I} \mathbf{h}_i \mathbf{h}_i^T, \tag{13.1}$$

where T stands for transposition, $\mathbf{h}_i$ represents the HRIR at spatial location L_i for $i = 1, ..., I$. Spatial location, on a spherical coordinate system, is defined by azimuth angle θ and elevation angle ϕ. Therefore, a HRIR at location L_i can be expressed as an N-by-1 vector $\mathbf{h}(\theta_i, \phi_i)$ where I denotes the total number of HRIR's in consideration. For example, on a 10-degree azimuth-elevation grid, $I = 614$, counting north and south poles as one location.

Although HRIRs measured at different locations are distinct, similarity and systematic changes exist between measured HRTFs at neighboring points. This fact leads to a speculation that these HRIRs are actually laid in a subspace with dimension of M, even when each HRIR is represented by a larger N-by-1 vector. Therefore, if $M << N$, then an M-by-1 vector may be used to represent the HRIR without significant error. This speculation is verified by applying eigen-analysis to the sample covariance matrix of measured HRIRs. Figure 13.4 (a) depicts the eigenvalues of the HRIR sample covariance matrix, that is, the variance projected on each eigenvector of HRIR sample covariance matrix. It is observed that the first few eigenvalues represent most of the variations contained in all the HRIRs. It was shown that doubling the density of HRIR sampling on the sphere, for example, using all HRIRs sampled on a 5-degree grid with a total of 2376 HRIRs, to construct the covariance matrix does not significantly change the distribution of this eigenvalue plot. This indicates a 10-degree sampling is adequate to represent the variations contained in the HRIRs on the whole sphere. Figure 13.4 (b) depicts the sum of first M eigenvalues as a function of M. It is shown that the subspace spanned by first 3 eigenvectors covers 95% of the variance contained in all 614 HRIRs, the first 10 covers 99.6%, and the first 16 eigenvectors cover 99.9%. This suggests that with a mean square error

$$\epsilon^2(M) = \sum_{i=M+1}^{N} \lambda_i,$$

an HRIR can be expressed as

$$\tilde{h}(n) = \sum_{i=1}^{M} w_i q_i(n). \tag{13.2}$$

Equation (13.2) is known as the *Karhunen-Loève* expansion (KLE) of the HRIR [17]. The vector set $q_i(n),\ i, n = 1, ..., N$, forms an orthonormal basis for the HRIR space. The first M basis vector are termed the most significant eigenvectors, or in a continues space, eigenfunctions (EFs). By back projecting all of the measured HRIRs to the first M most significant eigenvectors, M sets of coordinates or weights are obtained. Considering each set of these weights as discrete samples of a continuous function, M such functions can be synthesized by using a two-dimensional regression model, also known as regularization algorithm [25, 11]. The regularized continuous functions are termed spatial characteristic functions (SCFs) because they are functions of spherical spatial coordinates θ and ϕ only. Using (13.2), HRIRs at arbitrary locations (θ, ϕ) can be synthesized. Equation (13.2) is termed as the time domain spatial feature extraction and regularization (TDSFER) model of the HRIR. Note that (13.2) also implements a separation of time and spatial coordinates, showing an example in which a multivariate function HRIR can be represented as a linear combination of univariate and bi-variate functions.

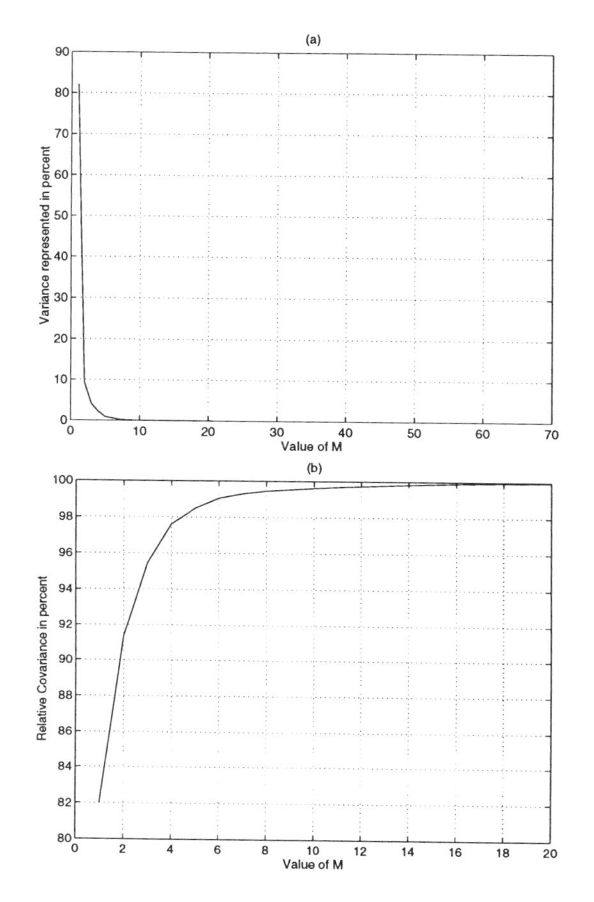

Figure 13.4 Covariance analysis: (a) Percent variance distribution. (b) The percentage variance of HRIR represented by first M eigenvalues.

Some advantages, as we show later, result from this separation. Compared with a frequency domain SFER model, the TDSFER model has all real-valued kernel functions (EFs and SCFs). This provides a very convenient and efficient programming model for DSP coding.

3.2 TDSFER MODEL FOR MULTIPLE 3D SOUND SOURCE POSITIONING

In this section we illustrate how the SFER model can be used as a highly efficient algorithm for virtual acoustic environment synthesis (VAES) where multiple sound sources and multiple reflections need to be positioned in 3D.

Let $s(n)$ represent the source to be positioned, $y(n)$ the output signal processed by the HRIR filter, and $h(n, \theta, \phi)$ be the HRIRs used to position the source at spatial location (θ, ϕ). Then,

$$y(n) = s(n) * h(n). \tag{13.3}$$

Plugging (13.2) into (13.3), we have,

$$y(n) = s(n) * \sum_{m=1}^{M} w_m(\theta, \phi) q_m(n). \tag{13.4}$$

It is obvious that (13.4) is M-times more computationally expensive than the direct convolution (13.3). Now consider the case when two sources at two different locations (θ_1, ϕ_1) and (θ_2, ϕ_2) are involved. Using (13.2), the output is

$$\begin{aligned} y(n) &= s_1(n) * h(n, \theta_1, \phi_1) + s_2(n) * h(n, \theta_2, \phi_2) \quad (13.5) \\ &= s_1(n) * \sum_{m=1}^{M} w_m(\theta_1, \phi_1) q_m(n) + s_2(n) * \sum_{m=1}^{M} w_m(\theta_2, \phi_2) q_m \\ &= \sum_{m=1}^{M} [w_m(\theta_1, \phi_1) s_1(n) + w_m(\theta_2, \phi_2) s_2(n)] * q_m(n), \quad (13.6) \end{aligned}$$

where $s_1(n)$ and $s_2(n)$ represent sources at two different spatial locations and $h_1(n)$ and $h_2(n)$ represent the corresponding HRIRs. The M-th order KLE is used to represent each HRIR. Compared with (13.4), (13.6) does not double the number of convolutions even though the number of sources and HRIRs are doubled, instead, it adds M multiplications and $M - 1$ additions. Since convolution involves N multiplications and $M - 1$ additions per sample base, if $M << N$, the number of instructions is certainly reduced.

The principle of (13.6) can be immediately extended to multiple sources. Suppose K independent sources at different spatial locations need to be rendered to form a one-ear input signal, $y(n)$. Then, $y(n)$ is the summation of each source convolved with its respective HRIR. Plugging (13.2) into this expression, we have,

$$\begin{aligned} y(n) &= s_1(n) * h(n, \theta_1, \phi_1) + ... + s_K(n) * h_K(n, \theta_K, \phi_K) \\ &= \sum_{k=1}^{K} s_k(n) * h(n, \theta_k, \phi_k) \quad (13.7) \\ &= \sum_{k=1}^{K} s_k(n) * \sum_{m=1}^{M} w_m(\theta_k, \phi_k) q_m(n) \\ &= \sum_{m=1}^{M} \left[\sum_{k=1}^{K} w_m(\theta_k, \phi_k) s_k(n) \right] * q_m(n). \quad (13.8) \end{aligned}$$

In (13.8), the inner sum takes MK multiplications and $M(K-1)$ additions for $m = 1, ..., M$. For a DSP processor utilizing multiplication-accumulation instructions, it takes MK instructions to finish the inner sum loop. If each $q_m(n)$ has N taps, then the outer sum including the convolution takes MN instructions to finish. Therefore, the total instructions needed will be $MN + MK = M(N + K)$. In contrast, the direct convolution will need KN instructions. The computation efficiency improvement ratio η is,

$$\eta = \frac{KN}{M(N+K)}. \tag{13.9}$$

For a moderate size of K, $10 < K < 1000$, η is a function of all the parameters M, N, and K. When $K \to \inf$, $\eta \to \frac{N}{M}$. Figure 13.5 depicts η as a function of K with both M and N as parameters. It is obvious that by using (13.8) there will be M convolutions regardless of how many sources are involved. Each source requires M multiplications and $M - 1$ additions. If $K \leq M$, (13.8) is less efficient than (13.7). However, if $K \geq M$, (13.8) is more efficient than (13.6). When K is significantly larger than M, the gain of using (13.8) to simulate multiple sound source and reflections is substantial.

To quantitatively illustrate this improvement, Table 13.1 shows the comparison between the direct convolution method and the TDSFER model method for different numbers of signal sources. Figure 13.5 (a) depicts the computation efficiency improvement ratio for $N = 128$ which is typically used when the sampling rate is 44.1 kHz. Figure 13.5 (b) is the case where $N = 64$ at a sampling rate of 22.05 kHz. Both cases of $M = 4$ and $M = 8$ are shown. The larger the M is, the higher the quality of the SFER model, that is, the synthesized HRIR will be closer to the measured one. Initial testing supports the idea that a value of M between 3 and 10 gives performance from acceptable to excellent.

For a typical example of authentic 3D sound environment simulation, several sources are considered with first- and second-order room reflections included. For example, four sources with second order reflections included, results in a total of $2 \times (4+4 \times (6+36)) = 344$ sources and reflections to be simulated for both ears. If direct convolution is used, 22016 instructions are needed for each sample at a sampling rate of 22.05 kHz, which is equivalent to 485 MIPS in computational requirements. This is beyond a single DSPs capacity by today's standard. However, if this is implemented using TDSFER model, only 3264 instructions are needed per sample, which is equivalent to 72 MIPS. Note that this is based on $M = 8$. If $M = 4$ is considered then only 36 MIPS are needed. Today, many off-the-shelf single DSPs have this computational capacity.

Table 13.1 Comparison of number of instructions for HRIR filtering between direct convolution and TDSFER model.

K	$N = 64$			$N = 128$		
	Direct Conv.	TDSFER		Direct Conv.	TDSFER	
		$M = 8$	$M = 4$		$M = 8$	$M = 4$
2	128	528	264	256	1040	520
10	640	592	296	1280	1104	552
100	6,400	1,312	656	12800	1,824	912
1,000	64,000	8,512	4,256	128,000	9,024	4,512
10,000	640,000	80,512	40,256	1,280,000	81,024	40,512
100,000	6,400,000	800,512	400,256	12,800,000	801024	400,512

4. COMPUTING ARCHITECTURES USING TDSFER MODEL

So far, we have only discussed the processing of HRIRs. As was mentioned earlier, a complete 3D system also needs to incorporate other important cues, including IIDs and ITDs. To implement the core algorithms effectively, several architectures and programming models are now proposed.

4.1 MULTIPLE SOURCES WITH MULTIPLE REFLECTIONS

VAES with multiple independent sources and their respective reflections, presents the most complicated case of implementation. Figure 13.6 illustrates the computing architecture for VAES. The top block consists of an image model using inputs of room specifications, source specifications and listener specifications to derive the distance and direction parameters for each virtual source with respect to the listener. The "Delay and Attenuation Calculation" block uses the distance data and ITD database to calculate delays and attenuations for each independent source and their images. Usually, delays are represented in terms of numbers of samples. The integer portion of the delay is implemented by delaying the source signal a number of samples while the fractional portion is implemented through simple interpolation of the signal. Note that there is a FIFO buffer associated with each independent source. Reflections which are simply a delayed and attenuated version of the source are generated by properly selecting the tap outputs of each source FIFO buffer.

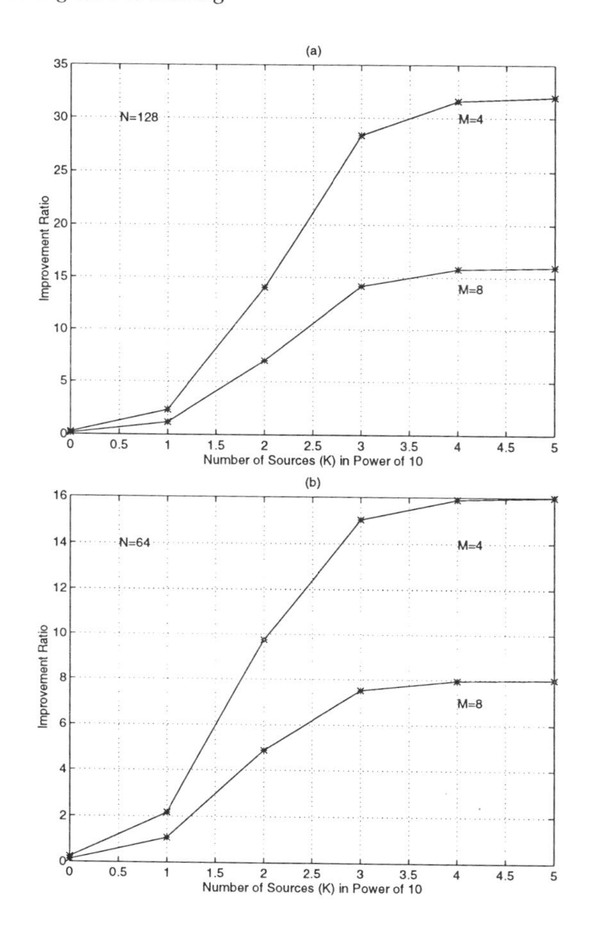

Figure 13.5 Computation efficiency improvement ratio of TDSFER model over direct convolution. (a) HRIR filter tap number $N = 128$. (b) HRIR filter tap number $N = 64$.

In Fig. 13.6 there are S_i, $i = 1, ..., K$, independent sources. Each source may have at most J reflections obtained from the tap outputs of the corresponding FIFO buffer. 3D positioning is done by passing these signals, attenuated and delayed, through the "SCF Weighting Matrix" in a method according to (13.8). The "SCF Weighting Matrix" obtains its values from SCF interpolation block, which is driven by the source direction data from the image model and the "SCF Sample Meshes." The "Sample Meshes" are obtained from the SFER model. The interpolation also can be a very simple algorithm based on two-dimensional linear interpolation to save computation.

On the right hand side of "SCF Weighting Matrix" block, there are two EF filter banks, one is for left ear channel and one for right. Each bank contains M EF filters. The input to each EF filter from "SCF Weighting Matrix" is

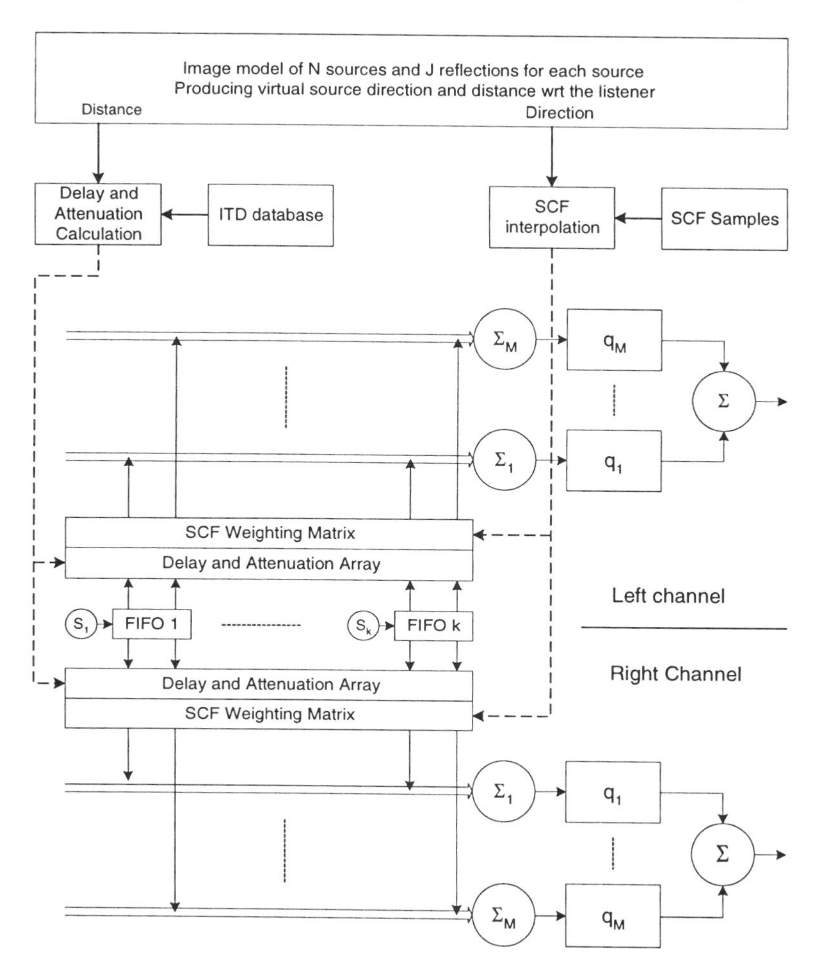

Figure 13.6 SFER computing model for multiple sound sources with multiple reflections.

the summation of its own weighted and attenuated input signals in a method according to (13.8). For example, the output signal of "SCF Weighting Matrix" to $\mathbf{q}_1$, t_{1l} is expressed as,

$$t_{1l} = \sum_{k=1}^{K} w_m(\theta_k, \phi_k) s_{l,k}(n). \tag{13.10}$$

The output of each EF filter is summed at summer S_1 and S_2 to form left and right channel output signals l' and r', respectively.

4.2 SINGLE SOURCE WITH MULTIPLE REFLECTIONS

When only one source is present in an acoustic enclosure, all of its images are delayed and attenuated versions of the source itself. An architecture that saves MIPS can be derived using this characteristic. The following derivation helps in understanding this approach.

Assuming $y(n)$ represents a monaural output signal to a headphone set, without making distinction of left and right,

$$\begin{aligned} y(n) &= s(n-\tau_0) * h(n, \theta_0, \phi_0) + \ldots + s(n-\tau_J) * h(n, \theta_J, \phi_J) \\ &= \sum_{j=0}^{J} s(n-\tau_j) * h(n, \theta_j, \phi_j), \end{aligned} \tag{13.11}$$

where $s(n-\tau_0)$ represents the source and $s(n-\tau_j)$, $j = 1, \ldots, J$, represents the images, τ_j, $j = 0, \ldots, J$, are the corresponding delays for the source and its images. Substituting $h(n, \theta, \phi)$ with its TDSFER model representation, we rewrite (13.11) as,

$$\begin{aligned} y(n) &= \sum_{j=0}^{J} s(n-\tau_j) * \sum_{m=1}^{M} w_m(\theta_j, \phi_j) q_m(n) \\ &= \sum_{j=0}^{J} \sum_{m=1}^{M} s(n-\tau_j) * q_m(n) w_m(\theta_j, \phi_j). \end{aligned} \tag{13.12}$$

Taking the Z-transform of (13.12),

$$\begin{aligned} Y(Z) &= \sum_{j=0}^{J} \sum_{m=1}^{M} S(Z) Z^{-\tau_j} Q_m(Z) w_m(\theta_j, \phi_j) \\ &= \sum_{j=0}^{J} \left[\sum_{m=1}^{M} S(Z) Q_m(Z) w_m(\theta_j, \phi_j) \right] Z^{-\tau_j} \qquad (13.13) \\ &= \sum_{j=0}^{J} R(Z, \theta_j, \phi_j) Z^{-\tau_j}, \qquad (13.14) \end{aligned}$$

where $S(Z)Z^{-\tau_j}$ is the Z-transform of $s(n-\tau_j)$ and $Q_m(Z)$ is the Z-transform of $q_m(n)$, and $R(Z, \theta_j, \phi_j) = \sum_{m=1}^{M} S(Z) Q_m(Z) w_m(\theta_j, \phi_j)$. Equation (13.14) suggests an alternative implementation in which only one set of EF filters is needed and thus the number of convolutions is further reduced.

5. SPECIFIC ISSUES FOR VAES IMPLEMENTATION

Several practical issues are involved in the application of 3D audio systems besides the ones discussed above. We are not able to discuss all of these issues in detail in this chapter and only list them for future research.

1. Headphone equalization. The signals fed into the left and right ears have to be rendered by electroacoustic devices. Headphones give the best control over what is to be delivered to each ear and hence they are the most suitable devices for VAES rendering. When a listener wears a set of headphones, an acoustic cavity forms between the headphone and the eardrum. The volume of the acoustic cavity, the resonance of this cavity, and the frequency response of the headphone form a transfer function that can significantly compromise the quality of the 3D audio rendering. An equalizer needs to be designed to compensate for this non-ideal transfer function.

2. Crosstalk cancellation and the sweet spot. When loudspeakers are used for rendering, acoustic crosstalk occurs between the loudspeaker and the ear on the opposite side. Crosstalk significantly blurs 3D positioning. Another problem associated with loudspeaker presentation is the listening sweet spot. The sweet spot is a small area where 3D effects are maximized. It constraints the movement of the listener and hence is undesirable. These problems are the focus of Chapter 14.

3. Head movement. In a natural listening situation listeners are free to move their head. The resulting dynamic changes in all the acoustic cues enhance localization. To simulate this, a corresponding mechanism has to be introduced into the VAES system. Typically, a head tracker has been used for headphone rendering to compensate for head movement. There has been no report on how to address this problem for loudspeaker presentation.

6. CONCLUSIONS

3D audio technology has received attention because the availability of computing power of DSP's and microprocessors has enabled real-time implementations. With the SFER model for measured HRTFs functional representation, dense measurement of the HRTF is unnecessary. The SFER model also provides a very efficient method for 3D audio synthesis when multiple source 3D positioning is required. Virtual acoustic environment synthesis can take advantage of the SFER model such that hundreds of sources and their images can be rendered using modest computing resources. For source numbers between 100 to 1000 the efficiency improvement can be more than a factor of 10. Several computing architectures have been proposed to integrate the SFER model with other important localization cues for implementation.

A number of issues, including crosstalk cancellation, head movement compensation, frontal image elevation, and others, remain unresolved and are the subject of ongoing research projects.

References

[1] J. B. Allen and D. A. Berkley, "Image method for efficiently simulating small room acoustics," *J. Acoust. Soc. Am.*, vol. 65, pp. 943-950, Apr. 1979.

[2] D. W. Batteau, "The role of the pinna in human localization," In *Proc. Roy. Soc.*, pp. 158-180, London, 1967.

[3] D. W. Batteau, *Listening with the naked ear: The neuropsychology of spatially oriented behavior.* Dorsey Press, Homewood, IL, 1968.

[4] J. Blauert, *Investigations of directional hearing in the median plane with the head immobilized.* Ph.D. thesis, Technische Hochschule, Aachen, 1969.

[5] J. Blauert, *Spatial Hearing.* MIT Press, Cambridge. MA, 1983.

[6] R. A. Butler, *Handbook of Sensory Physiology*, Vol. 2. Chapter "The influence of the external ear and middle ear on auditory discrimination." Springer-Verlag, Berlin, 1975.

[7] J. Chen, *Auditory Space Modeling and Virtual Auditory Environment Simulation.* Ph.D. thesis, University of Wisconsin-Madison, H6/573, 600 Highland Ave., Madison, WI 53792, 1992.

[8] J. Chen, B. D. Van Veen, and K. E. Hecox, "External ear transfer function modeling: a beamforming approach," *J. Acoust. Soc. Am.*, pp. 1933-1944, 1992.

[9] J. Chen, B. D. Van Veen, and K. E. Hecox, "A spatial feature extraction and regularization model for the head-related transfer function," *J. Acoust. Soc. Am.*, vol. 97, pp. 439-452, Jan. 1995.

[10] K. Genuit, "A description of the human outer ear transfer function by elements of communication theory," In *Proceedings of 12th International Congress on Acoustics*, Toronto, Canada, 1986.

[11] C. Gu, "Rkpack and its applications: Fitting smoothing spline models," Technical Report 857, Department of Statistics, University of Wisconsin-Madison, 1989.

[12] J. Hebrank and D. Wright, "Spectral cues used in the localization of sound source on the median plane," *J. Acoust. Soc. Am.*, pp. 1829-1834, 1974.

[13] D. J. Kistler and F. L. Wightman, "A model of head-related transfer functions based on principal components analysis and minimum-phase reconstruction," *J. Acoust. Soc. Am.*, vol. 91, pp. 1637-1647, 1992.

[14] W. L. Martens, "Principal components analysis and resynthesis of spectral cues to perceived direction," In J. Beauchamp, editor, *The International Computer Music Conference*, pp. 274-281, San Francisco, CA, 1987.

[15] S. Mehrgardt and V. Mellert, "Transformation characteristics of the external human ear," *J. Acoust. Soc. Am.*, vol. 61, pp. 1567-1576, 1977.

[16] J. C. Middlebrooks and D. M. Green, "Observations on a principal components analysis of head-related transfer functions," *J. Acoust. Soc. Am.*, pp. 597-599, 1992.

[17] R. A. Reale, J. Chen, J. E. Hind, and J. F. Brugge, *Virtual Auditory Space: Generation and Applications*. S. Carlile, editor. Chapter "An implementation of virtual acoustic space for neurophysiological studies of directional hearing." R. G. Landes Company, Austin, TX, 1996.

[18] E. A. G. Shaw, "Physical models of the external ear," In *Proc. 8th IntUl. Congr. Acoust.*, pp. 206, 1974.

[19] E. A. G. Shaw, "Wave properties of the human ear and various physical models of the ear," *J. Acoust. Soc. Am.*, pp. S3(A), 1974.

[20] E. A. G. Shaw, "The external ear: New knowledge," In S. Daalsgaard, editor, *Earmolds and associated problems, Scand. Audiol. Suppl.*, vol. 5, pp. 24-48, 1975.

[21] E. A. G. Shaw, "The elusive connection: 1979 Rayleigh Medal Lecture," In *Ann. Mtg. Institute of Acoustics (U. K.)*, Southampton, England, 1979.

[22] B. Shinn-Cunningham and A. Kulkarni, *Virtual Auditory Space: Generation and Applications*, S. Carlile, editor. Chapter "Recent developments in virtual auditory space." R. G. Landes Company, Austin, TX, 1996.

[23] A. N. Tikhonov and V. Y. Arsenin, *Solutions of Ill-Posed Problems*. John Wiley and Sons, New York, 1977.

[24] J. V. Tobias, editor, *Foundations of Modern Auditory Theory*. Academic Press, New York, 1972.

[25] G. Wahba, "Spline Models for Observational Data," Society for Industrial and Applied Mathematics, Philadelphia, Pennsylvania, 1990.

[26] A. J. Watkins, *Localization of Sound: Theory and Application*. Chapter "The monaural perception of azimuth: A synthesis approach." The Amphora Press, Groton, CT, 1979.

[27] F. L. Wightman and D. J. Kistler, "Headphone simulation of free-field listening: I: Stimulus synthesis," *J. Acoust. Soc. Am.*, pp. 858-867, 1989.

[28] F. L. Wightman and D. J. Kistler, "Headphone simulation of free-field listening: II: Psychophysical validation," *J. Acoust. Soc. Am.*, pp. 868-878, 1989.

[29] D. Wright, J. H. Hebbrank, and B. Wilson, "Pinna reflections as cues for localization," *J. Acoust. Soc. Am.*, pp. 957-962, 1974.

Chapter 14

VIRTUAL SOUND USING LOUDSPEAKERS: ROBUST ACOUSTIC CROSSTALK CANCELLATION

Darren B. Ward
University College, The University of New South Wales
darren.ward@adfa.edu.au

Gary W. Elko
Bell Laboratories, Lucent Technologies
gwe@research.bell-labs.com

Abstract In this chapter we describe how virtual sound may be delivered to a single listener using a small number of loudspeakers. We also outline the fundamental robustness problem inherent in all such systems, and explain how judicious choice of the loudspeaker locations can be used to help overcome this problem.

Keywords: Virtual Sound, Acoustic Crosstalk Cancellation, 3D Audio, Binaural Sound, Head-Related Transfer Function

1. INTRODUCTION

For multiple participant teleconferencing, providing multi-channel audio has been advocated as an alternative to a conventional single audio channel. The use of multi-channel audio provides a means of supplying listeners with spatialized sound cues. Using these cues, remote participants could be virtually placed around the listener. Not only would this greatly add to the realism of such a teleconferencing system, it would also provide the listener with the ability to easily differentiate between competing talkers based on their locations in the virtual sound space. This will become more important as telephony moves to IP networks: the unavoidable packet delays mean that with full duplex communication, it is inevitable that conference participants will be talking simultaneously (even if that is not necessarily their intention).

The system we will describe in this chapter is only practical for a single listener. In cases where virtual sound must be produced for several listeners simultaneously in a single room, *sound field* techniques (e.g., [3]) are more appropriate. These techniques aim to recreate an acoustical pressure field over a region of space, and typically require a large number of loudspeakers to reproduce sound originating from different directions.

To create virtual sound for a single listener it is necessary to recreate the acoustic pressures at the listener's ears that would result from the natural listening environment to be simulated. This requires two systems: a binaural synthesizer that creates the appropriate ear signals corresponding to the target scene; and a means of delivering these binaural signals to the listener. The previous chapter described the binaural synthesizer step. In this chapter we consider how to deliver these binaural signals to the listener. This can be trivially achieved using headphones, although studies show that headphone reproduction often suffers from front-back reversals and other localization errors [2], unless individualized HRTFs (head-related transfer functions) are used [11]. Requiring the listener to wear headphones is also inconvenient and cumbersome.

If loudspeakers are used, precise binaural control can be achieved using a *crosstalk cancellation system* (CCS), which consists of a particular filter network applied to the loudspeaker signals [1]. These filters are designed to equalize the transmission paths from loudspeakers to ears, requiring accurate knowledge of the relevant acoustic transfer functions. Although these transfer functions can be measured or modeled, in practice they can never be known exactly. This is because transmission depends on uncontrollable and unknown factors such as room reverberation (which varies significantly from room to room, and even from point to point within the same room), the listener's head shape (which can vary substantially from person to person), and head position (although this can be tracked and accounted for to some extent [7]).

The main disadvantage of any CCS is that it is critically dependent on the listener's head being in a fixed design position, the so-called "sweet-spot" [13, 14]. Studies show that lateral (sideways) movement away from the design position of as little as a few centimeters can result in loss of the three-dimensional audio effect.

Several authors have noted that moving the loudspeakers close together increases the size of the sweet spot. Specifically, the stereo-dipole [9] is based on using loudspeaker positions of $\pm 5^\circ$ (relative to the center of the listener's head), as opposed to commonly-used stereo loudspeaker positions of $\pm 30^\circ$.

The chapter is organized as follows. In the following section we describe the problem of designing a CCS. In Section 3 we discuss the fundamental robustness problem of these systems. Section 4 explains how the position of

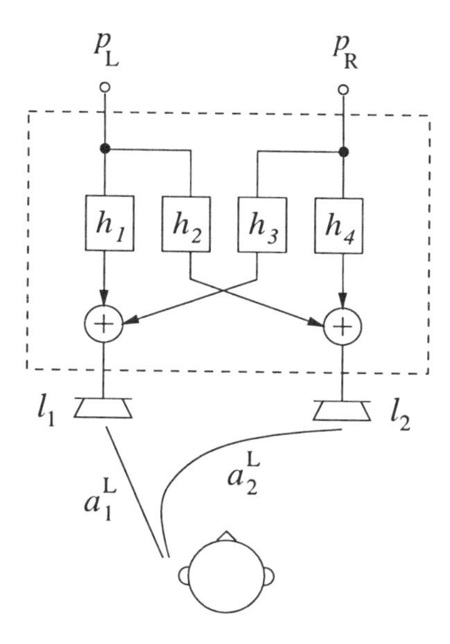

Figure 14.1 Schematic diagram of a crosstalk cancellation system.

the loudspeakers can be used to improve the system robustness, and describes a CCS that is more robust than conventional systems. Finally, in Section 5 we discuss new prospects and challenges in this field.

2. ACOUSTIC CROSSTALK CANCELLATION

Consider the classic Atal-Schroeder CCS [1] shown in Fig. 14.1, in which p_L and p_R are the left and right program signals respectively (i.e., these are the signals that, if delivered exactly to the listener's ears, would produce the desired virtual sound effect), l_1 and l_2 are the loudspeaker signals, and a_n^L, $n = 1, 2$, is the acoustic transfer function (TF) from the nth loudspeaker to the left ear (a similar pair of TFs for the right ear, denoted by a_n^R, are not shown). We assume the loudspeakers are placed at a distance x and an angle $\pm\theta$ relative to the center of the listener's head (where $\theta = 0°$ is directly in front).

2.1 PROBLEM STATEMENT

The objective is to find the filters h_1, h_2, h_3, h_4 such that: (i) the signals p_L and p_R are reproduced at the left and right ears respectively; and (ii) the crosstalk signals are canceled, i.e., none of the p_L signal is received at the right ear, and similarly for the p_R signal and the left ear.

Denoting the actual signals received at the left and right ears as $\widehat{p}_L$ and $\widehat{p}_R$ respectively, this block diagram may be described by

$$\begin{bmatrix} \widehat{p}_L \\ \widehat{p}_R \end{bmatrix} = \begin{bmatrix} a_1^L & a_2^L \\ a_1^R & a_2^R \end{bmatrix} \begin{bmatrix} h_1 & h_3 \\ h_2 & h_4 \end{bmatrix} \begin{bmatrix} p_L \\ p_R \end{bmatrix}$$

$$\widehat{\mathbf{p}} = \mathbf{A}\,\mathbf{H}\,\mathbf{p}. \tag{14.1}$$

Clearly, to reproduce the program signals identically at the ears requires $\mathbf{AH} = \mathbf{I}$.

Note that equation (14.1) describes the system at a single frequency. To simplify notation we will drop the explicit frequency dependence in the remainder.

2.2 SELECTION OF THE DESIGN MATRIX

If the TF matrix $\mathbf{A}$ is known, then designing the filters is straightforward (see, e.g., [10]). However, it is clearly impractical to know $\mathbf{A}$ exactly, and therefore $\mathbf{H}$ is designed using an assumed model for $\mathbf{A}$. We will refer to this as the *design matrix* and denote it by $\mathbf{A}_0$.

In choosing an appropriate $\mathbf{A}_0$, there are several parameters to be considered including:

(a) head position,

(b) TFs from loudspeakers to ears, and

(c) loudspeaker positions.

For a two-loudspeaker CCS, it is usual to assume the head is symmetrically positioned between the loudspeakers and a certain distance back. This distance is somewhat application-dependent, and a PC-based teleconferencing system might typically use a distance of $x = 0.5$ meters.

To model the TFs from loudspeakers to ears there are two main considerations. The first of these is the propagation from the loudspeakers to the head. In a real room this propagation will consist of multi-path reverberant propagation, which will differ significantly from room to room, and even from point to point within the same room. Accurately modeling this reverberant path is infeasible, and typically only the direct path (i.e., free field propagation to the head) is modeled. The second consideration is the effect of the head on the impinging sound wavefront. It is impractical to assume that individualized HRTFs are available, so the listener's exact HRTF will usually be unknown. For this reason, a generic head model will be most appropriate. Data obtained from a KEMAR mannequin is widely available (see, e.g., [6]) and will typically provide good results for a large number of users. It has also been reported that a CCS based on a simple spherical head model "produces immensely satisfying results for a wide range of listener's heads" [5].

The last aspect of $\mathbf{A}_0$ is the loudspeaker positions. This is one parameter over which the designer has complete control, and, as we will see, proves to be fundamentally important in designing a CCS that will work well in practice.

3. ROBUSTNESS ANALYSIS

Once $\mathbf{A}_0$ is chosen it is straightforward to design the CCS filters such that $\mathbf{A}_0\mathbf{H} = \mathbf{I}$. In practice, however, the actual TF matrix will differ from $\mathbf{A}_0$. This raises the following questions:

(a) To what extent will $\mathbf{AH} \approx \mathbf{I}$ for $\mathbf{A} \approx \mathbf{A}_0$?

(b) Is there a specific choice of design matrix for which the CCS will give better cancellation under real-world conditions?

It is the second question that we will answer here. In particular, we intend to determine $\mathbf{A}_0$ such that filters designed according to $\mathbf{A}_0\mathbf{H} = \mathbf{I}$ will also be effective when the TF matrix differs slightly from $\mathbf{A}_0$. In other words, we aim to determine $\mathbf{A}_0$ such that the resulting CCS is robust.

3.1 ROBUSTNESS MEASURE

To enable us to compare different design matrices, an appropriate robustness measure is required. For this purpose we will choose the matrix condition number, which reflects the robustness to perturbation of a linear system such as (14.1). For a complex-valued matrix, $\mathbf{X}$, the condition number is defined as

$$\text{cond}\{\mathbf{X}\} = \frac{\sigma_{\max}(\sqrt{\mathbf{X}\mathbf{X}^H})}{\sigma_{\min}(\sqrt{\mathbf{X}\mathbf{X}^H})}, \tag{14.2}$$

where $\sigma_{\min}(\cdot)$ and $\sigma_{\max}(\cdot)$ represent the smallest and largest singular values, respectively.

If $\text{cond}\{\mathbf{A}_0\}$ is small, then $\mathbf{A}_0$ is well-conditioned and the CCS will be inherently robust. On the other hand, if $\text{cond}\{\mathbf{A}_0\}$ is large, then $\mathbf{A}_0$ is ill-conditioned and the CCS will be very sensitive to perturbations from the assumed design matrix.

3.2 ANALYSIS OF THE DESIGN MATRIX

We have already seen that the design matrix should include head effects (at least generically). However, in performing a robustness analysis it is unnecessary to go to this level of detail. In fact, a very simple TF model can be used to perform a useful robustness analysis. Using a straightforward propagation model provides a closed-form expression for the conditioning of the design matrix under different conditions. These results may then be used to derive optimum loudspeaker positions.

Let the acoustic TF between the nth loudspeaker and the left ear be

$$a_n^{\mathrm{L}} = e^{-j2\pi\lambda^{-1}d_n^{\mathrm{L}}}, \quad n = 1, 2, \tag{14.3}$$

where λ is the wavelength, and d_n^{L} is the distance from the nth loudspeaker to the left ear (and similarly for the right ear and a_n^{R} and d_n^{R}). This model only considers the direct acoustic path, and ignores attenuation and diffraction and scattering effects from the head. Hence, it only models the interaural delay. While we do not claim that this model should be used to design the CCS filters, it does provide adequate information to allow us to draw conclusions about system robustness.

To verify that this is the case, Fig. 14.2 shows the conditioning (14.2) of $\mathbf{A}_0$ as a function of frequency for two different loudspeaker positions. The solid lines show results obtained from HRTF measurements of KEMAR [6], whereas the dashed lines show results of using the model (14.3). It is clear that the conditioning of the KEMAR design matrix has peaks and troughs at approximately the same frequency as the model. Since we are primarily concerned with finding cases where $\mathbf{A}_0$ is well-conditioned (troughs in Fig. 14.2), or vice versa, avoiding cases where $\mathbf{A}_0$ is ill-conditioned (peaks in Fig. 14.2), using the model will provide us with this information. Furthermore, the model provides us with the ability to obtain closed-from design equations, and the results of Fig. 14.2 verify that these equations will also be valid when real HRTFs are included in the design model.

As an aside, we note that the large peak at around 8.2 kHz in the KEMAR curves of Fig. 14.2 is due to a notch in the HRTFs at this frequency. This notch is caused by interaction of the incident sound wave with the external ear (specifically, concha reflections).

3.3 EXAMPLE OF EAR RESPONSES

The results of Fig. 14.2 indicate that with loudspeaker positions of $\pm 30°$, the CCS becomes non-robust at a frequency of around 2 kHz, whereas for loudspeaker positions of $\pm 10°$ the system is non-robust at around 5 kHz. To validate these conclusions, we simulated ear responses for a CCS designed using $\mathbf{A}_0$ with each set of loudspeaker positions used in Fig. 14.2. The results are shown in Fig. 14.3. In each case, the CCS was designed using a spherical head model [4], anechoic propagation, and assuming the head was positioned symmetrically between the loudspeakers at a radial distance of $x = 0.5$ meters.

The figure shows the acoustic TF between the left program input (p_{L} in Fig. 14.1) and each ear. Ideally the left ear response should be unity, and the right ear response should be zero. In each case, the thick lines are for the design listening position (i.e., directly between the loudspeakers), the thin lines

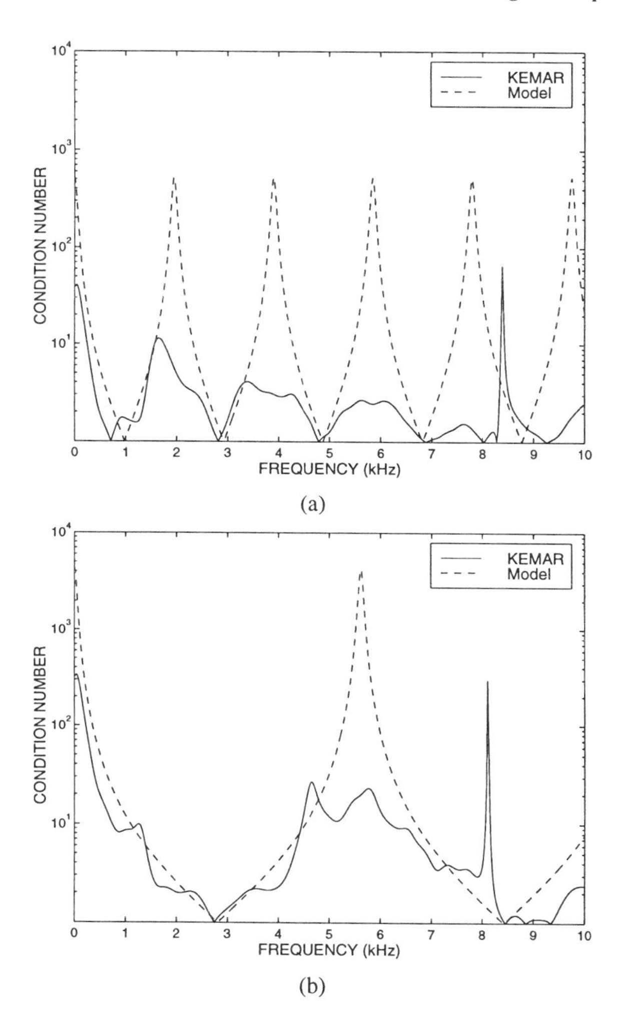

Figure 14.2 Conditioning of acoustic TF matrix versus frequency with loudspeaker positions of (a) $\pm 30^\circ$, and (b) $\pm 10^\circ$.

are with the head displaced laterally by 2 cm, and the dash-dot lines are with the head laterally displaced by 4 cm.

With the head in the design listening position, crosstalk cancellation of at least 20 dB is obtained at all frequencies between about 200 Hz and 7 kHz. As predicted by Fig. 14.2(a), loudspeaker positions of $\pm 30^\circ$ result in a system that is extremely sensitive to perturbation at around 2 kHz. This is verified by Fig. 14.3(a), which indicates that crosstalk cancellation is completely ineffective above about 2 kHz if the head moves by as little as 4 cm. Similarly, for a

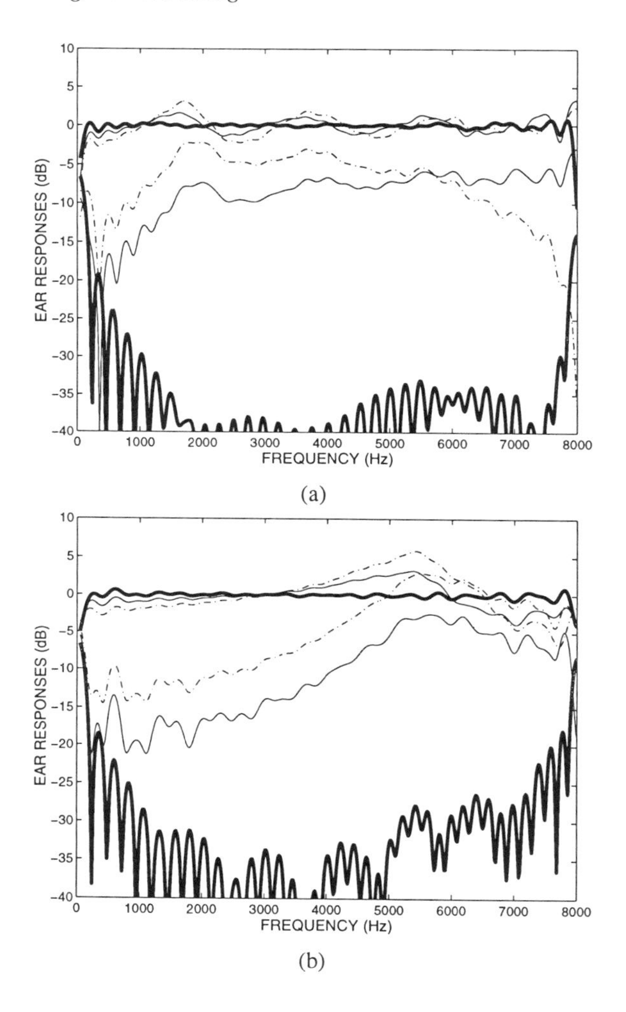

Figure 14.3 Example of ear responses for a CCS designed using loudspeaker positions of (a) $\pm 30^{\circ}$, and (b) $\pm 10^{\circ}$.

loudspeaker position of $\pm 10^{\circ}$, the CCS is ineffective above about 4 to 5 kHz if the head moves. These results verify the effectiveness of Fig. 14.2 in predicting the frequency at which a given CCS will become non-robust.

3.4 SPATIAL RESPONSES

To gain a better understanding of why the CCS suffers from robustness problems, we will now consider some example spatial responses.

Assume the CCS filters are designed to satisfy $\mathbf{A}_0\mathbf{H} = \mathbf{I}$, where $\mathbf{A}_0$ is composed of free-field acoustic TFs (14.3) with the head located symmetrically

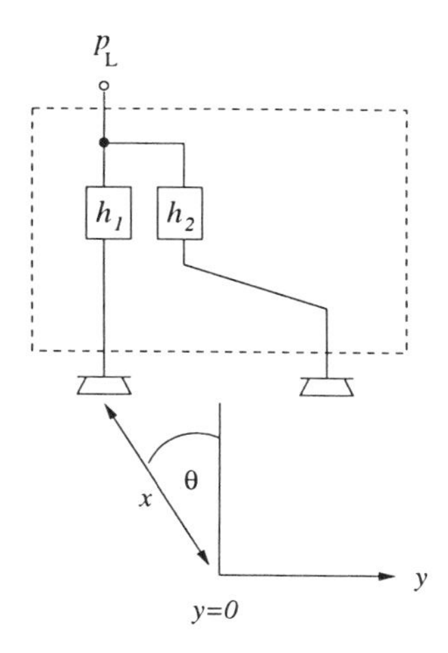

Figure 14.4 Block diagram for spatial responses.

between the loudspeakers at a radial distance of $x = 0.5$ m. Consider the response from the left program input p_L to a point in space y, as denoted in Fig. 14.4.

The resulting spatial response as a function of the position y is shown in Fig. 14.5 at a frequency of 2 kHz, for three different loudspeaker angles θ. Also shown are the nominal positions of the left and right ears (using a default head radius of 0.0875 m). Since we are considering the left program signal only, the CCS is designed to provide unity response for the left ear position and zero response for the right ear position. Thus, there are two design constraints: a flat frequency response to the left ear and a spatial null for the left channel signal at the right ear. Furthermore, in this case the CCS can be thought of as a two-channel beamformer, with filters h_1 and h_2 on the respective channels (see [12] for an excellent review of beamforming). As there are only two filter responses, all available degrees of freedom are used to impose the two design constraints. For a loudspeaker angle of $\theta = 10°$ (Fig. 14.5(a)), this does not unduly effect robustness, since the spatial response is close to unity in the vicinity of the left ear and is close to zero in the vicinity of the right ear.

However, it is a physical property of beamformers that as the elements (in this case the loudspeakers) move further apart, the spatial response becomes narrower. This is apparent in Fig. 14.5(b), where we observe that with $\theta = 20°$ the beam is compressed and the peak has moved accordingly to accommodate the design constraints. Finally, Fig. 14.5(c) shows that at a loudspeaker angle of $\theta = 28°$, the beam is almost precisely narrow enough that both ear positions fall into nulls of the spatial response. This is a situation in which the CCS is fundamentally non-robust, since any movement away from the nominal ear

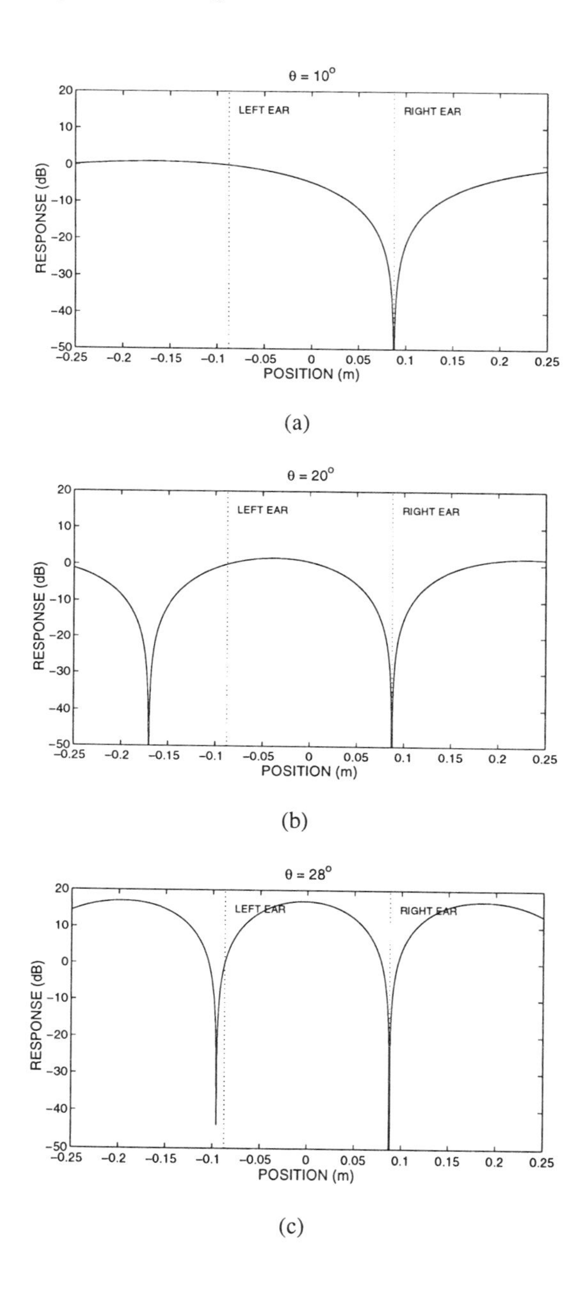

Figure 14.5 Spatial response at 2 kHz for the left program signal p_L with loudspeaker positions of: (a) $\theta = \pm 10^\circ$; (b) $\theta = \pm 20^\circ$; and (c) $\theta = \pm 28^\circ$.

positions will result in responses that are far from the ideal. For example, moving both ears about 1 cm to the left would result in the left ear response being zero and the right ear response being unity – the complete opposite of what is required for the left program signal!

We also note in Fig. 14.5(c) that a significant amount of power is directed to locations away from the head. This is likely to cause problems due to reflections in the listening room. It may be avoided through judicious choice of the loudspeaker angle θ (which we consider in the following section), or by adding more loudspeakers and thereby providing additional degrees of freedom. Alternatively, it may be more practical to limit the output power of the loudspeakers at the expense of the amount of crosstalk cancellation achieved. This particular approach has been considered to some extent in [10], although it requires further analysis and psychoacoustic investigation.

4. EFFECT OF LOUDSPEAKER POSITION

In the previous section we demonstrated that loudspeaker position plays a critical role in determining CCS robustness. In this section we will derive expressions for the optimum loudspeaker positions to use in a given situation.

For the acoustically symmetric system shown in Fig. 14.1, we have $a_2^{\mathrm{L}} = a_1^{\mathrm{R}}$ and $a_2^{\mathrm{R}} = a_1^{\mathrm{L}}$. Let $d_1^{\mathrm{R}} = d_1^{\mathrm{L}} + \Delta$, where Δ represents the *interaural path difference*. Substitution into (14.1) yields:

$$\mathbf{A}_0 = a_1^{\mathrm{L}} \begin{bmatrix} 1 & e^{-j2\pi\lambda^{-1}\Delta} \\ e^{-j2\pi\lambda^{-1}\Delta} & 1 \end{bmatrix}.$$

Solving for the roots of the characteristic equation, it is straightforward to show that the eigenvalues of $\mathbf{A}_0\mathbf{A}_0^H$ are:

$$\sigma_s = 1 \pm \cos(2\pi\lambda^{-1}\Delta).$$

There are two cases of interest:

$$\mathrm{cond}\{\mathbf{A}_0\} = \begin{cases} \infty, & \text{if } \cos 2\pi\lambda^{-1}\Delta = \pm 1 \\ 1, & \text{if } \cos 2\pi\lambda^{-1}\Delta = 0. \end{cases} \tag{14.4}$$

If the interaural distance Δ is such that $\cos(2\pi\lambda^{-1}\Delta) \approx \pm 1$, then the resulting CCS will be extremely sensitive to any variations from the assumed model. On the other hand, if Δ is such that $\cos(2\pi\lambda^{-1}\Delta) \approx 0$, then the CCS will be robust to variations from the assumed model. The value of the parameter Δ is thus of paramount importance in determining how robust the system will be. We must now relate Δ to the loudspeaker positions.

Assume that loudspeaker 1 is placed at a distance x and angle θ relative to the head center, and loudspeaker 2 is at $(x, -\theta)$. Straightforward geometry dictates that the distance from loudspeaker 1 to the left ear is

$$\begin{aligned} d_1^{\mathrm{L}} &= x\sqrt{1 - \frac{2r_{\mathrm{H}}\sin\theta}{x} + \frac{r_{\mathrm{H}}^2}{x^2}} \\ &\approx x\left(1 - \frac{r_{\mathrm{H}}\sin\theta}{x} + \frac{r_{\mathrm{H}}^2}{2x^2}\right), \end{aligned} \tag{14.5}$$

where r_{H} is the radius of the head. Similarly,

$$d_1^{\mathrm{R}} \approx x\left(1 + \frac{r_{\mathrm{H}}\sin\theta}{x} + \frac{r_{\mathrm{H}}^2}{2x^2}\right). \tag{14.6}$$

Substituting $d_1^{\mathrm{R}} = d_1^{\mathrm{L}} + \Delta$ gives

$$\Delta = 2r_{\mathrm{H}}\sin\theta. \tag{14.7}$$

Relating this to (14.4) we find

$$\mathrm{cond}\{\mathbf{A}_0\} = \begin{cases} \infty, & \text{if } \sin\theta = i\frac{\lambda}{0.35}, \\ 1, & \text{if } \sin\theta = \frac{\lambda}{0.7} + i\frac{\lambda}{0.35}, \end{cases} \tag{14.8}$$

where $i \in \mathbb{Z}$, and we have used the commonly-cited average adult head radius of $r_{\mathrm{H}} = 0.0875$ m.

Since $\lambda = c/f$, where f is the frequency of operation and c is the speed of wave propagation (approximately 340 m/s for sound waves in air), it follows from (14.8) that:

(a) Crosstalk cancellation is inherently non-robust at low frequencies, i.e., $\mathrm{cond}\{\mathbf{A}_0\} \to \infty$ as $f \to 0$.

(b) For a given loudspeaker angle, there are limited frequency bands within which the system is stable.

(c) The best loudspeaker position is frequency dependent.

The loudspeaker positions predicted by (14.8) are shown in Fig. 14.6. Positions indicated by the solid curve correspond to cases where the CCS will be robust, whereas those shown dashed should be avoided since they produce an inherently non-robust CCS. For verification, we also performed numerical calculations on KEMAR data, and obtained the results indicated by the crosses and circles. These results agree very well with the model predictions, verifying the validity of our analysis.

These results indicate that to obtain a robust system, it may be necessary to use four loudspeakers, with a wide spacing at low frequencies, and a narrow spacing at high frequencies. The use of close loudspeaker spacing to improve robustness has been previously proposed in [9].

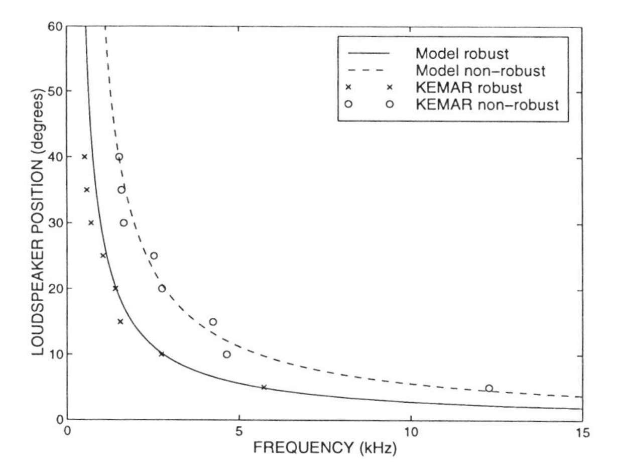

Figure 14.6 Loudspeaker positions versus frequency.

4.1 A ROBUST CCS

In summary, we make the following recommendations to obtain a robust (i.e., practical) CCS:

(a) At frequencies of around 500 Hz to 1.5 kHz, one should use a pair of widely-spaced loudspeakers, i.e., $\theta \approx \pm 30°$.

(b) At frequencies of around 1.5 kHz to 5 kHz, one should use a pair of closely-spaced loudspeakers, i.e., $\theta \approx \pm 7°$.

(c) At frequencies below about 500 Hz, CCS is inherently non-robust and should not be used. In this case, the left and right program signals should be sent directly to the outside pair of loudspeakers.

(d) At frequencies above about 5kHz, there is substantial variation between the HRTFs of different people, so it is almost impossible to implement a generic CCS for these high frequencies. At high frequencies, the shadowing effect of the head also comes into play and helps separate the left and right channels. Hence, at high frequencies the left and right program signals should also be sent directly to the outside pair of loudspeakers.

The block diagram of such a CCS is shown in Fig. 14.7. There are actually three independent systems implemented here, separated on each channel by a filter bank consisting of a bandstop filter (BSF, which attenuates frequencies between 500 Hz and 5 kHz), and two bandpass filters (BPF1, which passes frequencies between 500 Hz and 1.5 kHz, and BPF2, which passes frequencies between 1.5kHz and 5kHz). These systems are: (i) a conventional stereo arrangement that is active at very low and very high frequencies; (ii) a CCS using

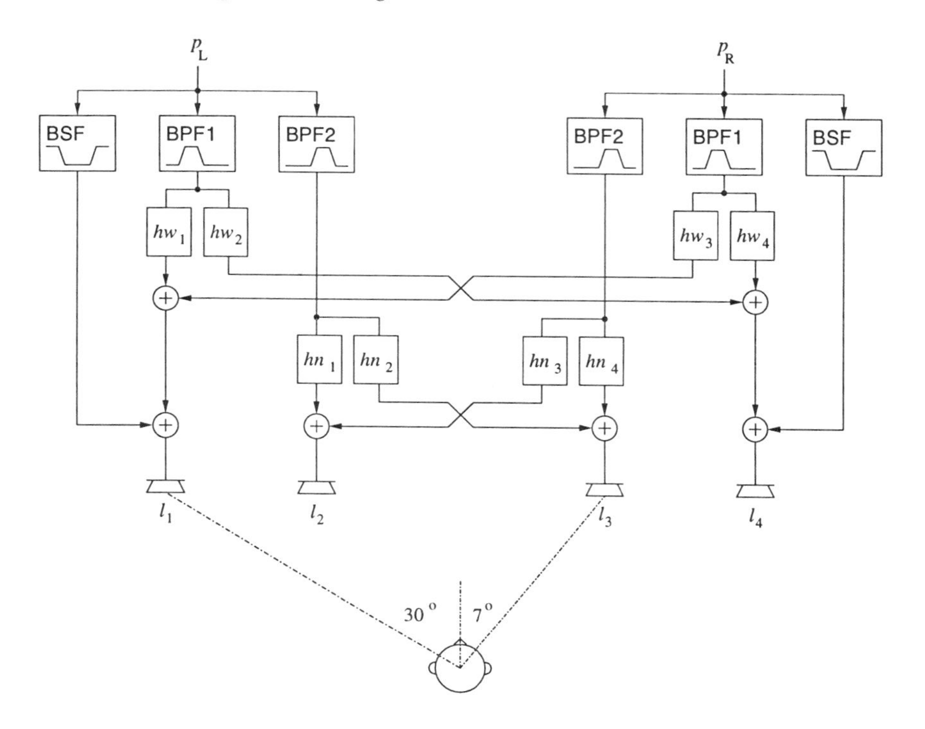

Figure 14.7 Block diagram of a robust CCS (not drawn to scale).

wide loudspeaker positions that is active at low to mid frequencies, denoted by filters hw_n, $n = 1, \ldots, 4$; and (iii) a CCS using narrow loudspeaker positions that is active at mid to high frequencies, denoted by filters hn_n, $n = 1, \ldots, 4$.

5. DISCUSSION AND CONCLUSIONS

We have described the design of single-user virtual sound systems using loudspeakers. In particular, we focused on the fundamental robustness problem with such systems, and showed that the loudspeaker positions play a critical role in determining the sensitivity of the system to modeling errors.

Although using the loudspeaker positions we propose will greatly improve the system robustness, its performance will still be poor if the listener moves too far from the design position. Thus, it will be necessary to track the position of the listener's head in real time and update the CCS filters accordingly. This idea is particularly well-suited to video-conferencing applications in which a video camera will be available to provide this tracking capability. In particular, the availability of inexpensive multimedia PCs incorporating loudspeakers, microphones, and video cameras, means that conferencing systems incorporating virtual audio interfaces are becoming feasible. Combining passive head tracking with adaptive updating of the CCS filters will ensure that virtual sound can be delivered to the listener without requiring headphones. Additional research is required to develop computationally-efficient tracking and updating algorithms so that an adaptive CCS may be implemented in real-time on a native PC without requiring the processing power of an external DSP.

References

[1] B. S. Atal and M. R. Schroeder, "Apparent sound source translator," U.S. Patent 3,236,949, Feb. 1966.

[2] D. R. Begault, "Challenges to the successful implementation of 3-D audio," *J. Audio Eng. Soc.*, vol. 39, pp. 864-870, 1990.

[3] A. J. Berkhout, D. de Vries , and P. Vogel, "Acoustic control by wave field synthesis," *J. Acoust. Soc. Amer.*, vol. 93, pp. 2764-2778, May 1993.

[4] C. P. Brown and R. O. Duda, "A structural model for binaural sound synthesis," *IEEE Trans. Speech Audio Processing*, vol. 6, pp. 476-488, Sep. 1998.

[5] D. H. Cooper and J. L. Bauck, "Prospects for transaural recording," *J. Audio Eng. Soc.*, vol. 37, pp. 3-19, Jan. 1989.

[6] W. G. Gardner and K. D. Martin, "HRTF measurements of a KEMAR," *J. Acoust. Soc. Amer.*, vol. 97, pp. 3907-3908, 1995.

[7] W. G. Gardner, "Head tracked 3D audio using loudspeakers," in *Proc. 1997 IEEE Workshop on Applicat. of Signal Processing to Audio and Acoust.*, New Paltz, NY, USA, Oct. 1997.

[8] G. H. Golub and C. F. Van Loan, *Matrix Computations.* The Johns Hopkins University Press, 3rd edition, 1996.

[9] O. Kirkeby, P. A. Nelson, and H. Hamada, "The stereo dipole – A virtual source imaging system using two closely spaced loudspeakers," *J. Audio Eng. Soc.*, vol. 46, pp. 387-395, May 1998.

[10] O. Kirkeby, P. A. Nelson, H. Hamada, and F. Orduna-Bustamante, "Fast deconvolution of multichannel systems using regularization," *IEEE Trans. Speech Audio Processing*, vol. 6, pp. 189-194, Mar. 1998.

[11] H. Møller, M. F. Sørenson, C. B. Jensen, and D. Hammershøi, "Binaural technique: Do we need individual recordings?" *J. Audio Eng. Soc.*, vol. 44, pp. 451-469, June 1996.

[12] B. D. Van Veen and K. M. Buckley, "Beamforming: A versatile approach to spatial filtering," *IEEE ASSP Mag.*, vol. 5, pp. 4-24, Apr. 1988.

[13] D. B. Ward and G. W. Elko, "Effect of loudspeaker position on the robustness of acoustic crosstalk cancellation," *IEEE Signal Processing Lett.*, vol. 6, pp. 106-108, May 1999.

[14] D. B. Ward and G. W. Elko, "A robustness analysis of 3D audio using loudspeakers," in *Proc. 1999 IEEE Workshop on Applicat. of Signal Processing to Audio and Acoust.*, New Paltz, NY, USA, Oct. 1999, pp. 191-194.

VI

BLIND SOURCE SEPARATION

Chapter 15

AN INTRODUCTION TO BLIND SOURCE SEPARATION OF SPEECH SIGNALS

Jacob Benesty
Bell Laboratories, Lucent Technologies
jbenesty@bell-labs.com

Abstract Blind source separation attempts to recover independent sources given sensor outputs in which the sources have been mixed by an unknown channel. A very important approach for blind source separation is based on maximum entropy (ME). Different learning rules based on ME can be derived and one of them, the natural gradient, is a major contribution to this topic because its performance is independent of the mixing matrix. However, no normalization is made with respect to the source signals. We propose a normalized natural gradient algorithm that is able to track the nonstationarities of the signals. Extensive simulations with speech signals as sources and in the instantaneous mixing case show that this new normalized algorithm converges faster than without normalization.

Keywords: Maximum Entropy, Blind Source Separation, Natural Gradient

1. INTRODUCTION

Blind source separation was first introduced by Herault and Jutten in 1986 [1]. It attempts to find a set of unobservable signals (sources) from a set of observed mixtures without knowing anything about the sources and the mixing process, except the number of sources. Let us consider N unknown zero-mean source signals $s_i(n)$, $i = 1, ..., N$, which are mutually independent at any fixed time n. We assume that at most one source is Gaussian, that we have as many observations as sources, and that all the sources have unit variance (the last assumption is only a normalization convention since the amplitude of each source can be incorporated in the mixing matrix). The model for the sensor outputs is

$$\mathbf{x}(n) = \mathbf{A}\mathbf{s}(n), \tag{15.1}$$

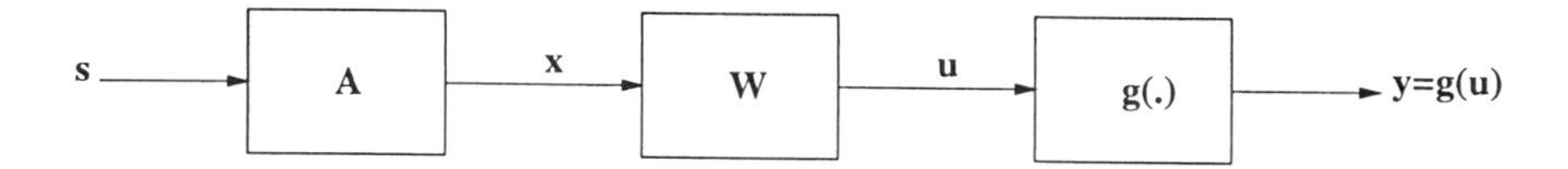

Figure 15.1 Instantaneous mixing, unmixing, and nonlinear transformation.

where $\mathbf{A}$ is an $N \times N$ unknown nonsingular mixing matrix,

$$\mathbf{s}(n) = \begin{bmatrix} s_1(n) & s_2(n) & \cdots & s_N(n) \end{bmatrix}^T$$

is an $N \times 1$ vector containing the source signals,

$$\mathbf{x}(n) = \begin{bmatrix} x_1(n) & x_2(n) & \cdots & x_N(n) \end{bmatrix}^T$$

is an $N \times 1$ vector of the observations (which is the only information that is available), and ^{T} denotes the transpose of a vector or a matrix. The task (using only the observations) is to find a demixing square matrix $\mathbf{W}$ such that

$$\mathbf{u}(n) = \mathbf{W}\mathbf{x}(n) = \mathbf{W}\mathbf{A}\mathbf{s}(n) \qquad (15.2)$$

is an estimate of the source signals (see Fig. 15.1) [2, 3], where

$$\mathbf{u}(n) = \begin{bmatrix} u_1(n) & u_2(n) & \cdots & u_N(n) \end{bmatrix}^T.$$

Note that, without additional information, it is not possible to estimate the original sources $s_i(n)$ in an exact order and amplitude because of the indeterminacy of permutation and scaling of $\{s_i\}$ due to the product of two unknowns: the mixing matrix $\mathbf{A}$ and the source vector $\mathbf{s}(n)$ [4]. We define the following matrix

$$\mathbf{P} = \mathbf{W}\mathbf{A} \qquad (15.3)$$

as the performance matrix so that if $\mathbf{P}$ is normalized and reordered, a perfect separation leads to the identity matrix.

This chapter introduces a new way to normalize the natural gradient algorithm in order to better track the nonstationarities of speech. The organization of the chapter is as follows. Section 2 gives a brief review of the information maximization principle. In Section 3, we discuss different learning rules including the one that we propose. Section 4 presents some simulations where three different learning rules with speech signals as sources are compared. Finally, we summarize our conclusions in Section 5.

2. THE INFORMATION MAXIMIZATION PRINCIPLE

Bell and Sejnowski were the first to apply the *infomax* (information maximization) principle to source separation [5]. They showed that maximizing the joint entropy $H(\mathbf{y};\mathbf{W})$ of the output of a neural processor (see Fig. 15.1) can approximately minimize the information among the output components $y_i = g_i(u_i)$, $i = 1, ..., N$ (so when the mutual information shared among the outputs is zero, the variables are statistically independent). Nonlinear transformations $g_i(u_i)$ are necessary for bounding the entropy in a finite range. In the following, we suppose that all the $y_i = g_i(u_i)$ have a unique inverse $u_i = g_i^{-1}(y_i)$. The joint entropy of $\mathbf{y}$ is

$$\begin{aligned} H(\mathbf{y};\mathbf{W}) &= -\int p(\mathbf{y};\mathbf{W}) \log p(\mathbf{y};\mathbf{W}) d\mathbf{y} \\ &= -E\{\log p(\mathbf{y};\mathbf{W})\}, \end{aligned} \tag{15.4}$$

where $p(\mathbf{y};\mathbf{W})$ is the joint probability density function (pdf) of $\mathbf{y}$ determined by $\mathbf{W}$ and $\{g_i\}$, and $E\{\cdot\}$ denotes mathematical expectation. Equation (15.4) can be re-written

$$H(\mathbf{y};\mathbf{W}) = \sum_{i=1}^{N} H(y_i;\mathbf{W}) - I(\mathbf{W}), \tag{15.5}$$

where

$$H(y_i;\mathbf{W}) = -E\{\log p(y_i;\mathbf{W})\} \tag{15.6}$$

are the marginal entropies and $I(\mathbf{W})$ is their mutual information.

The nonlinear mapping between the output density $p(y_i;\mathbf{W})$ and source estimate density $p(u_i;\mathbf{W})$ can be described by the absolute value of the derivative of y_i with respect to u_i [6]:

$$\begin{aligned} p(y_i;\mathbf{W}) &= \frac{p(u_i;\mathbf{W})}{|\partial y_i/\partial u_i|} \\ &= \frac{p(u_i;\mathbf{W})}{|g_i'(u_i)|}, \end{aligned} \tag{15.7}$$

which can be substituted in (15.6), so that (15.5) becomes

$$H(\mathbf{y};\mathbf{W}) = -I(\mathbf{W}) + \sum_{i=1}^{N} E\{\log \frac{p(u_i;\mathbf{W})}{|g_i'(u_i)|}\}. \tag{15.8}$$

We can see from (15.8) that if

$$p(u_i;\mathbf{W}) = |g_i'(u_i)|, \tag{15.9}$$

i.e. the density function of the estimated source u_i is the derivative of the nonlinear function g_i (or, equivalently, the nonlinearity g_i is the cumulative distribution function –cdf– of the source estimate u_i), then maximizing the joint entropy of the output is equivalent to minimizing the mutual information among the components at the outputs. We have to keep in mind that in general it is very hard to find a nonlinearity that is exactly the cdf of the unknown sources, so by using the maximum entropy (ME) principle, the second term in (15.8) may interfere during the convergence of any adaptive algorithm. Very good discussions on maximizing entropy versus minimizing mutual information are given in [5] and [4]. In the following, we suppose that we can always find a nonlinear function such that $p(u_i; \mathbf{W}) \approx |g_i'(u_i)|$. For super-Gaussian distributions (like speech), i.e. distributions with positive Kurtosis, a good choice for the nonlinear function is $g(u) = \tanh(u)$.

3. DIFFERENT STOCHASTIC GRADIENT ASCENT RULES BASED ON ME

In this section, we first derive the infomax learning rule [5]. Then we give the natural gradient [7, 8], or equivalently, the relative gradient [2]. Finally, we introduce a new normalized natural gradient algorithm.

3.1 THE INFOMAX STOCHASTIC GRADIENT ASCENT LEARNING RULE

The infomax learning rule can be derived by maximizing the output entropy

$$\begin{aligned} H(\mathbf{y}; \mathbf{W}) &= -E\{\log p(\mathbf{y}; \mathbf{W})\} \\ &= -E\{\log \frac{p(\mathbf{x})}{|J(\mathbf{x}, \mathbf{y})|}\}, \end{aligned} \tag{15.10}$$

where

$$J(\mathbf{x}, \mathbf{y}) = \det(\mathbf{W}) \prod_{i=1}^{N} |g_i'(u_i)| \tag{15.11}$$

is the Jacobian of the transformation from $\mathbf{x}$ to $\mathbf{y}$ [6]. Now, the output entropy can be written as

$$H(\mathbf{y}; \mathbf{W}) = H(\mathbf{x}) + \log|\det(\mathbf{W})| + \sum_{i=1}^{N} E\{\log|g_i'(u_i)|\}. \tag{15.12}$$

It can be easily shown that

$$\frac{\partial \log|\det(\mathbf{W})|}{\partial \mathbf{W}} = (\mathbf{W}^{-1})^T \tag{15.13}$$

and

$$\frac{\sum_{i=1}^{N} E\{\log|g_i'(u_i)|\}}{\partial \mathbf{W}} = -E\{\Phi(\mathbf{u})\mathbf{x}^T\}, \tag{15.14}$$

where

$$\Phi(\mathbf{u}) = \left[\begin{array}{ccc} -\frac{g_1''(u_1)}{g_1'(u_1)} & \cdots & -\frac{g_N''(u_N)}{g_N'(u_N)} \end{array} \right]^T.$$

Hence, the stochastic gradient of $H(\mathbf{y}; \mathbf{W})$ with respect to $\mathbf{W}$ is

$$\frac{\partial \hat{H}}{\partial \mathbf{W}} = (\mathbf{W}^{-1})^T - \Phi(\mathbf{u})\mathbf{x}^T \tag{15.15}$$

and the adaptive algorithm as proposed in [5] is

$$\mathbf{W}(n+1) = \mathbf{W}(n) + \mu_{\mathrm{i}}\{[\mathbf{W}^{-1}(n)]^T - \Phi[\mathbf{u}(n)]\mathbf{x}^T(n)\}, \tag{15.16}$$

where the subscript i on μ stands for "infomax." An adaptive algorithm whose performance is independent of the mixing matrix $\mathbf{A}$ is said to be equivariant [2]. Unfortunately, the original infomax learning rule does not have this property [4]. As a result, the convergence of the algorithm can be greatly affected.

3.2 THE NATURAL GRADIENT ALGORITHM

The *natural* gradient algorithm was first proposed in [7] and later studied and justified in [8]. The natural gradient rescales the entropy gradient by post-multiplying the entropy gradient by $\mathbf{W}^T\mathbf{W}$ giving

$$\frac{\partial \hat{H}}{\partial \mathbf{W}}\mathbf{W}^T\mathbf{W} = [\mathbf{I} - \Phi(\mathbf{u})\mathbf{u}^T]\mathbf{W}, \tag{15.17}$$

where $\mathbf{I}$ denotes the identity matrix. The adaptive algorithm based on (15.17) is [8], [9]

$$\mathbf{W}(n+1) = \mathbf{W}(n) + \mu_{\mathrm{r}}\{\mathbf{I} - \Phi[\mathbf{u}(n)]\mathbf{u}^T(n)\}\mathbf{W}(n), \tag{15.18}$$

(the subscript r on μ stands for "relative") and has two important properties: the equivariant property [2] and the property of keeping $\mathbf{W}(n)$ from becoming singular [4]. To summarize, we can say that the matrix $\mathbf{W}^T\mathbf{W}$ rescales the gradient, simplifies the original infomax learning rule, and speeds up considerably the convergence of the algorithm.

3.3 A NORMALIZED NATURAL GRADIENT ALGORITHM

While the natural gradient rescales the entropy gradient and makes the performance of the algorithm independent of $\mathbf{A}$, it does nothing to track the non-stationarities of signals like speech. In [2] an NLMS-type normalization is proposed but it falls short in practice with highly correlated signals. Here, we propose an RLS-type normalization. Let us consider the following matrix

$$\mathbf{R} = E\{\Phi[\mathbf{u}(n)]\mathbf{u}^T(n)\}, \tag{15.19}$$

since $\Phi(\mathbf{u})$ is in general an odd function [i.e. $\Phi(-\mathbf{u}) = -\Phi(\mathbf{u})$], $\mathbf{R}$ can be seen as the covariance matrix of the estimated source signals. Thus, we propose to rescale the entropy gradient by post-multiplying the entropy gradient by $\mathbf{W}^T\mathbf{R}^{-1}\mathbf{W}$ giving

$$\frac{\partial \hat{H}}{\partial \mathbf{W}}\mathbf{W}^T\mathbf{R}^{-1}\mathbf{W} = [\mathbf{I} - \Phi(\mathbf{u})\mathbf{u}^T]\mathbf{R}^{-1}\mathbf{W}. \tag{15.20}$$

Intuitively, $\mathbf{R}^{-1}$ will update the elements of $\mathbf{W}$ individually (instead of globally without this normalization, so each element of $\mathbf{W}$ has its own adaptation step), and after convergence, when the estimated sources become independent, $\mathbf{R}$ becomes a diagonal matrix and then has no effect on the estimation of $\mathbf{W}$. During convergence of the adaptive algorithm, matrices $\mathbf{R}$ and $\mathbf{W}$ are naturally linked in the sense that when $\mathbf{R}$ converges to a diagonal matrix, $\mathbf{W}$ converges to the true solution and vice versa. For nonstationary signals, we replace (15.19) by its recursive estimate

$$\begin{aligned} \mathbf{R}(n) &= \sum_{l=1}^{n} \lambda^{n-l}\Phi[\mathbf{u}(l)]\mathbf{u}^T(l) \\ &= \lambda\mathbf{R}(n-1) + \Phi[\mathbf{u}(n)]\mathbf{u}^T(n), \end{aligned} \tag{15.21}$$

where λ $(0 < \lambda \leq 1)$ is an exponential forgetting factor. The proposed adaptive algorithm is now

$$\mathbf{W}(n+1) = \mathbf{W}(n) + \mu_{\mathrm{n}}\{\mathbf{I} - \Phi[\mathbf{u}(n)]\mathbf{u}^T(n)\}\mathbf{R}^{-1}(n)\mathbf{W}(n), \tag{15.22}$$

where the subscript n on μ stands for "normalized" and where $\mathbf{R}^{-1}(n)$ can be computed recursively by using the matrix inversion lemma which gives

$$\mathbf{R}^{-1}(n) = \frac{1}{\lambda}\{\mathbf{R}^{-1}(n-1) - \frac{\mathbf{R}^{-1}(n-1)\Phi[\mathbf{u}(n)]\mathbf{u}^T(n)\mathbf{R}^{-1}(n-1)}{\lambda + \mathbf{u}^T(n)\mathbf{R}^{-1}(n-1)\Phi[\mathbf{u}(n)]}\}. \tag{15.23}$$

4. SIMULATIONS

Extensive simulations were performed on 4.8-second speech segments from ten different speakers (five females and five males) to compare the three learning rules explained in Section 3. All signals were sampled at 8 kHz and were normalized such that their variance is equal to 1. For the nonlinearity, we used

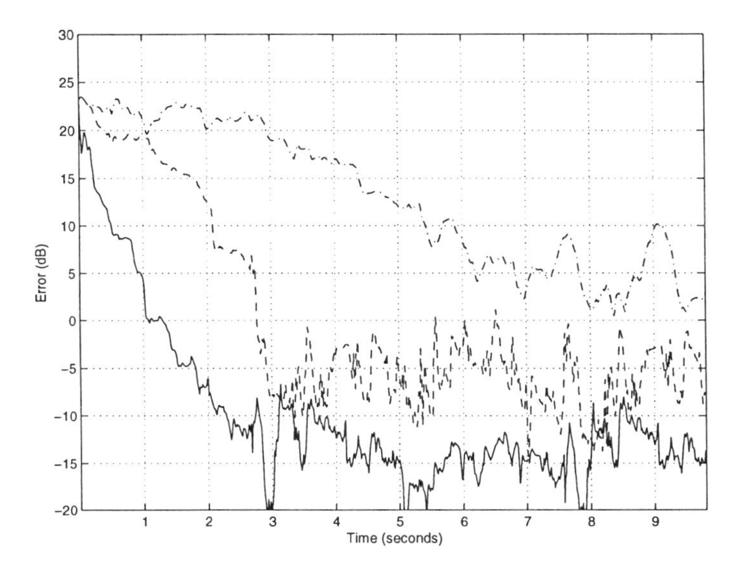

Figure 15.2 Performance (using error measure ϵ in dB) of the original infomax learning rule (-.), the natural gradient (- -), and proposed algorithm (-) with four speech signal sources.

$g_i(u_i) = \tanh(u_i)$ so that $\Phi(\mathbf{u}) = 2\tanh(\mathbf{u})$. In all the simulations, a nonsingular random mixing matrix $\mathbf{A}$ was generated with values uniformly distributed between -1 and 1 to make the mixed time series $\mathbf{x}$ from the original sources $\mathbf{s}$. We show here the performance of the algorithms only for two examples: four sources (Fig. 15.2) and ten sources (Fig. 15.3). To measure the performance of the algorithms, we use the cross-talking error [4]

$$\epsilon = \sum_{i=1}^{N}(\sum_{j=1}^{N} \frac{|p_{ij}|}{\max_k |p_{ik}|} - 1) + \sum_{j=1}^{N}(\sum_{i=1}^{N} \frac{|p_{ij}|}{\max_k |p_{kj}|} - 1), \tag{15.24}$$

where $\mathbf{P} = (p_{ij}) = \mathbf{WA}$. The parameter settings chosen for simulation of Fig. 15.2 are: $N = 4$; $\mathbf{W}(0) = \mathbf{I}$; $\mu_{\mathrm{i}} = 0.0005$, $\mu_{\mathrm{r}} = 0.0005$, $\mu_{\mathrm{n}} = 0.008$; $\mathbf{R}(0) = \mathbf{I}$, $\lambda = 0.991$; Number of passes: 2 (2×4.8 seconds). The parameter settings chosen for simulation of Fig. 15.3 are: $N = 10$; $\mathbf{W}(0) = \mathbf{I}$; $\mu_{\mathrm{i}} = 0.0002$, $\mu_{\mathrm{r}} = 0.0002$, $\mu_{\mathrm{n}} = 0.008$; $\mathbf{R}(0) = \mathbf{I}$, $\lambda = 0.995$; Number of passes: 3 (3×4.8 seconds). We can see from Figs. 15.2 and 15.3 that the proposed learning rule outperforms the two others and listening confirms that there is almost perfect separation of the sources. We could have taken larger adaptation steps to speed up the convergence but the separation would then not have been that good.

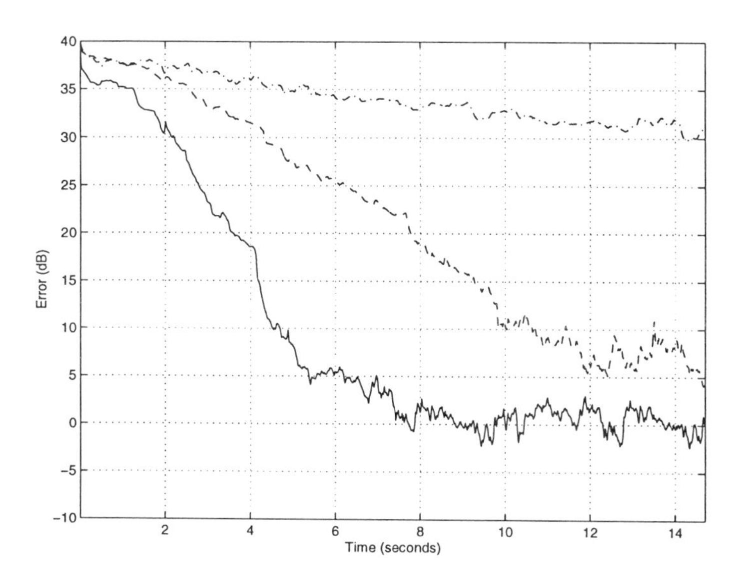

Figure 15.3 Performance (using error measure ϵ in dB) of the original infomax learning rule (–.), the natural gradient (– –), and proposed algorithm (–) with ten speech signal sources.

5. CONCLUSIONS

With a large number of sources, it becomes very hard with the original infomax learning rule to separate the signals in a reasonable amount of time. That is why it is very important to continue to investigate on the use of adaptive filtering in the context of blind source separation. The natural gradient is a great improvement and the normalized version is another step forward. However, in the convolutive mixing case (to deal with acoustic impulse responses for example), time-domain adaptive algorithms are not suitable because of their high level of complexity and the difficulty of the problems to solve. (The length of room impulse responses is typically of the order of a couple of thousands taps.) It goes without saying that a frequency-domain implementation of these algorithms is the approach to take. But, because of the non-linearities involved and the permutation problem, such translation is not obvious. The question is how can we approximate time-domain algorithms to deduce fast versions without worsening significantly their performances?

References

[1] J. Herault and C. Jutten, "Space or time adaptive signal processing by neural network models," in *Neural Networks for Computing: AIP Conference Proceedings 151*, J. S. Denker, ed. American Institute of Physics, New York, 1986.

[2] J.-F. Cardoso and B. H. Laheld, "Equivariant adaptive source separation," *IEEE Trans. Signal Processing*, vol. 44, pp. 3017-3030, Dec. 1996.

[3] T.-W. Lee, *Independent component analysis: theory and applications*. Boston: Kluwer Academic Publishers, 1998.

[4] H. H. Yang and S. Amari, "Adaptive online learning algorithms for blind separation: maximum entropy and minimum mutual information," *Neural Comput.*, vol. 9, pp. 1457-1482, 1997.

[5] A. J. Bell and T. J. Sejnowski, "An information-maximization approach for blind separation and blind deconvolution," *Neural Comput.*, vol. 7, pp. 1129-1159, 1995.

[6] A. Papoulis, *Probability, random variables, and stochastic processes*. New York: McGraw-Hill, 1991.

[7] A. Cichocki, R. Unbehauen, L. Moszczynski, and E. Rummert, "A new online adaptive learning algorithm for blind separation of source signals," in *Proc. ISANN*, Taiwan, 1994, pp. 406-411.

[8] S. Amari, A. Cichocki, and H. H. Yang, "A new learning algorithm for blind signal separation," in *Advances in neural information processing systems*, 8, Cambridge, 1996, pp. 757-763.

[9] S. Amari, S. C. Douglas, A. Cichocki, and H. H. Yang, "Multichannel blind deconvolution and equalization using the natural gradient," in *Proc. IEEE Workshop Signal Proc. Adv. Wireless Comm.*, Paris, France, 1997, pp. 101-104.

Index